XV International Seminar on Physics and Chemistry of Solids 2009

(ISPCS15)

Journal of Physics: Conference Series Volume 289

Szklarska Poreba, Poland
7-10 June 2009

ISBN: 978-1-61782-910-9
ISSN: 1742-6588

Printed from e-media with permission by:

Curran Associates, Inc.
57 Morehouse Lane
Red Hook, NY 12571

CURRAN ASSOCIATES INC. proceedings .com

Some format issues inherent in the e-media version may also appear in this print version.

Printed by Curran Associates, Inc. (2011)

Institute of Physics
Dirac House, Temple Back
Bristol BS1 6BE UK

Phone: 44 1 17 929 7481
Fax: 44 1 17 920 0979

techtracking@iop.org

Additional copies of this publication are available from:

Curran Associates, Inc.
57 Morehouse Lane
Red Hook, NY 12571 USA
Phone: 845-758-0400
Fax: 845-758-2634
Email: curran@proceedings.com
Web: www.proceedings.com

Table of contents

Journal of Physics: Conference Series

Volume 289

2011

15th International Seminar on Physics and Chemistry of Solids
7–10 June 2009, Szklarska Poreba, Poland

Preface

Papers

PREFACE

The XV International Seminar on Physics and Chemistry of Solids (ISPCS15) was held from 7–10 June 2009 in Szklarska Poręba. Over eighty participants joined together in this mountain resort, at the foot of Szrenica peak, in the Polish part of the Sudety range. The majority or participants, in accordance with Seminar tradition, were from Ukraine and Poland.

The pleasant and warm atmosphere created by the organizers was conducive to fruitful discussions, making new contacts and to joyful gatherings with friends. Even the rainy weather could not change that.

Lectures and communications mainly covered the fields of solid state physics and chemistry, and possible applications. This time, however, a new section was introduced – one devoted to modern topics in liquid chemistry. Sometimes such a look over the borders of scientific specialties leads to interesting insights and original research solutions.

Some of the papers presented during ISPCS15 are collected in this volume. Their diversity is representative of both the scope and character of this Seminar. The majority of the papers are research reports, but a review article and a paper focussed on problems connected with environmental protection are also included.

This Conference has functioned for over a decade due to the permanent support of the rectors of both co-organizing universities: Professor Ivan Vakarchuk from Ivan Franko National University of Lviv and Professor Zygmunt Bąk from Jan Dlugosz University in Częstochowa. It is our pleasure, on behalf of the all participants of the past Seminars, to express our gratitude for this assistance.

We would also like to thank all the invited speakers who kindly accepted our invitation, namely Professors Roman Gladyshevskii (Ivan Franko National University, Lviv, Ukraine), Mihaela Gulea (Laboratoire de Chimie Moleculare et Thioorganique, CAEN, France), Osama I Abd El-Salam (National Research Centre Dokki, Cairo, Egypt), Jarosław Lewkowski (University of Łódź, Łódź, Poland), Jean Ebothe (University of Reims Champagne-Ardenne, France), Alain Gibaud (University du Maine, Le Mans, France), and Zbigniew Czapla (University of Wrocław, Wrocław, Poland).

Finally, it is both our duty and pleasure to express our gratitude for the work done by the referees. Without their efforts many of the papers in this volume would not have been improved.

Bogdan Kotur, Lviv
Piotr Brągiel, Częstochowa

Journal of Physics: Conference Series **289** (2011) 012001 doi:10.1088/1742-6596/289/1/012001

Synthesis and Optical Investigations of the Guest-Host Nanostructures Alumina-SiC and Alumina-In_2O_3

A Bouifoulen[1,2], A Kassiba[1], A Outzourhit[2], M Edely[1], M Makowska-Janusik[3], J Szade[4], A Oueriagli[2]

[1] Laboratoire de Physique de l'Etat Condensé, UMR-CNRS 6087, Institut de recherche IRIM2F-FR-CNRS 2575-Université du Maine, Avenue Olivier Messiaen, 72085 Le Mans Cedex 9, France
[2] Laboratoire de Physique du Solide et Couches Minces, Faculté des Sciences Semlalia, Université Cadi Ayyad, B. P. 2390, Marrakech 40000, Maroc
[3] Institute of Physics, Al. Armii.Krajowej, 13/15, J. Dlugosz University, 42-200 Czestochowa, Poland
[4] A.Chełkowski Institute of Physics - University Śląski in Katowice - 40-219 Katowice, Poland

Email: kassiba@univ-lemans.fr

Abstract. Several strategies were developed to synthesise two classes of nanostructured thin films with nanocrystals of SiC (nc-SiC) or In_2O_3 (nc-In_2O_3) confined in alumina. The syntheses were performed by using Rf-sputtering and co-pulverisation process of the suitable reactants. Thus, Al_2O_3/nc-SiC and Al_2O_3/nc-In_2O_3 composite thin films were obtained and their structural and optical features analyzed respectively by XRD, XPS and UV-VIS absorption. The deposition conditions and the post-synthesis treatments were optimized in order to improve the crystalline character of confined nanocrystals. The optical properties were compared in the range 200 nm-1200 nm for bare alumina films or nanostructured ones with the semiconducting nanocrystals. The direct and indirect band band gaps were evaluated and discussed with regard to the stoechiometry and morphologies of the nanocomposite films

1. Introduction

Nanostructured films composed by semi-conducting nanocrystals confined in background host matrices are attractive architectures with regard to expected original electronic and optical behaviours. The key parameters which monitor the quality of the nanostructure consist in narrow size and shape distributions of the nanocrystals in the matrix as well as their homogeneous crystalline order. Among the considered materials, silicon carbide nanocrystalline (SiC) is a wide bandgap semiconductor with particular characteristics such as high breakdown field strength, high thermal conductivity, and high saturation drift velocity [1]. These excellent physical properties make SiC a promising semiconductor material for the fabrication of electronic devices [2]. Moreover, as an important wide band gap semiconductor, cubic silicon carbide (β-SiC) nanostructures would be favourable for applications in high temperature, high power, and high frequency nanoelectronic devices [3]. From another hand, besides the interest in applications in optoelectronics, In_2O_3 nanostructures have also a key role from a fundamental to applications in several areas such as solar cells, sensors or flat panel displays [4]. As main characteristics of indium oxide, the direct band gap is around 3.75 eV and the indirect bandgap at about 2.61 eV [5] lead to high optical transparency in the visible light region [6].

In another respect, alumina (Al_2O_3) is the most important dielectric material suitable for optical coatings in large spectral region [7]. Therefore, it is considered to be the most promising candidates to realize composites with improved optical properties taking benefit from SiC and In_2O_3 nanocrystals whose the size and crystalline polytypes can modulate the band gap in a large extent.

For the practical achievement of the synthesis, although DC magnetron sputtering is more preferred for industrial applications, rf-sputtering is more suited for our purpose. The experimental versatility allows modulating the structural and optical properties of alumina-SiC and alumina-In_2O_3 through the nature of substrates, deposition conditions such as temperature of substrate during deposition, annealing and also the electrical power of the deposition process.

Thus, this work is dedicated to realize composites such as alumina-SiC and alumina-In_2O_3 films deposited on (100) silicon or borosilicate (BK7) substrates at different temperatures and to investigate the variations of structural and optical properties. The morphology, microstructure and the composition of alumina-SiC and alumina-In_2O_3 films were characterized by methods such as XRD and XPS. In addition, optical transmittance spectra were compared between the as-deposited and post-annealed composites films associating alumina and SiC or In_2O_3 nanocrystals. The evaluation of direct and indirect band gap was performed as function of the compositions of nanostructured films.

2. Experimental

2.1. Materials

Alumina-SiC and alumina-In_2O_3 thin films were deposited onto crystalline (100) silicon wafer and borosilicate glass substrates by rf-sputtering method using co-deposition from Al_2O_3 target and pellets of SiC or In_2O_3.The rf sputtering power was varied in the range 50 – 100 Watt. The SiC pellets were hot-pressed while the In_2O_3 pellets were sintered at 1300°C [8]. Typically, the base vacuum was about $1.4\ 10^{-6}$ mbar, and the pressure was stabilized at about $10.98\ 10^{-2}$ mbar during the deposition process with the flow rate of argon about 100 sccm. After cleaning, the float substrates were placed in the vacuum chamber. The substrates were heated from room temperature (20°C) to 600 °C. The film thickness of about 310 nm was achieved. In this study, only the substrate temperature was varied, while all deposition times presented here were 60 min and all other parameters were fixed.

2.2. Methods

The film thickness was measured using an interferometer (FilmeTrics Model 205-0070). The phase composition and structure of the samples was determined by X-ray diffraction using a Philips X-pert diffractometer with a Cu-Kα radiation. Optical transmittance spectra of the film structure were measured with a UV–VIS–NIR scanning spectrophotometer (DH-2000-BAL). High resolution transmission electron microscopy (TEM) images were obtained on the composite samples based on SiC and Al_2O_3. XPS investigations were performed with a Mg Kα X-ray excitation source (1253.6 eV). The residual pressure in the analysis chamber was maintained below 10^{-9} torr during data acquisition. The various XPS spectra with high resolution were recorded for the elements Si 2p, C 1s and Al 2p.

3. Results and discussion

3.1. Structural and morphology features

Figure 1(a) shows XRD patterns of as-deposited and 800°C annealed films. The annealing is performed in a furnace under Ar atmosphere during 3 hours. The high temperature annealing (800°C) favours the development of a large diffraction background from of small (111) SiC peak at about $2\theta=33.6°$. Previous experimental results [9] showed that this structure appears when the annealing is made at 900°C.Figure 1(b) reports the XRD pattern of alumina-In_2O_3 samples which were deposited onto silicon substrate and prepared at temperature of substrates at about 20°C, 250°C and 400°C. It can be seen that the main growth directions are (2 1 1), (2 2 2), (4 0 0) and (4 4 0), which are related to the cubic structure of In_2O_3 [10]. Under the same conditions, depositions of these samples were also made on a borosilicate glass. This procedure allow to resolve better the main diffraction line related for the In_2O_3 main phase (Figure 1(c)). As-deposited films at RT and the one with the substrate temperature fixed at 250°C exhibit amorphous features. The films were crystallized only for the temperature as high as 400°C leading to a XRD (222) peak enhancement. Finally, None of the diffraction patterns bring the mark of any characteristic peaks of alumina. This is indicative of amorphous nature of the alumina matrix.

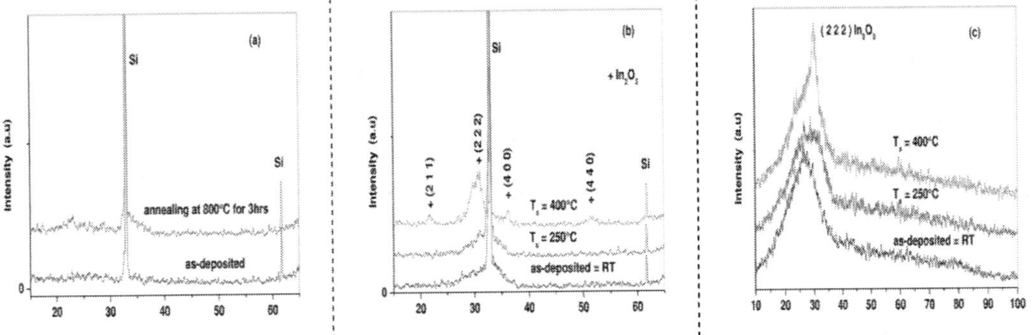

Figure 1. (a) XRD spectrum for alumina-SiC deposited on a silicon substrate at RT and annealed at 800°C for 3 hours. (b) XRD patterns of the as-synthesized alumina-In_2O_3 on a silicon substrate formed at different substrate temperatures. Indices of the peaks are marked above the peaks. (c) XRD spectra of the as-synthesized alumina-In_2O_3 nanocomposites on a BK7 substrates maintained during the synthesis at different temperatures.

The XPS experiments give valuable information on the composition of the films. XPS spectra were recorded and deconvolved for precise identification of the involved chemical species. Figure 2(a) shows the XPS spectra of Si 2p core level peak. The deconvolution of this spectrum reveals two components. The one located at a binding energy of 103.17 eV is associated to environments such as Si-O_2 [11]. The second contribution located at energy about 99.60 eV is assigned to bonding which can be attributed to Si-C [12]. Such involved bonding with silicon seems sound with respect to the coexistence of SiC clusters and the expected formation of SiO_2 during the co-pulverization of Alumina.

The spectrum of the C 1s level in figure 2(b) seems to reveal three bonding features. The peak position located at a binding energy of 284.81 eV corresponds to C-C [11] bonding. The second contribution located around the energy of 283.49 eV can be assigned to C-Si [11] bond while the features at the binding energy of 286.45 eV correspond to C-H bond [13]. It is worth noting that, the involved hydrogen bonding occurs from the composition of the SiC pellets made from powders which contain hydrogen due to the used synthesis process [14].

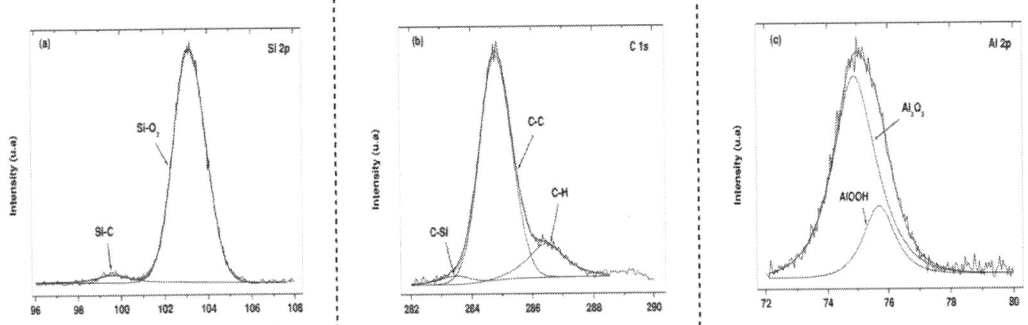

Figure 2. (a) Typical deconvoluted XPS spectrum of Si 2p energy region for the alumina-SiC. (b) Typical deconvoluted XPS spectrum of C 1s energy region for the alumina-SiC. And (c) Typical deconvoluted XPS spectrum of Al 2p energy region for the alumina-SiC.

Figure 2(c) presents the XPS spectra of the Al 2p core level peak and its decomposition reveals two superimposed components with a small difference in the binding energy. Both are most likely associated to alumina [15]. The low intensity line at about 75.7 eV may be related to a non-stoechiometric compound.

3.2. Optical features

The transparency of the synthesized films for all temperatures exhibits similar features shown in Figure 3(a,b). At the opposite of SiC based composites, the more pronounced optical effects were obtained after the incorporation of In_2O_3. Indeed at given wavelength about 650 nm, the transmission percent of alumina-In_2O_3 films changes from 84% (RT deposited), to 89% for deposition with the substrate at (250°C) and to 93% for substrate temperature about (400°C). The ripples in the spectrum resulted from the interference fringes due to suitable film thikness for resolving interferences effects [16].

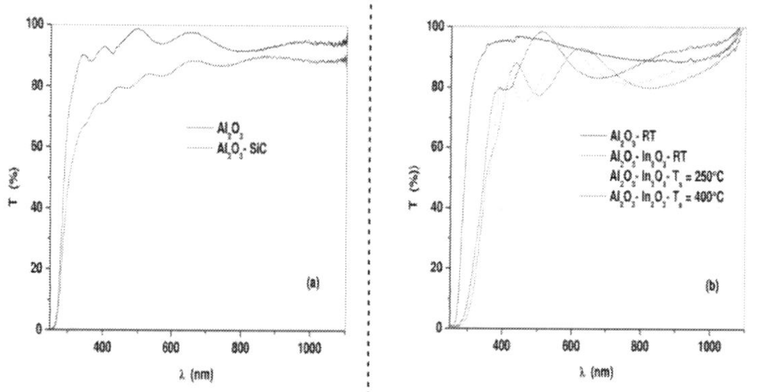

Figure 3. (a) Optical transmittance of alumina and alumina-SiC film deposited on borosilicate glass substrate at RT. (b) Optical transmittance spectra of alumina and alumina-In_2O_3 films with different BK7 substrate temperatures.

The type of the optical transitions, whether direct or indirect, can be determined by studying the dependence of the absorption coefficient $\alpha = [\ln(1/T)/d]$, as function of the photon energy (hv) and the thickness (d) of the films. For direct/indirect optical transition between presumably parabolic bands, the relevant optical band gaps parameters can be deduced as follows [17].

$$(\alpha.h.v)^n = A.(h.v - E_g) \quad (1)$$

Where A is a constant and Eg is the optical band gap. The direct and indirect bandgaps of the composites films may roughly be estimated by plotting $(\alpha.hv)^n$ versus hv (n=2 for allowed direct and n=1/2 for allowed indirect optical transitions). By extrapolating the linear region of the plot toward low energies [18], we may deduce the optical band-gaps. Figure 4(a,b) and figure 5(a,b) represents these plots for the alumina (RT), alumina-SiC (RT) and alumina-In_2O_3 (RT, 250, 400°C). Table 1 summarizes all bandgaps obtained by extrapolating the linear parts of the $(\alpha hv)^n$ versus hv plots.

Table 1. Variation of the direct and indirect band-gap of nanocomposite films.

			Al_2O_3-In_2O_3		
n	Al_2O_3	Al_2O_3-SiC	RT	250°C	400°C
1/2 (eV) (indirect)	4.04	3.94	3.09	3.07	3.20
2 (eV) (direct)	4.51	4.48	3.92	4.05	4.09

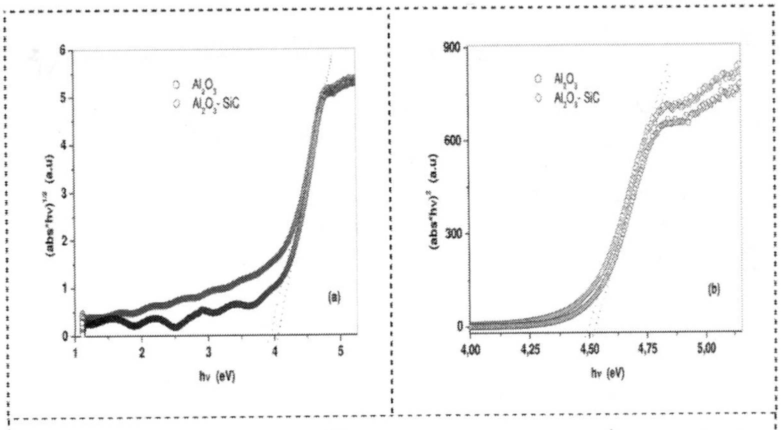

Figure 4. Plots of (a) $(\alpha h\nu)^{1/2}$ versus hν and (b) $(\alpha h\nu)^2$ versus hν for alumina and alumina-SiC.

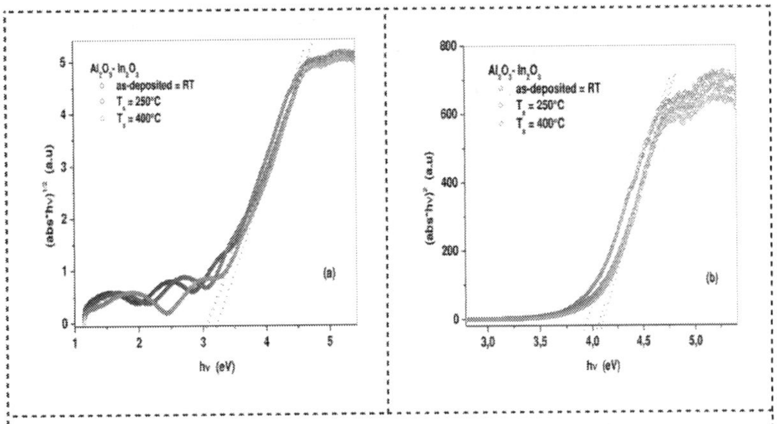

Figure 5. Plots of (a) $(\alpha h\nu)^{1/2}$ versus hν and (b) $(\alpha h\nu)^2$ versus hν for alumina-In_2O_3 deposited at different substrate temperatures.

The obtained band gap, direct and indirect, seem relevant to monitor the composition of the film alumina-nanocrystals.
The optical properties traduce the change in the film composition through the modification of the caractéristiques optical band-gaps. The effects are more importnat on the alumina-In_2O_3 compared to composites with SiC.

4. Conclusions

Innovative guest-host nanostructures can be realized from alumina as host matrix and nanocrystals incorporated in the amorphous media. Two main wide band gap semiconductors (SiC, In_2O_3) were used because their gap can be finely tuned as function of nanocrysal dimensions and their cristalline polytypes. The XRD experiments showed that the crystallinity of alumina-In_2O_3 nanocomposites was improved by using appropriate substrates maintained at temperatures between RT and 400°C. From optical aspects, high quality films with a transmittance 93% were obtained and the allowed-indirect and allowed-direct optical bandgap were found to vary respectively from 3.07 to 3.20 eV and from 3.92 to 4.09 eV, depending on the temperature of the substrates. The effects of In_2O_3 incorportaion are more important on optical behavior compared to composites based on SiC. The carried out synthesis procedures are currently under development to improve the film quality, cristalline features and morphologies.

Acknowledgements
A part of the work was realized under the France-Morocco cooperation program VOLUBILIS with the financial support from the "Comité inter-universitaire Franco-Marocain".

Journal of Physics: Conference Series **289** (2011) 012001 doi:10.1088/1742-6596/289/1/012001

References

[1] Kang B C et al. *Thin Solid Films* 464-465 (2004) 215-219
[2] Fisher A, Schroter B, Richter W, *Appl. Phys. Lett.* 66 (23) 1995
[3] Li J C, Lee C S, Lee S T, *Chem. Phys. Lett.* 355 (2002)
[4] Chen et al. *J. Phys. Chem. C*, Vol. 111, No. 49, 2007
[5] Mazzera M et al. *Nanotechnology* 18 (2007) 355707
[6] Mohan Babu P, *J. Opt. Adv. Mat.* Vol. 6, No. 1, March 2004, p. 205 – 210
[7] Zhang F et al. *Applied Surface Science* 254 (2008) 3045–3048
[8] Xu H, Yu A, *Materials Letters* 61 (2007) 4043-4045
[9] Yoon D H and Suh S J, *J. Korean Phy Soc,* Vol. 42, February 2003, pp. S943-S946
[10] Gheidari A M et al. *Materials Science and Engineering* B 136 (2007) 37-40
[11] Santoni A, Frycek R, Castrucci P, Scarselli M, Cres M D, *Surf. Sci.* 582 (2005) 125
[12] Shimoda K, Park J –S, Hinoki T and Kohyama A, *Appl. Surf. Sci.* 253 (2007) 9450
[13] Dikov Y P, Ivanov A V, Wlotzka F, M Galimov E, Wanke G, *Solar System Research* 36, 1(2002)1 and references therein
[14] Bouclé J, Kassiba A, Makowska-Janusik M, Herlin-Boime N, Reynaud C, Desert A, Emery J, Bulou A, Sanetra J, Pud A A, and Kodjikian S, *Phys. Rev.* B 74 (2006) 205417
[15] NIST standard XPS reference database 20 version 3.5 (2003)
[16] Lin H, Kumon S, Yoko T, Kozuka H, *Thin Solid films* 1998, 315, 266–272
[17] K Subba Ramaiah et al. *Semicond. Sci. Technol.* 15 (2000) 676–683
[18] Novkkovski N and Tanusevski A, *Semicond. Sci. Technol.* 23 (2008) 095012

Modelling noncollinear spin structure in $YNiO_3$ perovskite: a GGA+U APW approach

Malgorzata Szwaja[1,2] **and Florent Calvayrac**[2]

[1]Faculty of Mathematical-Natural Sciences, Jan Długosz University, Częstochowa, Poland
[2]Laboratoire de Physique de l'Etat Condensé LPEC UMR CNRS 6087 FR 2575 Université du Maine, PRES UNAM, Av Olivier Messiaen Le Mans, France

E-mail: Florent.Calvayrac@univ-lemans.fr

Abstract. In this work we compute the magnetic spin structure of the YNiO3 perovskite using the noncollinear APW implementation of WIEN2K. We compare LDA+U, GGA+U results to experimental results and find an improvement in comparison to previous theoretical work supposing an antiferromagnetic collinear alignment of the magnetic momenta.

1. Introduction

The perovskite system $RNiO_3$ (where R is a trivalent rare earth) is a very attractive family of compounds around which intense research activity has been recently developed, see [1-3] for instance. This family of materials is particularly interesting because it exhibits structural and metal-insulator phase transitions around room temperature, especially when doped with some rare earths. This allows for instance to modulate the amount of infrared radiation transmitted as a function of temperature, with potential applications in defense (infrared stealth) or building (smart windows which let heat in winter and out in summer).

In this work we will compute *ab initio* the magnetic structure of $YNiO_3$. Due to nonmagnetic character of Y, the magnetic properties of the Ni lattice can be studied in $YNiO_3$ without interference with other magnetic species.

Magnetic properties of transition metals compounds, especially oxides, have become a subject of intensive research, both from the experimental and theoretical point of view. One of the most challenging and interesting aspects of that field is the subtle interplay between magnetic ordering and geometric structure. Finite size effects and a reduced dimensionality often lead to a significantly different magnetic behavior.

Noncollinear magnetism is manifested in a wide variety of substances. Those compounds can be ferromagnetic (FM), where the spins are aligned in parallel directions; antiferromagnetic (AFM), where the spins are antiparallel; or noncollinear, where the spins are not parallel and may be disordered. Noncollinear magnetism can arise naturally due to geometric frustration of AFM interactions. Other effects also giving rise to noncollinear magnetism include magnetic anisotropy, which arises due to a parallel direction of magnetization. Finally, the competition between exchange interaction and magnetic anisotropy also contributes to noncollinear magnetism [4].

In this paper we study the noncollinear spin structure of this nickelate and compare it to experimental data. This structure can serve as a benchmark to check previous computations restricting to collinear configurations.

2. Previous results

2.1. Experimental facts

At high temperatures, $YNiO_3$ is orthorhombic and metallic but below $T_{MI} = 582K$ it changes to a monoclinic insulator due to a charge disproporation $2Ni^{3+} \rightarrow Ni^{2+} + Ni^{4+}$, that develops at the opening of the gap. Crystallographic studies combining X-ray diffraction and neutron diffraction [5] have shown the presence of two alternating NiO_6 octahedra, with expanded Ni◖‾ and contracted Ni◖‾ $Ni-O$ bonds and a magnetic structure $\vec{k} = \left(\frac{1}{2}, 0, \frac{1}{2}\right)$, with unequal moments at Ni◖‾ and Ni◖‾ octahedra. $YNiO_3$ systems contains four Ni ions in its crystallographic unit cells and each locates on the inversion center. Four Ni sites are labeled as Ni◖‾ $= \left(\frac{1}{2}, 0, 0\right)$, Ni◖‾ $= \left(0, \frac{1}{2}, 0\right)$, Ni◖‾ $= \left(\frac{1}{2}, 0, \frac{1}{2}\right)$, Ni◖‾ $= \left(0, \frac{1}{2}, \frac{1}{2}\right)$. In $YNiO_3$, no symmetry operations transfer a position Ni◖‾ or Ni◖‾ to that of Ni◖‾ or Ni◖. Therefore, there are two inequivalent positions of Ni ions in $YNiO_3$. The structure of $YNiO_3$ is shown on Fig. 1 and Fig. 2 (graphics were generated by *XCrySDen* [6]).

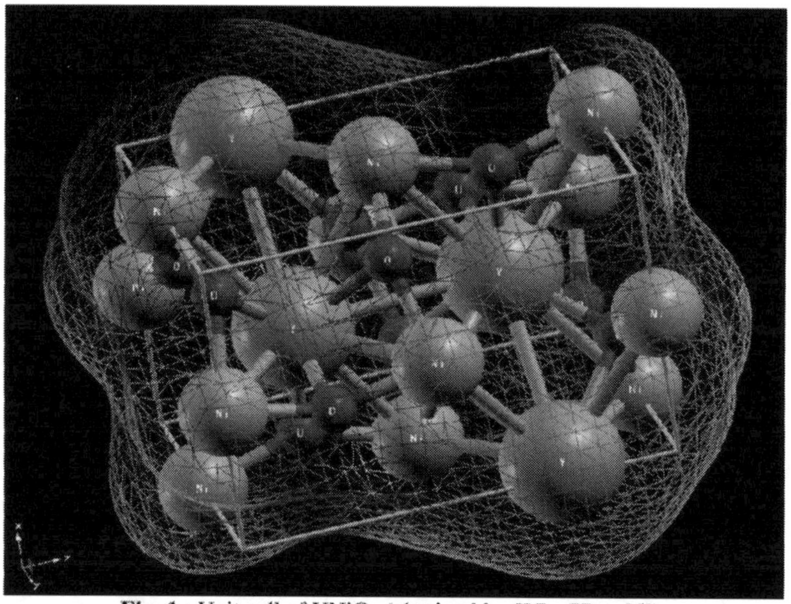

Fig. 1.: Unit cell of $YNiO_3$ (obtained by *XCrySDen* [6]).

Fig. 2.: Structure of $YNiO_3$ (obtained by *XCrySDen* [6]).

Finally, we shall focus on the magnetic structure. The experimentally observed magnetic moments are $\pm 1.4\mu_B$ for Ni^{2+} ions and $\pm 0.7\mu_B$ for Ni^{4+} ions [5]. The charge disproportionation is mainly due to the crystal field effect. $YNiO_3$ are antiferromagnetic insulators in their low temperature phases. The spin alignment in these two nickelates is quite unique one such that the symmetrical coexistence of ferro- and antiferromagnetic couplings along the three pseudocubic axes:

$$\cdots Ni\uparrow - \overset{FM}{O} - Ni\uparrow - \overset{AFM}{O} - Ni\downarrow - \overset{FM}{O} - Ni\downarrow \cdots.$$

In $YNiO_3$, the charge disproportionation $2Ni^{3+} \rightarrow Ni^{2+} + Ni^{4+}$ and the symmetrical coexistence of ferro- and antiferromagnetic couplings are observed, though the Ni^{4+} ions have small magnetic moments.

2.2. *Previous theoretical work*

Previous work on the present class of materials has been performed using typically the LSDA+U approach in an antiferromagnetic setting, using a supercell, and the LMTO-ASA approximation. This calculation results in a band structure with insulating and metallic characteristics for monoclinic and orthorhombic phases, respectively [7,8] with gap $E_g = 1.03eV$ for the insulating phase. This corresponds more or less to experimental results [3] obtained on closely related compounds, exhibiting however some dispersion due to the measuring method (four points methods or optical methods). In those works, however, the magnetic moments for one of the nickel species come out as zero, in contradiction with experimental results. Some authors have also studied the effect of cell parameters on the conducting characteristics of related materials in order to understand thin film or pressure effects, or optimized the oxygen positions within the LDA or LDA+U picture, getting magnetic momenta closer to experimental results of [5].

3. Theoretical choices

An essential difference between collinear and noncollinear magnetism is that the latter lacks a global spin quantization axis. In the case of noncollinear magnetism a solution is to follow Von Barth and Hedin [10] formulation of DFT and Kohn-Sham equations: adopt the spinor formalism and work with 2×2 matrixes for the operators as done in [11,12] for instance.

3.1. Noncollinear Density Functional Theory

The one-electron Hamiltonian-like Kohn-Sham operator for magnetic systems can be approximated from Dirac's equation in the non relativistic case by:

$$h_i = -\frac{\hbar^2}{2m}\nabla^2 + V_{eff} + \mu_B\vec{\sigma}\cdot\vec{B}_{eff} + \zeta(\vec{\sigma}\cdot\vec{l}),$$

where the scalar product $\mu_B\vec{\sigma}\cdot\vec{B}_{eff}$ expresses the interaction of spins with effective magnetic field, $\zeta(\vec{\sigma}\cdot\vec{l})$ express the spin-orbit coupling.

In this case, the Schrödinger-like equation appears as:

$$h_i\begin{pmatrix}\psi_{i1}\\ \psi_{i2}\end{pmatrix} = \varepsilon_i\begin{pmatrix}\psi_{i1}\\ \psi_{i2}\end{pmatrix},$$

where the one-electron Kohn-Sham wavefunction is a $2-$component vector (spinor): ψ_{i1} $-$spin up component, ψ_{i2} $-$spin down component.

Effective potential is sum of external and exchange-correlation potential:

$$V_{eff} = V_{ext} + V_H + V_{xc},$$

where V_H is the Hartree potential.

Effective magnetic field is sum of external and exchange-correlation fields:

$$\vec{B}_{eff} = \vec{B}_{ext} + \vec{B}_{xc}.$$

In the density functional formalism, the exchange-correlation potential and magnetic exchange-correlation vector field are defined as functional derivatives:

$$V_{xc} = \frac{\partial E_{xc}(n,\vec{m})}{\partial n},$$

and

$$\vec{B}_{xc} = \frac{\partial E_{xc}(n,\vec{m})}{\partial \vec{m}},$$

where n is the a spin density and $\vec{m}$ is a charge densities vector.

In a noncollinear case $\vec{m}$ varies in space. Furthermore, all three components of the scalar products $\mu_B\vec{\sigma}\cdot\vec{B}_{eff}$ have to be taken into account. Note that $\sigma_x = \begin{pmatrix}0 & 1\\ 1 & 0\end{pmatrix}$, $\sigma_y = \begin{pmatrix}0 & -i\\ i & 0\end{pmatrix}$ and the potential matrix is of the form:

$$\hat{V} = V_{eff}I + \vec{B}_{eff}\vec{\sigma} = \begin{pmatrix} V_{eff} + \mu_B B_z & \mu_B(B_x - iB_y)\\ \mu_B(B_x + iB_y) & V_{eff} - \mu_B B_z\end{pmatrix}.$$

The diagonalization can not be performed for up and down spins, separately. The total matrix in spin space (2×2) must be setup and locally diagonalized. In this case, complex basis functions are necessary.

We can write the Schrödinger-like equation as:

$$\begin{pmatrix} -\frac{\hbar}{2m}\nabla^2 + V_{eff} + \mu_B B_z + \ldots & \mu_B(B_x - iB_y)\\ \mu_B(B_x + iB_y) & -\frac{\hbar}{2m}\nabla^2 + V_{eff} + \mu_B B_z + \ldots\end{pmatrix}\psi_i = \varepsilon_i\psi_i.$$

Here we have noncollinear magnetic moment and solutions can be non-pure spinors:

$$\psi_i = \begin{pmatrix}\psi_{i1}\\ \psi_{i2}\end{pmatrix},\ \psi_{i1}, \psi_{i2} \neq 0.$$

The local spin density matrix in a noncollinear system has the form:

$$\rho_{\sigma\sigma'} = \frac{1}{2} n I_z + \sigma \vec{m} = \frac{1}{2}\begin{pmatrix} n+m_z & m_x - i m_y \\ m_x + i m_y & n - m_z \end{pmatrix}.$$

The space dependent spin density matrix contains the whole information necessary to describe the magnetic state of the considered material. In a collinear calculation, m_x and m_y are zero, thus $\rho_{\sigma\sigma'}$ is diagonal. In the non magnetic case m_z is zero thus $\uparrow\uparrow$ and $\downarrow\downarrow$ parts are the same [12]. Magnetic moments are estimated as the quantum mechanical average over all electrons summed up in a sphere of arbitrary radius centered around each atom.

3.2. *Numerical choices*

Ab initio calculations were performed by the *WIEN2k* and *WIENncm* program package [13,14], based on the density functional theory in the local density approximation (LDA) and the generalized-gradient approximation (GGA), and on-site U corrections. In this case, we can note that for instance the presumably faster. *QuantumESPRESSO* pseudopotential program package has not been used because noncollinear calculations can not be combined with LDA+U. However, *WIEN2k* and *WIENncm* program packages work perfectly in this case, although quite slowly because all electrons are taken into account. The interest of *WIEN2k* consists in the high accuracy of the results and the capacity to compute effects linked to the core electrons, such as NMR or Mössbauer parameters. Calculations were based on the crystallography parameters reported by Alonso and al [5,15]: monoclinic space group $P2_1/n$, $a = 9.7875(1)\, bohr$, $b = 10.4224(1)\, bohr$, $c = 14.0152(3)$ bohr, $\beta = 90.081(1)^{\circ}$. It is interesting to note that the $P2_1/n$ symmetry there are two crystallographyically independent Ni positions ($Ni(1)$ and $Ni(2)$), as well as three nonequivalent oxygen atoms ($O(1)$, $O(2)$ and $O(3)$) all in general (x, y, z) positions. Parameters used in the calculation are presented in Table 1:

Table 1. Atomic parameters for $YNiO_3$ from [5].

Atom	Site	x	y	z
Y	*4e*	0.9816(2)	0.0729(1)	0.2502(4)
Ni(1)	*2d*	0.5	0	0
Ni(2)	*2c*	0.5	0	0.5
O(1)	*4e*	0.0998(3)	0.4705(3)	0.2457(5)
O(2)	*4e*	0.6973(7)	0.3080(7)	0.0467(5)
O(3)	*4e*	0.1882(6)	0.2038(7)	0.9465(5)

In addition, we adopted in the calculations: lattice type: P, number of inequivalent atoms in unit cell: 12, number of equivalent atoms: 2 for Y, 1 for Ni, 2 for O, number of radial mesh points: 781, first radial mesh point: $R0 = 0.000010000$ for Y, $R0 = 0.000050000$ for Ni, $R0 = 0.000100000$ for O, atomic sphere radius: $RMT = 2.23000$ for Y, $RMT = 1.89000$ for Ni (the same radius used to compute the magnetic momenta), $RMT = 1.68000$ for O.

We asked for simultaneous charge convergence to 0.00001 units and force convergence to $1.0 mRy/bohr$. SCF was run with $RK\max = 6$ and 8 k-points in irreducible Brillouin zone.

In order to describe the strong electronic correlation of the material, we use as other authors a phenomenological on-site Hubbard like correction within the LDA+U or GGA+U formalism. [12-14]. The values of U and J were fixed close to received value from the literature for this family of compounds, namely at $U = 5eV$ and $J = 0eV$. We could check after the SCF a resulting estimated U of 4.7 eV, so that the procedure is almost self-coherent. A trial doubling of the number of k-points in the irreducible Brillouin zone did not change this result significantly, so we kept it at 8 to limit computational time. All the calculations were performed on the 10 core Linux computational cluster of LPEC laboratory.

4. Results and discussion

Results are summarized on Table 2, where we compare the results of using the LDA+U and GGA+U methods as implemented in the *WIENncm* package:

Table 2. Magnetic moments on nickel atoms in $YNiO_3$.

Ions	Experiment	LDA+U	GGA+U
Ni^{2+}	$\pm 1.4\mu_B$	$\pm 1.6\mu_B$	$\pm 1.6\mu_B$
Ni^{4+}	$\pm 0.7\mu_B$	$\pm 0.2\mu_B$	$\pm 0.8\mu_B$

Experimental data from Ref. 5.

Analyzing the results, it can be concluded, that the results closer to experiments are given by the GGA+U method, which is not surprising due to the fact that GGA+U has been developed to depend explicitly on the gradient of the density in addition to the density itself. However, the LDA+U method is the least computationally demanding approach.

One can note that the results are better than the ones of [15], which were using an AFM collinear modeling of the system with a large supercell, and could only reproduce the momenta of one of the nickel species (Ni^{2+}), finding a null momentum for the other species, and closer to experiment than the ones of [9] which find momenta of 1.25 and 0.28 μ_B .

Alltogether, the magnetic structure of $YNiO_3$ that we computed can be observed on Fig. 3 and 4.

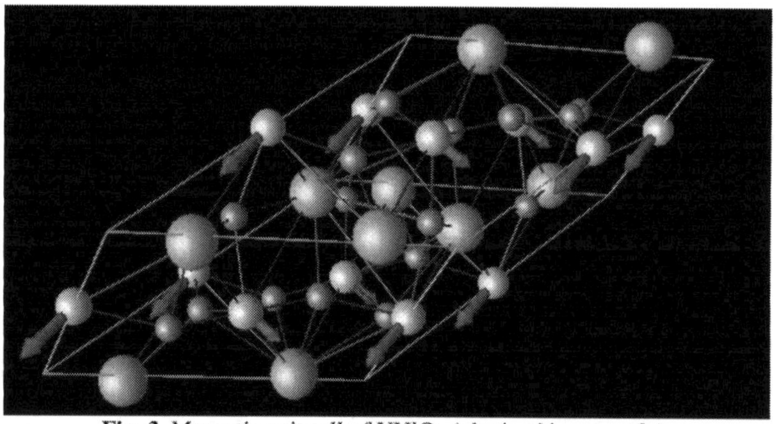

Fig. 3. Magnetic unit cell of $YNiO_3$ (obtained by *struct2x*).

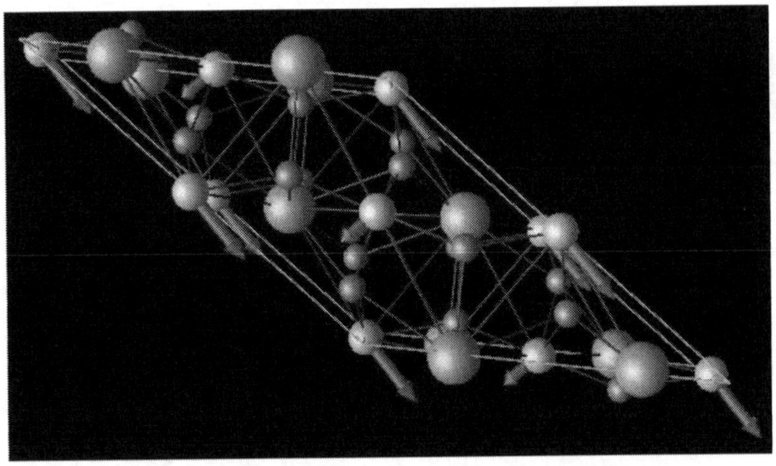

Fig. 4. Magnetic unit cell of $YNiO_3$ (obtained by *struct2x*).

The compounds can be ferromagnetic (FM), where the spins are aligned in parallel directions; antiferromagnetic (AFM), where the spins are antiparallel. This system is noncollinear, where the spins are not parallel and are partially disordered.

In large distorting perovskite such as $YNiO_3$, the charge disproportionation $2Ni^{3+} \rightarrow Ni^{2+} + Ni^{4+}$ occurs. The charge ordering mechanism can explain the stabilization of the local magnetic moments around each nickel site.

5. Conclusion

In summary, we have studied the magnetic properties of the Ni lattice in the insulator $YNiO_3$. The GGA+U calculations gives the best results for magnetic moments, close to experimental results and improving previous studies supposing antiferromagnetic collinear order. Further work could be done to study the effects of cell distortion on the magnetic order, to simulate epitaxied thin films effects or pressure effects; a fully self-consistent GGA+U procedure could also be tried. The whole family of compounds could be studied and compared to experimental results for the Nd substitution for instance where experimentalists using resonant soft x-ray diffraction also suspect a noncollinear spin structure [16].

References:

1. Laffez P., Zaghioui M., Retoux R. and Lacorre P., 2000 *Journal of Magnetism and Magnetic Materials* **211** 111.
2. Alonso J. A. et al. 2001 *Phys.Rev.*B **64** 094102.
3. Capon et al., 2005 *J.Phys : Condens.Matter* **17** 1137.
4. Hobbs D., Kresse G. and Hafner J., 2000 *Phys.Rev.*B 62, 11556.
5. Alonso J. A. et al, 1998 *Phys.Rev.Lett.* **82**, 3871.
6. Kokalj A., Comp. Mater. Sci., 2003, Vol. 28, p. 155.
7. Yamamoto Y. and Fujiwara T, 2002 *J.Phys.Soc.Jpn.*, **71**, 1226.
8. Yamamoto S. and Fujiwara T., 2002 *J.Phys.Chem.Sol*, **63**, 1347.
9. Mazin I. I. et al, 2007 *Phys.Rev.Lett.* **98,** 176406.
10. Barth U. and Hedin L., 1972 *J.Phys.*C. **5**, 1629.
11. Kohl C. and Bertsch G., 1999 *Phys.Rev.*B **60** 4205.
12. Calvayrac F., *Comp.Mat.Sci.* **17**, 164 (2000).
13. Blaha P., 2001 *WIEN2k*, Technische Universität Wien, Austria.
14. Laskowski R., 2006 *Non-collinear version of WIEN2k package*, November 16.
15. http://www.crystallography.net/
16. Scagnoli V. et al, 2008 *Phys.Rev.*B **77**, 115318.

Optical absorption and photoluminescence properties of ZnO/PMMA nanocomposite films

B Kulyk[1], V Kapustianyk[1], O Krupka[2], B Sahraoui[3]

[1]Department of Physics, Scientific and Educational Center "Fractal", Scientific-Technical and Educational Center of Low Temperature Studies, Ivan Franko National University of L'viv, 50 Dragomanova Str., L'viv, Ukraine
[2]Department of Chemistry, Kyiv Taras Shevchenko National University, 60 Volodymyrska Str., Kyiv, Ukraine
[3]Department of Physics, University of Angers, 2 Lavoisier Av., Angers, France

E-mail: bohdan_kulyk@yahoo.com

Abstract. The ZnO nanocrystals (ZnO NCs) with particle size, less than 100 nm, have been blended with polymethylmethacrylate (PMMA) by solution mixing to prepare PMMA/ZnO nanocomposite films. The structure of ZnO/PMMA nanocomposite films was characterized using X-ray diffractometry. The prepared nanocomposite films are highly transparent and a clear excitonic peak is observed in their absorption spectra. Measurements of temperature evolution of the photoluminescence (PL) spectra show intensive UV emission peak corresponding to the donor-bound excitons with binding energy of 51 meV and green emission band related to the intrinsic defects in ZnO. The temperature evolution of the emission peaks energy position, intensity and integral intensity in ZnO/PMMA nanocomposite films were examined.

1. Introduction

Semiconductor nanoparticles are receiving much attention owing to their novel physical and chemical properties. In comparison with the bulk semiconductor they possess many specific properties such as ultrafast optical nonlinear response, photoelectricity switch and attractive piezoelectric properties. The recent high-speed development of biotechnology, energy sources, environment and advanced manufacture technology urgently require considerable material miniaturization, high integration and high-density transmission, that imply a broad application for semiconductor nanoparticles.

It is expected that ZnO nanoparticles should possess some characteristics close to those for the bulk zinc oxide and the others to be quite another in nanoscale system. First of all this concerns a large exciton binding energy (nearly 60 meV) and excellent stability. At the same time, ZnO nanoparticles have attracted much attention due to the strong commercial desire for photocatalysis, photoelectrochemistry, blue and UV light emitters and detectors based on their specific properties. Indeed, it is well known that small particles have the large surface-to-volume ratio that strongly modifies the physical processes in the nanocomposites. The PL spectra of the ZnO nanoparticle samples usually show a nearband-edge (NBE) UV line accompanied by a strong visible luminescence, which can result in decrease of the carrier/exciton lifetime and higher emission efficiency of the UV light emitting devices. Therefore, the most crucial aspects of high luminescence efficiencies from ZnO nanoparticles are the surface texture and an efficient surface passivation. Fortunately, polymer capping can effectively passivate the surface of nanoparticles and reduce the surface-related visible emission [1,2].

Generally, the UV emission is attributed to the exciton radiative recombination while that of the visible emission still remains uncertain. Various mechanisms have been proposed, in fact, most probable origin of the visible PL is caused by the intrinsic defects. The density of these defects has to be higher on the surface of the sample, first of all due to the zinc and oxygen vacancies. In the case of nanoparticles the influence of surface is expected to be more pronounced due to the increased surface-to-volume ratio.

In this paper we report the method of ZnO/PMMA nanocomposite film preparation as well as the investigation of optical absorption and temperature evolution of PL in order to compare the features of ZnO/PMMA composite film with the bulk or nanostructured ZnO.

2. Experimental details

The investigated samples were manufactured on the basis of ZnO nanocrystals embedded into PMMA polymeric matrix. For preparation of the nanocomposite films the nanopowder from Sigma-Aldrich (<100 nm in size) have been used. The glass substrates used were carefully cleaned in a commercial surfactant using ultrasonication. The cleaning procedure was ended by baking in a 200°C oven for 60 minutes. Spin coating technique was used to fabricate nanocomposite films with ZnO nanocrystals dispersed in PMMA. The solution of 1,2,2-trichloroethane containing PMMA 100 g/L and ZnO was coated on BK7 glass slides. The weight concentration of ZnO nanocrystals in PMMA were 5, 10 and 15%. The principle of deposition (of the mixture with certain viscosity) is based on a homogeneous spreading out of the solution on the rotating substrate with an angular speed of 1000 rpm during 60 s. Immediately after the deposition, the thin films were cured in an 60°C oven for 60 minutes in order to eliminate any remaining solvent. The thickness of nanocomposite films was verified by a Dektak surface stylus profiler (Veeco) to be about 1 μm.

The XRD measurements were carried out using STOE STADI P diffractometer with linear position sensitive detector in transmission Bragg-Brentano geometry (CuKα1 radiation, Ge(111) monochromator, 2θ range: 20-90°, step - 270 seconds).

The optical absorption spectra in visible and near UV region were measured using a Lambda 19 spectrometer from PerkinElmer. The temperature evolution of the PL spectra near the intrinsic absorption edge and in a visible region was studied under excitation with a LGI-21 nitrogen laser at 337 nm emission wavelength with10 ns pulse duration and the power of 1 kW.

3. Results and discussion

3.1. *XRD data analysis.* Figure 1 shows X-ray diffraction pattern of the ZnO/PMMA (15 wt.% of ZnO NCs) nanocomposite film. The observed set of sharp diffraction peaks corresponds to the hexagonal wurtzite structure of ZnO. Detailed description of the observed diffraction peaks, their angular position and relative intensity towards most intensive (101) peak are given in Table 1. Intensive and sharp diffraction peaks of XRD pattern in ZnO/PMMA nanocomposite film are similar to that in ZnO bulk. This confirms that we deal with the nanoparticles of quite large sizes as compared with ZnO quantum dots. The large amount of diffraction peaks corresponding to various crystalline orientations indicates that ZnO NCs within PMMA are disordered and do not have preferred axis orientation.

The average size of the particles can be calculated by the Debye-Scherrer formula, i.e.:

$$D = \frac{0.9\lambda}{B\cos\theta_B}, \tag{1}$$

where D is the average size of the particle, λ is the X-ray wavelength (1.54 Å), B is the FWHM width of the diffraction peak, and θ_B is the corresponding diffraction angle of the diffraction peak. According to the data presented in Fig. 1 the average particle size of the ZnO nanocrystals calculated on the basis of Eq.(1) for the (002), (101) and (100) diffraction peaks was found to be about 60÷65 nm.

Journal of Physics: Conference Series **289** (2011) 012003 doi:10.1088/1742-6596/289/1/012003

Table 1 Angular position and relative intensity of the diffraction peaks in XRD pattern of ZnO/PMMA (15 wt.% of ZnO NCs) nanocomposite film.

Diffraction peak	Peak position 2θ (deg)	Peak intensity (%)
(100)	31.79	55.2
(002)	34.44	51.0
(101)	36.27	100
(102)	47.57	23.8
(110)	56.62	35.4
(103)	62.91	32.8
(200)	66.43	6.1
(112)	68.00	26.2
(201)	69.13	14.2
(004)	72.62	3.3
(202)	77.01	5.0
(104)	81.45	2.9
(203)	89.67	10.3

Table 2 Parameters of Varshni formula for the donor-bound excitons in single crystals and nanocrystals of ZnO.

	$E(0)$ [eV]	α [meV/K]
ZnO bulk [7]	3.36	0.67
ZnO NCs	3.32	1.1

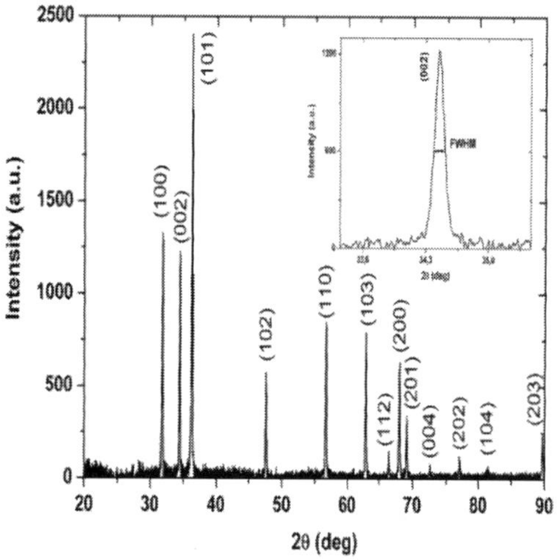

Figure 1 X-ray diffraction pattern of ZnO/PMMA (15 wt.% of ZnO NCs) nanocomposite film. In inset – the FWHM of (002) diffraction peak.

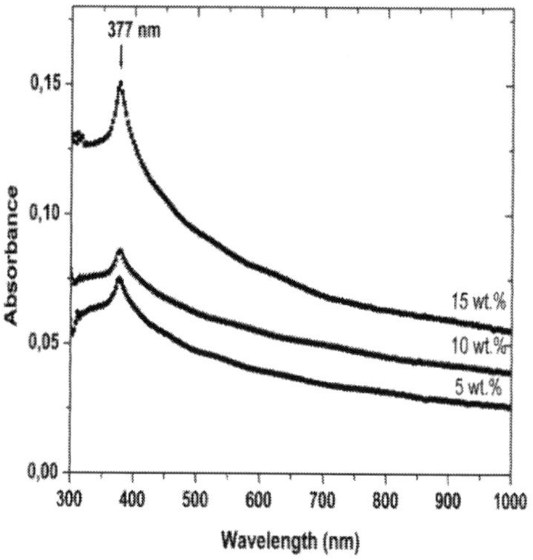

Figure 2 UV-vis spectra of ZnO/PMMA nanocomposite films with different ZnO NCs concentrations: 5, 10, and15 wt.%. The thickness of all films is around 1 μm.

3.2 *Optical absorption measurements.*

The absorption spectra of ZnO/PMMA nanocomposite films shown in Fig. 2 characterize by overall nonzero absorbance in visible region connected with partial light scattering on agglomerated particles. This circumstance does not allow observing the sharp absorption edge as for the bulk or thin film ZnO. The spectra clearly demonstrate the absorption peaks at 377 nm (3.29 eV) corresponding to the exciton state inherent to the bulk ZnO [3]. Since the size of ZnO NCs is nearly 100 nm that is much more than the exciton Bohr radius in ZnO (a_B=2.34 nm [4]) we do not observe in our NCs system any shift of the corresponding peak as manifestation of the quantum confinement effect. The intensity of the peaks enhances with increasing of ZnO NCs concentration.

3.3. *PL spectra temperature evolution.*
The Fig. 3 shows the temperature evolution of PL spectra in ZnO nanocrystals embedded into polymeric matrix PMMA in the temperature range from 85 to 290 K. Analysis of the low-temperature PL in ZnO NCs and its comparison with PL in bulk ZnO permits to define the nature and characteristics of the recombination processes in the nanoscale ZnO crystals near the absorption edge. The PL spectra are characterized by intensive emission in the near-UV region and lower emission in the green region of spectrum (507 nm) connected with the presence of intrinsic defects [5,6]. The recombination processes in ZnO NCs differ from that in bulk ZnO mainly, because of the larger surface-to-volume ratio and, therefore, larger amount of defects on the surface. Basing on the previous investigation of PL in NCs [7], the obtained experimental results assume that the UV PL in ZnO NCs in the temperature range 85-290 K emerges in consequence of recombination of donor-bound excitons (D,X). The maximum of UV peak emission correspond to 382 (3.25 eV) nm at 290 K and its intensity grows with temperature decreasing meanwhile it shifts to shorter wavelength. Besides, UV band has an asymmetric shape with a clear shoulder on the high-energy side, which indicates the presence of the free excitons emission peak with the energy of 3.30 eV at 290 K extracted from spectrum using the method of derivative spectrophotometry.

The UV emission band from the side of lower energy has the step-like shape with the clearly defined equidistant peaks. Energy distant between these peak is approximately equal to 70 meV, which corresponds to the energy of LO-phonon in ZnO [8,9]. The presence of the LO-phonon peaks define a large exciton-phonon interaction in ZnO NCs, which enhances with temperature decreasing and appears in temperature broadening of UV emission band.

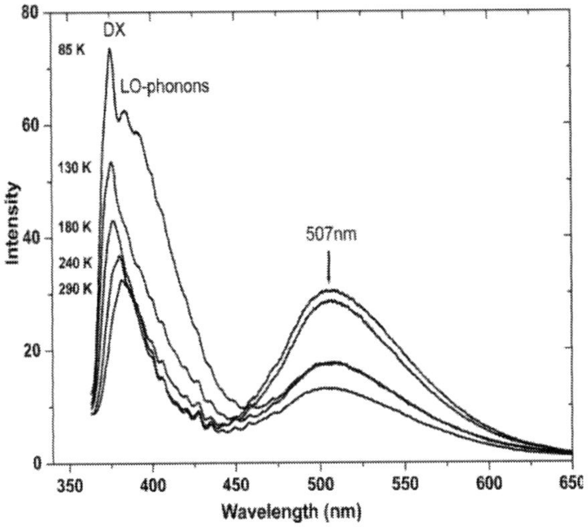

Figure 3 PL spectra of ZnO NCs embedded into PMMA matrix at 15 wt.% obtained in temperature range 85-290 K.

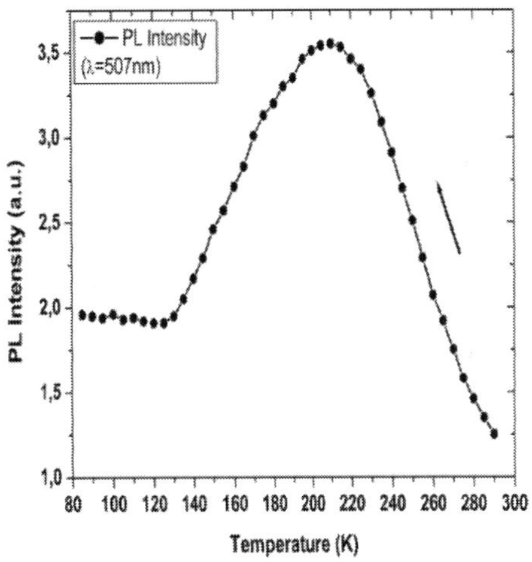

Figure 4 Temperature evolution of green emission band at 507 nm in ZnO NCs.

The temperature evolution of green emission band in ZnO NCs is shown in Fig. 4. Increasing of the full integral intensity of the band at cooling is observed due to the suppression of quenching mechanism – thermal activation of nonemitting centers and charge carrier recombination on localized center [10]. The energy position of the band does not depend on temperature in the investigated range. At the same time it is necessary to mention the peculiar temperature dependence of this band intensity with the clear maximum at 210 K, which is not inherent for ZnO. Such a behavior can be attributed to the temperature dependant properties of PMMA, first of all with nonlinear behavior of its electret field. As it is known from the thermally stimulated discharge data in PMMA at the 220-230 K the relaxation processes causing a partial orientational relaxation of dipoles take place. The dipoles of the polar ester side groups (—$COOCH_3$) reorient by local motion around the C-C bond [11,12]. The nature of the "green" band is connected with the deep centers of emission caused by zinc V_{Zn} [13,14] and oxygen vacancies V_O [15,16] and interstitial atoms of zinc (Zn_i) [17].

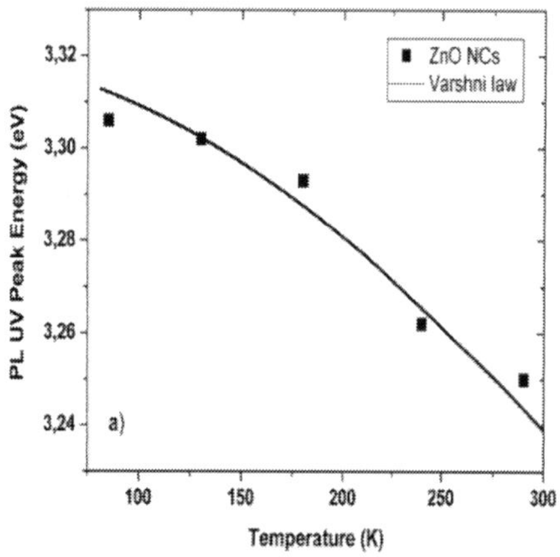

Figure 5a Energy position with approximation by Varshni law

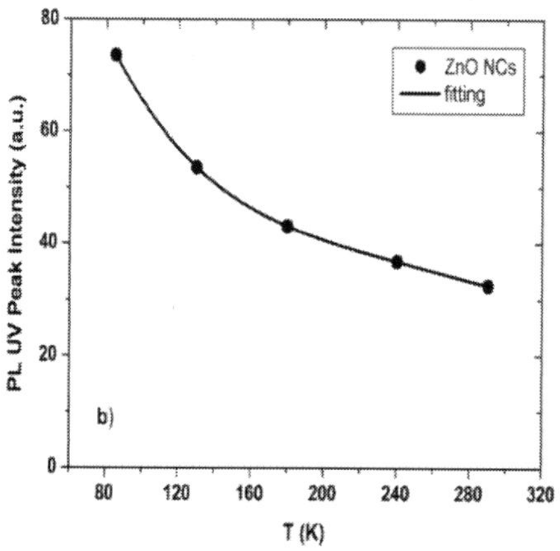

Figure 5b intensity of donor-bound exciton emission peak as a function of temperature.

In bulk ZnO the UV emission peak corresponding to the recombination of the free excitons was found to be at higher energy (about 3.3 eV) in comparison with ZnO NCs, where UV PL is caused by recombination of bound excitons. The difference in 51 meV between bound exciton peak in ZnO NCs and free exciton peak in ZnO bulk is close to the bound exciton binding energy – 50 meV [18-21].

In the Fig. 5 the UV emission peak energy position and its intensity are given depending on temperature. It can be seen that the emission intensity of donor-bound excitons and their energy diminish with heating. Depending on the temperature, the localization energy of donor-bound exciton varies from 50 to 100 meV in comparison with free exciton energy in bulk ZnO [7].

Significant diminution of the donor-bound exciton energy with temperature should be related to the presence of different kind of donors in ZnO NCs, for example, oxygen vacancies, interstitial zinc atoms. The width of UV peak also depends on the concentration of the oxygen vacancies responsible for a set of donor levels. Their energy should be different for the bulk of nanoparticle and its surface.

In the Fig. 5(a) the temperature dependence of the excitonic peak energy position was approximated by semiempirical formula of Varshni [22]:

$$E(T) = E(0) - \frac{\alpha T^2}{T + \theta_D}, \qquad (2)$$

where $E(0)$ is energy at T=0 K, α – parameter, θ_D – Debye temperature.

Table 2 presents the parameters obtained from approximation by Eq.(2), where θ_D=920 K for the recombination processes [23]. The calculated values of $E(0)$ and parameter α for the UV peak correlate well with the values obtained for single crystals and thin films of ZnO [7,21,24].

The integral intensity of UV emission of PL decreases with temperature increasing due to the effect of thermal quenching and is described by expression [25]:

$$I(T) = \frac{I_0}{1 + A\exp\left(-\frac{E_a}{k_B T}\right)}, \tag{3}$$

where I_0 is the integral intensity at T=0 K, A is the constant, E_a is the activation energy in the process of thermal quenching, k_B is the Boltzmann constant. Eq. 3 clarifies influence of binding energy of exciton on the emission intensity of PL, which grows almost exponentially with increase of binding energy at permanent temperature.

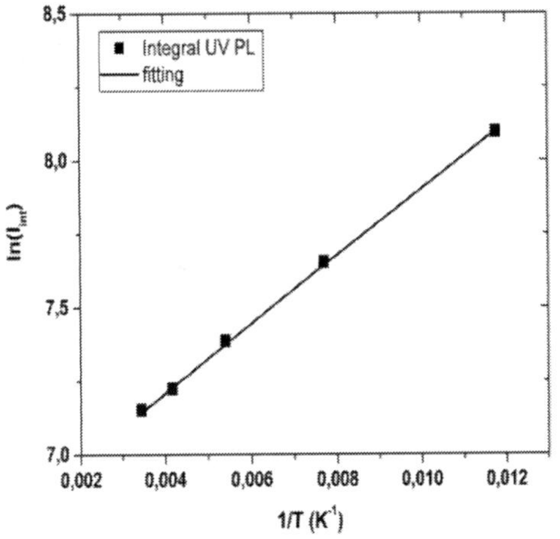

Figure 6 Log plot of the integral UV PL intensity as a function of the inverse temperature (from 290 to 85 K). Solid line gives linear fitting.

Fig.6 depicts the dependence of the integral PL intensity of UV band in logarithmic scale on the inverse temperature is shown. The solid line corresponds to fitting of results and possesses a linear character in the investigated temperature region.

4. Conclusions

In the present work, the high-quality ZnO/PMMA nanocomposite films were prepared by embedding of ZnO NCs into PMMA polymeric matrix. The films show high optical transmittance in near UV-vis region and presence of the near-band excitonic peak. The temperature study of PL in ZnO/PMMA nanocomposite films demonstrates the intensive UV emission, which was connected with donor-bound excitons and lower green emission due to the presence of internal intrinsic defects. When the ZnO structure changes from single crystal to the nanostructures (nanorods, nanotubes etc.) and then to the nanocomposite film the intensity of UV exciton related band enhance in respect to the intensity of the green band and its energy position could be changed due to the variation of the localised excitonic states corresponding to the different kinds of defects in these three types of materials. The unusual behaviour of green band temperature dependence was explained by nonlinear properties of electret field in PMMA caused by reorientation of dipoles within PMMA polymer. Besides, the temperature characteristics of UV band were studied, the binding energy of donor-bound excitons was found to be 51 meV at 290 K.

Acknowledgments
The XRD measurements were performed in "Interdepartmental scientific-educational laboratory of XRD analysis" of Ivan Franko National University of L'viv.

References

[1] Yang C L, Wang J N, Ge W K, Guo L, Yang S H and Shen D Z 2001 *J. Appl. Phys.* **90** 4489.
[2] Tong Y H, Liu Y C, Lu S X, Dong L, Cheng S J and Xiao Z Y 2004 J. *Sol-Gel Sci. Technol.* **30** 157-161.
[3] Özgür Ü, Alivov Ya I, Liu C, Teke A, Reshchikov M A, Doğan S, Avrutin V, Cho S-J and Morkoç H 2005 *J. Appl. Phys.* **98** 1-103.
[4] Lin K F, Cheng H M, Hsu H C, Lin L J and Hsieh W F 2005 *Chem. Phys. Lett.* **409** 208-211.
[5] Meyer B K, Alves H, Hofmann D M, Kriegseis W, Forster D, Bertram F, Christen J, Hoffmann A, Strassburg M, Dworzak M, Haboeck U and Rodina A V 2004 *Phys. Status Solidi* B **241** 231.
[6] Cho S 2006 *J. Korean Phys. Soc.* **49** (3) 985-988.
[7] Fonoberov V A, Alim K A, Balandin A A, Xiu F and Liu J 2006 *Phys. Rev.* B **73** 165317.
[8] Calleja J M and Cardona M 1977 *Phys. Rev.* B **16** 3753-3761.
[9] Alim K A, Fonoberov V A and Balandin A A 2005 *Appl. Phys. Lett.* **86** 053103.
[10] Li Q, Xu S J, Xie M H and Tong S Y 2005 *Europhys. Lett.* **71** 994-1000.
[11] Mazur K 1997 *J. Phys.* D: Appl. Phys. **30** 1383-1398.
[12] Kubon U, Schilling R, Wendorff J H 1988 *Colloid. Polym. Sci.* **266** 123-131.
[13] Kohan A F, Ceder G, Morgan D and Van de Walle C G 2000 *Phys. Rev.* B **61** 15019-15027
[14] Guo B, Qiu Z R and Wong K S 2003 *Appl. Phys. Lett.* **82** (14) 2290-2292.
[15] Studenikin S A, Golego N and Cocivera M 1998 *J. Appl. Phys.* **84** (4) 2287-2294.
[16] Leiter F H, Alves H R, Hofstaetter A, Hoffmann D M and Meyer B K 2001 *Phys. Stat. Sol.* (b) **226** (1) R4-R5.
[17] Korsunska N O, Borkovska L V, Bulakh B M, Khomenkova L Yu, Kushnirenko V I and Markevich I V 2003 *J. Lumin.* **102-103** 733-736.
[18] Zhang X T, Liu Y C, Zhi Z Z, Zhang J Y, Lu Y M, Shen D Z, Xu W, Fan X W and Kong X G 2002 *J. Lumin.* **99** 149.
[19] Chen J and Fujita T 2002 *Jpn. J. Appl. Phys.* Part 2 **41** L203.
[20] Gorla C R, Emanetoglu N W, Liang S, Mayo W E, Lu Y, Wraback M and Shen H 1999 *J. Appl. Phys.* **85** 2595.
[21] Ko H J, Chen Y F, Zhu Z, Yao T, Kobayashi I and Uchiki H 2000 *Appl. Phys. Lett.* **76** 1905.
[22] Varshni Y P 1967 *Physica* (Amsterdam) **34** 149.
[23] Hutson A R 1959 *J. Phys. Chem. So*lids **8** 467.
[24] Loughin S, French R H, Ching W Y, Xu Y N and Slack G A 1993 *Appl. Phys. Lett.* **63** 1182.
[25] Jiang D S, Jung H and Ploog K 1988 *J. Appl. Phys.* **64** 1371.

Journal of Physics: Conference Series **289** (2011) 012004 doi:10.1088/1742-6596/289/1/012004

Dielectric Mismatch at Finite Barrier Cubic Quantum Dots

V I Boichuk[1], I V Bilynsky[1], I O Shakleina[1] and I P Kogoutiouk[2]

[1]Drohobych State Pedagogical University, 24 Ivan Franko St., Drohobych, 8210, Ukraine

[2]Minnesota State University-Mankato, Mankato, MN 56001, USA

e-mail: ioshak@mail.ru

Abstract. We study the exciton states in the cubic quantum dots with finite potential barrier at the presence of dielectric mismatch. The dependence of electron and hole ground states as well as exciton binding energy and oscillator strength are found as a function of quantum dot size and dielectric constant of quantum dot and surrounding matrix. The developed theoretical approach has been applied to analyze the peculiarities of exciton spectra in *CdTe* quantum dots embedded in a variety of dielectric matrices.

1. Introduction

For the last two decades quantum dots (QDs), as well as other nanometer size structures have been attracting the increasing interest of experimental and theoretical researchers. From one side, such an interest is fueled by the possession by QD systems of unique physical properties that still require the thorough investigation and adequate explanation, and on the other hand by the perspectives of practical use of such materials, especially in the optoelectronic devices [1-4]. A variety of techniques was successfully developed to fabricate quantum dots: colloidal growth [5-7], molecular beam epitaxy [8], embedding into glass matrix by melting, quenching and subsequent nanocrystal growing [9]; twofold cleaved edge overgrowth technique [10] and some others. Great progress was also achieved in investigation of absorption and emission spectra of QDs and their behavior in electric and magnetic fields [11-14]. A number of studies have also been performed in order to explain the observed electronic and optical properties and theoretically model the physical processes in quantum dots [15-25].

Most of the existing theoretical investigations of excitons in QDs are based on some of the few simplifying approaches. Among them are:

1. An effective mass approach, which allows the use of theoretical techniques developed for the bulk semiconductor materials.
2. An infinite barrier potential approximation, in which particle and exciton wave functions are localized inside QD and vanish beyond its boundary.

3. Strong confinement approximation (SCA) that treats electron-hole interaction as a perturbation to their kinetic energy and permits the presentation of exciton wave functions as the product of electron and hole parts. SCA validity range is limited to the size of QDs less than exciton Bohr radius.
4. Assumption of the same dielectric constants for QD and surrounding matrix, that neglects creation of induced image charges at boundaries.
5. Assumption of spherical shape of QDs, which simplifies calculations and allows presentation of the solution in the analytical form.

In a number of papers researchers tried to consider more realistic models and get beyond the indicated simplifications, but overcoming even a few of them simultaneously is not an easy procedure. In Ref. [15], for example, the finite potential barriers for electrons and holes were introduced while studying cubic QDs in strong confinement regime with the same dielectric constants for QD and matrix. The authors of Ref. [16] went beyond SCA by using Ritz variational principle for investigating disc shape QD with no dielectric mismatch taken into account. The mismatch of dielectric constants between QD and surrounding matrix was accounted for in Ref. [17], where spherical and cubic shape QDs were considered in strong confinement regime and infinite barrier approximation.

It should be noted that the effective mass approximation (EMA) was questioned, taking into account the limited size of QD, and violation of the long range periodicity in it [18]. Evidently, it should be some small size limit for validity of EMA. By comparison of EMA results with data obtained through tight binding model and some other theoretical methods as well as with the experimental measurements, it was shown in Ref. [18,19], that EMA in the finite potential barrier model, can be applied to QDs as small as 15-20 Å. Partially it can be explained by the fact that exciton properties are obtained by averaging envelope functions and, therefore, are not so sensitive to small details of one-particle energy spectrum. There were also a few attempts to improve EMA by considering energy and position dependent electron and hole effective masses [20], or by introducing to QD effective masses much heavier than for bulk crystals that are anisotropic in case of large aspect ratio [21].

In this paper we consider a combined effect of finite barrier, dielectric mismatch, beyond the strong confinement approximation approach, and cubic shape on electronic and optical properties of QDs.

2. Theory

Let's consider cubic shape QD of size b with dielectric constant ε_Q that is embedded in dielectric matrix with dielectric constant ε_M. In this paper we consider quantum dots that are large enough (≥ 20 Å) to assume the validity of effective mass approximation, as well as to assign the QD the same value of the dielectric constant as for the appropriate bulk material. Such an approach was questioned in Ref. [18], and authors have shown that the effect of the permittivity change due to the limited volume of QD, and formation of image charges on the size dependent properties of QD is not significant. Following Ref. [21], we also neglect the complex nature of the electron and hole energy spectrum (in the case of *CdTe*, for example, only dominating heavy holes contribution is taken into consideration) and employ the single band effective mass approach. The considered system is described by the electron-hole Hamiltonian

$$H = H_e(\vec{r}_e) + H_h(\vec{r}_h) + W(\vec{r}_e, \vec{r}_h)$$
$$= -\frac{\hbar^2}{2}\nabla_e\left(\frac{1}{m_e(\vec{r}_e)}\nabla_e\right) + V_e(\vec{r}_e) - \frac{\hbar^2}{2}\nabla_h\left(\frac{1}{m_h(\vec{r}_h)}\nabla_h\right) + V_h(\vec{r}_h) + W(\vec{r}_e, \vec{r}_h), \qquad (1)$$

where $\vec{r}_e$, $\vec{r}_h$; $m_e(\vec{r}_e)$, $m_h(\vec{r}_h)$; and $V_e(\vec{r}_e)$, $V_h(\vec{r}_h)$ are coordinates, coordinate-dependent effective masses and finite confinement potentials for electrons and holes, respectively:

$$m_{e,h}(\vec{r}_{e,h}) = \begin{cases} m_{Q,e,h}, & |x_{1,2,3}| \le b/2,\ x_1 = x,\ x_2 = y,\ x_3 = z \\ m_{M,e,h}, & otherwise \end{cases},$$
$$V_{e,h}(\vec{r}_{e,h}) = \begin{cases} -V^0_{e,h}, & |x_{1,2,3}| \le b/2 \\ 0, & otherwise \end{cases}. \tag{2}$$

In some works on excitons in QD electron-hole interaction potential $W(\vec{r}_e,\vec{r}_h)$ was chosen in the Coulomb form [16, 22] which can be justified in the strong confinement approximation when dot size is considerably smaller than exciton Bohr radius: $b \ll a_B$. When QD size becomes comparable with a_B (for $CdTe$, for example, $a_B = 69$ Å), the electrostatic interaction potential is affected by the image charges induced at the boundary of QD and surrounding matrix, which leads to the appearance of the self-polarization correction term $W_S(\vec{r}_e,\vec{r}_h)$:

$$W(\vec{r}_e,\vec{r}_h) \equiv W_C(\vec{r}_e,\vec{r}_h) + W_S(\vec{r}_e,\vec{r}_h) = -\frac{e^2}{\varepsilon_Q|\vec{r}_e-\vec{r}_h|} + W_S(\vec{r}_e,\vec{r}_h). \tag{3}$$

It should be mentioned that in the case of common dielectric mismatch, when matrix permittivity is considerably less then the one of QD, the electron-hole interaction in QD increases substantially compared to the bulk crystal [23].

The electron-hole interaction potential can be obtained from the Poisson equation with given boundary conditions. Such an analytical solution for cubic QD with infinite confinement potential was found in Ref. [17], but it appeared to be divergent with increase of QD to matrix permittivity ratio, so authors of [17] had to mend it with the addition of "correcting function", which appears to have an approximately constant value for the set of parameters considered there. In addition, electrostatic potential found in [17] is divergent at QD interfaces, which creates a problem for performing integration in the case of finite potential barriers.

Another way to proceed with analysis would be to follow the challenging numerical calculations using, for example, finite difference method [24], but results achieved in such a way are limited to the specific material parameters.

In our current work we propose to approximate the potential term $W_S(\vec{r}_e,\vec{r}_h)$ for the cubic QD by the composition of three potentials for mutually orthogonal quantum wells:

$$W_S(\vec{r}_e,\vec{r}_h) = W_{S,xy}(\vec{r}_e,\vec{r}_h) + W_{S,yz}(\vec{r}_e,\vec{r}_h) + W_{S,xz}(\vec{r}_e,\vec{r}_h). \tag{4}$$

The analytical solution for single nanolayer (quantum well) located in X-Y plane and surrounded by the dielectric matrix was found in [23, 25] to have a form:

$$W_{S,xy}(\vec{r}_e,\vec{r}_h)=e^2\left(1-\frac{\varepsilon_M}{\varepsilon_Q}\right)\times$$

$$\int_0^\infty dk\left[J_0(k\rho)\frac{e^{-k(z_e+z_h)}\left(\left(e^{2kz_e}+e^{2kz_h}\right)\left(-\varepsilon_M+\varepsilon_Q\right)+\left(e^{bk}+e^{k(b+2(z_e+z_h))}\right)\left(\varepsilon_M+\varepsilon_Q\right)\right)}{e^{2bk}\left(\varepsilon_M+\varepsilon_Q\right)^2-\left(\varepsilon_M-\varepsilon_Q\right)^2}\right],$$

(5)

where $\rho=\sqrt{(x_e-x_h)^2+(y_e-y_h)^2}$; J_0 - Bessel function of the first kind. Potential $W_S(\vec{r}_e,\vec{r}_h)$ appears due to the dielectric mismatch which leads to the presence of connected interface charges, so $W_S(\vec{r}_e,\vec{r}_h)_e\approx 0$, if $\varepsilon_Q\approx\varepsilon_M$. Similar expressions can be written for two other planes by interchanging the appropriate coordinates.

There is no exact solution for the Schrödinger equation in the case of cubic QD and finite confinement potential, even for one-particle Hamiltonian, so we will employ the variational method. We start from finding the approximate solution of one-particle Schrödinger equation:

$$\hat{H}_{e,h}\,\phi_{e,h}\left(\vec{r}_{e,h}\right)=E_{e,h}\,\phi_{e,h}\left(\vec{r}_{e,h}\right), \tag{6}$$

and then consider one-particle wave functions $\phi_e(\vec{r}_e)$ and $\phi_h(\vec{r}_h)$ as a basis for finding the connected electron-hole wave function $\Psi(\vec{r}_e,\vec{r}_h)$.

Following Ritz variational procedure, the trial ground state one-particle (electron or hole indexes are dropped for brevity) wave function has been chosen in the form:

$$\phi(\vec{r})=\prod_{i=1}^{3}\phi_i(x_i), \tag{7}$$

where for cubic crystal:

$$\phi_i(x_i)=\begin{cases}C_1\cos(\beta x_i),|x_i|\le b/2\\ C_2\exp(-\gamma|x_i|),|x_i|>b/2\end{cases}, \tag{8}$$

β and γ, $(\gamma>0)$ are variational parameters. The total one-particle function consists from QD and matrix parts contributing with the weight coefficients C_1 and C_2, respectively, that can be found from the conditions of normalization and continuation.

Using connection rules, i.e. continuity of the wave function and component of the velocity normal to the interface, we obtain the system of uniform equations. It can be shown that such a system has a nontrivial solution if variational parameters satisfy the condition:

$$\gamma = \begin{cases} \frac{m_M}{m_Q}\beta\, tg(\frac{\beta b}{2}), & \frac{\pi}{b}n_i \le \beta \le \frac{\pi}{b}(n_i+1), \quad n_i = 0,1,2...,i=1,2,3 \\ -\frac{m_M}{m_Q}\beta\, ctg(\frac{\beta b}{2}), & otherwise \end{cases} , \tag{9}$$

Performing minimization of the energy of electron (hole) system we found the appropriate sets of variational parameters β and γ, as well as constants C_1 and C_2. It should be mentioned that there is a difference between found parameters for electrons and holes in view of different effective masses and confinement potentials for them. At the same time, because of the cubic symmetry of the system, found variational parameters and constants have the same value for all of the directions of coordinate axes.

The ground state of bound electron-hole pair can be found from Schrödinger equation:

$$\hat{H}\Psi(\vec{r}_e,\vec{r}_h) = E\Psi(\vec{r}_e,\vec{r}_h), \tag{10}$$

We again apply Ritz variational theory to search for the approximate solution. The trial wave function has been chosen in the form:

$$\Psi(\vec{r}_e,\vec{r}_h) = \phi(\vec{r}_e)\,\phi(\vec{r}_h)\exp\{-\alpha|\vec{r}_e-\vec{r}_h|\}, \tag{11}$$

where α – variational parameter, and electron-hole interaction is accounted by the exponential term.

The exciton energy can be determined by minimizing the functional:

$$E(\alpha) = \frac{\langle\Psi|\hat{H}|\Psi\rangle}{\langle\Psi|\Psi\rangle}. \tag{12}$$

Substituting an envelope function (11) into functional (12) and performing a minimization, we obtain the set of six- and seven-fold integrals that include Bessel functions. The multiplicity of such integration can be reduced at least by the order of two by taking into consideration the symmetry of the system, but even after such reduction the numerical minimization procedure still requires the challenging calculations, that are explained below.

For illustration and comparison purposes we will first consider the case of an infinite barrier that requires relatively simpler calculations. In order to simplify the computations, we introduce the "reduced" variables and constants, related to QD:

$$\tilde{\alpha} = \alpha\frac{b}{\pi}, \quad \tilde{b} = \frac{b}{a_0}, \quad \tilde{k} = \frac{kb}{\pi}, \quad \xi_{ie,h} = \frac{\pi}{b}x_{ie,h}, \quad \xi_i = \xi_{ie} - \xi_{ih},$$

$$a_0 = \frac{\varepsilon_Q\hbar^2}{\mu e^2}, \quad \frac{1}{\mu} = \frac{1}{m_e} + \frac{1}{m_h}, \quad R_0 = \frac{e^2}{2\varepsilon_Q a_0}.$$

After straightforward, but tedious, calculations we obtain the exciton energy expressed as a function of renormalized variational parameter $\tilde{\alpha}$:

$$E(\tilde{\alpha}) = T_0 + T_1(\tilde{\alpha}) + T_2(\tilde{\alpha}) + P_c(\tilde{\alpha}) + P_p(\tilde{\alpha}), \qquad (13)$$

The first three terms in expression (13) give the expectation value of the kinetic energy of the particles:

$$T_0 = \frac{3\pi^2\hbar^2}{2\mu\tilde{b}^2},\ T_1 = -\left(\frac{\pi}{\tilde{b}}\tilde{\alpha}\right)^2,\ T_2 = \frac{\pi^2}{\tilde{b}^2} B_0^{-1}(\tilde{\alpha})K(\tilde{\alpha}), \qquad (14)$$

where:

$$B_n(\tilde{\alpha}) = \int_0^\infty d\xi_1 \int_0^\infty d\xi_2 \int_0^\infty d\xi_3 f(\xi_1) f(\xi_2) f(\xi_3) t^n e^{-2\tilde{\alpha}t},\quad n = 0, -1,$$

$$f(\xi_i) = 2(\pi - \xi_i)(2 + \cos 2\xi_i) + 3\sin 2\xi_i;\ t = \sqrt{\xi_1^2 + \xi_2^2 + \xi_3^2}\ ;$$

$$K(\tilde{\alpha}) = \tilde{\alpha} B_0(\tilde{\alpha}) \frac{\exp(-2\tilde{\alpha}\sqrt{\xi_1^2 + \xi_2^2 + \xi_3^2}}{512(\xi_1^2 + \xi_2^2 + \xi_3^2)^{3/2}} \sum_{\substack{i,j,k=1 \\ (i \neq j \neq k)}}^{3} L_{ijk}(\tilde{\alpha}),$$

and

$$L_{ijk} = -f(\xi_i) f(\xi_j) \Big\{ \left[2\xi_i^2 + (\pi + \xi_i)(\xi_j^2 + \xi_k^2) \right] \cos(2\xi_i) + 3/2 (\xi_j^2 + \xi_k^2) \sin(2\xi_i) -$$
$$-2\left[2\xi_i^2 - (\pi - 2\xi_i)(\xi_j^2 + \xi_k^2) + \xi_i(\pi - \xi_i)(\xi_i^2 + \xi_j^2 + \xi_k^2) \sin(2\xi_i) \right].$$

The electron-hole Coulomb interaction is accounted by the fourth term in expression (13). It can be shown that due to cubic symmetry, all three terms of the potential (4) contribute equally to the fifth term of the potential energy $P_p(\alpha)$ that is related to the polarization effect. After some calculations we obtain:

$$P_p(\tilde{\alpha}) = P_1(\tilde{\alpha}) + P_2(\tilde{\alpha}), \qquad (15)$$

where:

$$P_1(\tilde{\alpha}) = \frac{1}{512} \int_0^\pi d\xi_1 \int_0^\pi d\xi_2 \int_0^\pi d\xi_3 \int_0^\infty d\tilde{k} f(\xi_1) f(\xi_2) f(\xi_3) J_0(\tilde{k}\sqrt{\xi_1^2 + \xi_2^2}) e^{-2\tilde{\alpha}t} \times$$
$$\times \left[\frac{2ch(\tilde{k}|\xi_3|)}{\exp\left[2(2\eta + \pi\tilde{k})\right] - 1} sh^{-1}(2\eta + \pi\tilde{k}) \right]$$

$$P_2(\tilde{\alpha}) = \frac{1}{64} \int_0^\pi d\xi_1 \int_0^\pi d\xi_2 \int_{-\pi/2}^{\pi/2} d\tilde{z}_e \int_{-\pi/2}^{\pi/2} d\tilde{z}_h \int_0^\infty d\tilde{k} f(\xi_1) f(\xi_2) J_0(\tilde{k}\sqrt{\xi_1^2 + \xi_2^2}) \times$$
$$\times \left[\frac{ch\left[\tilde{k}(\tilde{z}_e + \tilde{z}_h)\right]}{sh(2\eta + \pi\tilde{k})} \right].$$

For the finite confinement potential, the calculations become more complex and are reviewed in the Appendix. The detailed analysis of exciton energy dependence upon the properties of the QD and surrounding dielectric matrix is performed in the next section.

3. Numerical results and discussion

The numerical calculations were performed for *CdTe* QD with the parameters: $m_{eQ}=0.11m_0$; $m_{hQ}=0.35m_0$; $\varepsilon_Q=10.9$, where m_0 is the mass of free electron. In the experimental study [6] the QD was embedded in a variety of materials often with not precisely known electron and hole effective masses and magnitudes of the potential barrier, therefore in our calculations we consider a variety of values for the matrix confinement potentials as well as electron and hole effective masses (Fig.1-Fig.4). In order to study the properties of the system, we consider initially the electron and hole ground state energies (Fig.1), and later analyze the properties of exciton ground state.

As can be seen from Fig.1, electron and hole energies are decreasing functions of the QD size b. Because of the smaller electron effective mass, the corresponding energy dependence $E(b)$ has a steeper slope than for a hole. Such behavior is observed for both: an infinite (lines 1 and 4) and finite (lines 2, 3, 5 and 6) confinement potentials. We have considered two cases with the effective mass of electrons in the matrix to be $m_{eM}=0.11m_0$ (line 2) and $m_{eM}=m_0$ (line 3), and effective mass of the hole $m_{hM}=0.35m_0$ (line 5) and $m_{hM}=m_0$ (line 6), and as we see, the increase in the effective mass leads to decrease in the ground state energy of particles. It can be seen from Fig.1, that there is a substantial difference between results obtained in the infinite confinement potential and finite barrier models for $b<100\,\mathring{A}$ in the case of electrons, and $b\leq 40\,\mathring{A}$ for holes.

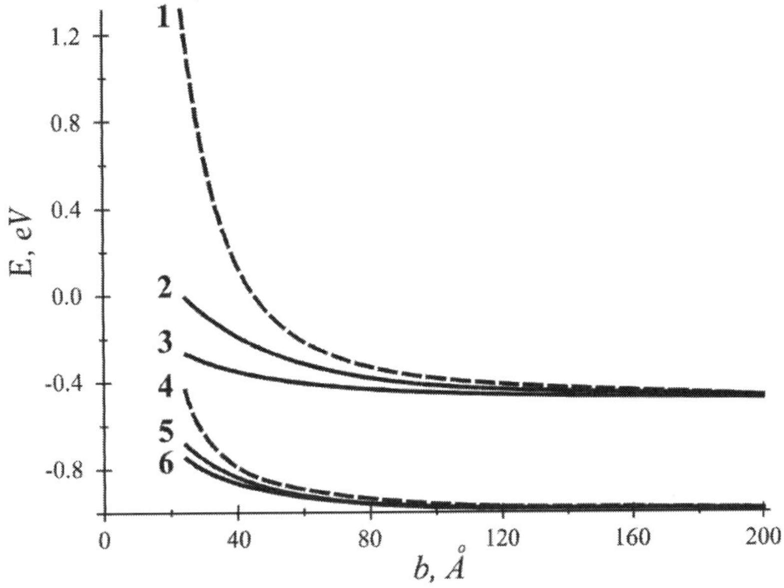

Figure 1. Electron (1,2,3) and hole (4,5,6) ground state energy versus QD size b. Dashed lines (1,4) represent an infinite confinement potential, solid lines (2,3,5,6) correspond to $V_e^0=0.5\,eV$, $V_h^0=1.0\,eV$; $2-m_{eM}=0.11m_0$; $3-m_{eM}=m_0$; $5-m_{hM}=0.35m_0$ and $6-m_{hM}=m_0$.

Journal of Physics: Conference Series **289** (2011) 012004 doi:10.1088/1742-6596/289/1/012004

Figure 2 illustrates dependence of exciton binding energy $E_c = E - T_0$ upon cubic QD size *b*.

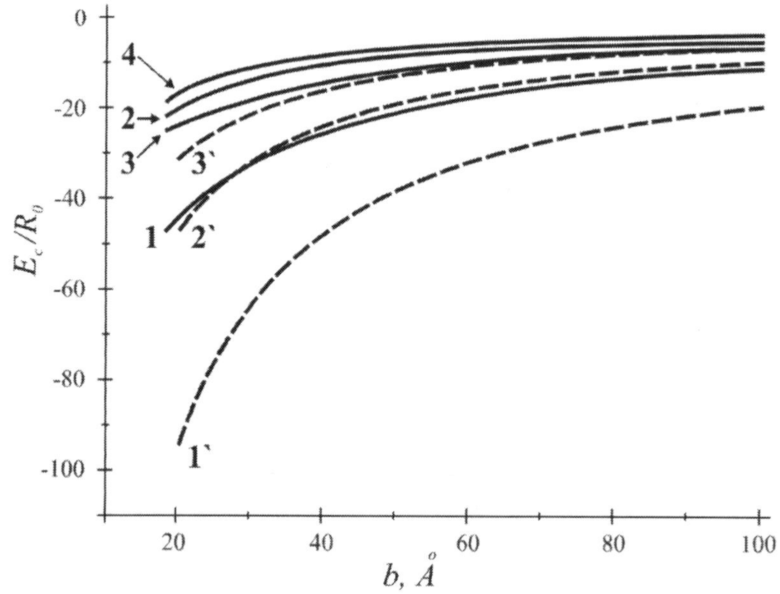

Figure 2. Exciton binding energy as a function of QD size b for various matrix dielectric constants: $1,1' - \varepsilon_M = 1$; $2,2' - \varepsilon_M = 4$; $3,3' - \varepsilon_M = 7$; $4,4' - \varepsilon_M \approx \varepsilon_Q; \varepsilon_Q = 10.9$.
Dashed lines represent an infinite confinement potential, solid lines correspond to $V_e^0 = 0.5\ eV$, $V_h^0 = 1.0\ eV$.

It should be noted that for *CdTe* QD of size $b = 20\ \mathring{A}$ magnitude of exciton binding energy exceeds the one for the bulk crystal more then five times for the finite barriers and more than ten times in case of infinite confinement potential. Binding energy is a monotonically increasing function of the QD size *b* and decreasing function of the matrix dielectric constant. At the absence of the dielectric mismatch (line 4 in Fig.2) the dependence of E_c upon QD size and its dielectric properties qualitatively remains similar to the described above, but its magnitude has a lower value. For example, at $b = 20\ \mathring{A}$, $|E_c| \approx 20 meV$ which is only two times larger than the appropriate value for the bulk crystal. In other words, accounting for the dielectric mismatch leads to the increase in differences of exciton binding energy between QD and bulk crystal, which is similar to the results obtained for the quantum wells [23, 25].

The other important characteristic of the exciton state is the oscillator strength *f*, which can be expressed by formula [22]:

$$\frac{f}{f_{ex,1s}} = \frac{\pi a_0^3}{V} \left| \int d^3 r \Psi(\vec{r}, \vec{r}) \right|^2 \frac{E_{ex,1s}}{E_c}, \qquad (16)$$

Journal of Physics: Conference Series **289** (2011) 012004 doi:10.1088/1742-6596/289/1/012004

where $f_{ex,\,1s}$, $E_{ex,1s}$ – oscillator strength and *1s* exciton binding energy for the bulk crystal of the volume *V*; E_c – exciton binding energy for QD, $\Psi(\vec{r},\vec{r}) \equiv \Psi(\vec{r}_e,\vec{r}_h)\big|_{\vec{r}_e=\vec{r},\ \vec{r}_h=\vec{r}}$ – exciton wave function for electron and hole located at the same point $\vec{r}$.

Figure 3 illustrates the ratio of oscillator strengths for QD and bulk *CdTe.*

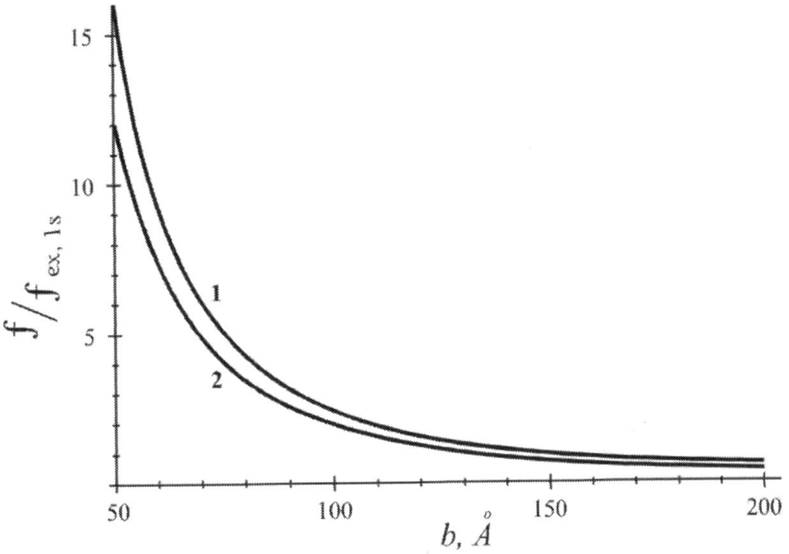

Figure 3. Oscillator strength versus QD size for various QD dielectric constants
$1 - \varepsilon_M = 1$; $2 - \varepsilon_M \approx \varepsilon_Q$.

Obtained according to formula (16) oscillator strength monotonically increases with decrease of QD size (Fig.3). At the absence of dielectric mismatch $f/f_{ex,\,1s} = 10$ for $b = 55$ Å. Consideration of the image boundary charges leads to increase in the oscillatory strength. For the same QD size $b = 55$ Å and $\varepsilon_M = 1$ we obtain $f/f_{ex,\,1s} = 13$.

One of the explanations of the Stocks shift in exciton absorption and photoluminescence spectra is based on the introduction of the bright and dark exciton states [26]. The difference between the energy of those states equals to the exchange interaction energy between electron and hole. If only short-range interaction is taken into account, then QD exciton exchange energy is expressed by formula [22]:

$$E_{c,exch} = E_{exch}\, \pi\, a_0^3\, K_0\,, \tag{17}$$

where $K_0 = \int d^3r \; |\Psi(\vec{r}, \; \vec{r})|^2$; E_{exch} – exchange energy for *1s*- exciton in the bulk crystal. The dependence of the ratio of QD exchange energy to the one of bulk crystal upon QD size at few values of QD dielectric constant is presented by Figure 4.

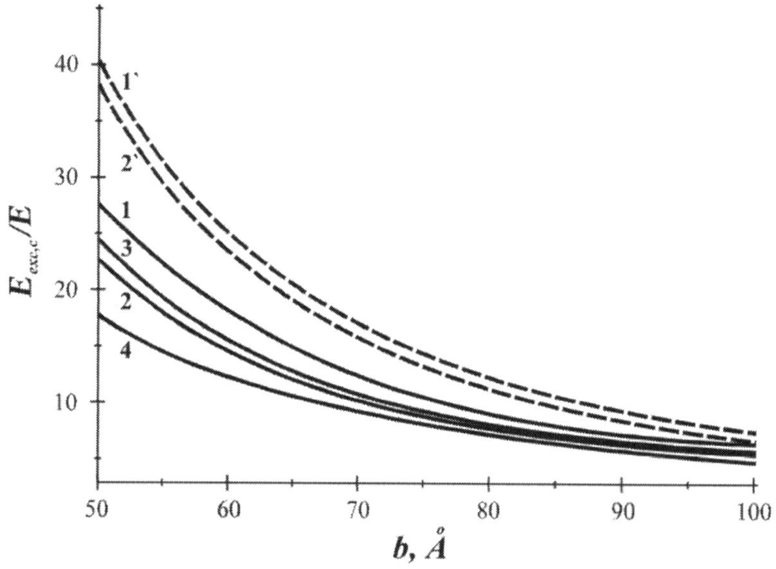

Figure 4. Exchange energy as a function of QD size, magnitude of confinement potential and matrix dielectric constant: $1,1' - \varepsilon_M = 1$; $2,2' - \varepsilon_M \approx \varepsilon_Q$; $3 - \varepsilon_M = 4$; $4 - \varepsilon_M = 81 \; (H_2O)$; dotted lines $1', 2'$ correspond to the infinite barrier, solid line 1-4 to the confinement potential $V_e^0 = 0.5eV, \; V_h^0 = 1.0eV$.

As it can be seen from the Fig.4, there is a drastic change in the exchange energy as QD decreases in size, especially at small b. Such behavior of $E_{c,exch}$ was experimentally confirmed for *CdTe* in [6] and our theoretical results show a qualitative agreement with the experimental data of the cited reference.

If we assume for bulk *CdTe* crystal Stocks shift equal to *0.15 meV* [6, 22], then for small-size QD we obtain the next results (Table 1).

Table 1. Small-size QD and Stocks shift parameters.

QD size (b), Å	$E^{\infty}_{c,exch}(V_{0e} = V_{0h} = \infty)$	$E_c(V_{0e} = 0.5eV, V_{0h} = 1eV)$	Stocks shift [6]
30	21 meV	10 meV	170 meV
25	45 meV	20 meV	180 meV
20	75 meV	35 meV	200 meV

For comparison purpose the last column of this table contains the experimental results [6]. As we can see, the experimental observation of Stocks shift provides higher values of $E_{c,exch}$ than were obtained in our calculations. In addition, the analysis of the theoretical results shows that in the framework of the current model, $E_{c,exch}$ has only a weak dependence upon the matrix permittivity. In our opinion it can be caused by the fact, that current theory underestimates the probability of charge localization near the QD boundary. It decreases, in turn, the matrix influence on the exciton and leads to a smaller magnitude of exciton exchange energy as well as its weak dependence upon ε_M .

Acknowledgments
The helpful discussions with D.V. Korbutyak are gratefully acknowledged.

References

[1] Mendes D R, Junior F Qu, Alcalde A M and Dantas N O 2003 Photoluminescence and optical absorption in CdSe nanocrystals *Microel. J.* **4** 787-801
[2] Cozzoli P D, Manna L, Curri M L, Kudera S, Giannini C, Striccodi M and Agostiano A. 2005 Shape and Phase Control of Colloidal ZnSe Nanocrystals *Chem. Mater.* **17** 1296-1306
[3] Bryant G W and Jaskolski W 2005 Surface Effects on Capped and Uncapped Nanocrystals *J. Phys Chem.* B. **109** 19650-56
[4] Woggon U and Gasponenko S V 1995 Excitons in Quantum Dots *Phys. Stat. Sol.* (b). **189** 285-343
[5] Wuister F S, Swart I, Driel F, Hickey S G and Donega C de M 2003 Highly Luminescent Water-Soluble CdTe Quantum Dots *Nano Lett.* **3** 503-507
[6] Kalytchuk S M, Korbutyak D V, Shcherbak L P, Feichuk P I 2008 Producing technology and photoluminescence properties of CdTe nanocrystals in colloid solutions and polymer matrixes *New Technologies.* **2** 135-140
[7] Liu Y, Chen W, Joly A G, Wang Y, Pope C, Zhang Y, Bovin J-O and Sherwood P 2006 Comparison of Water-Soluble CdTe Nanoparticles Synthesized in Air and in Nitrogen *J. Phys. Chem.* B **110** 16992-17000
[8] Munoz M, Guo S, Zhou X, Tamargo M C, Huang Y S, Trallero-Giner C and Rodriguez A H 2003 Contactless electroreflectance of CdSe/ZnSe quantum dots grown by molecular-beam epitaxy *Appl. Phys. Lett.* **83** 4339-4345
[9] Masumoto Y, Sonobe K 1997 Size-dependent energy levels of CdTe quantum dots *Phys. Rev. B* **56** 9734-9737
[10] Wegscheider W, Schedelbeck G, Abstreiter G, Rother M and Bichler M 1997 Atomically Precise GaAs/AlGaAs Quantum Dots Fabricated by Twofold Cleaved Edge Overgrowth *Phys. Rev. Lett.* **79** 1917-1920
[11] Besombes L, Marsal L, Kheng K, Charvolin T, Dang Le Si, Wasiela A and Mariette H 2000 Few-particle effects in single CdTe quantum dots *J. Cryst. Growth* **214-215** 742-746
[12] Prado S J, Trallero-Giner C, Alcalde A M, Lopez-Richard Vand Marques G E 2003 Influence of quantum dot shape on the Landé g-factor determination *Phys. Rev.* B **68** 235327-31
[13] Nirmal M, Norris D J, Kuno M, Bawendi M G, Efros Al L and Rosen M 1995 Observation of the "Dark Exciton" in CdSe Quantum Dot *Phys. Rev. Lett.* **75** 3728-31
[14] Prado S J, Trallero-Giner C, Alcalde A M, Lopez-Richard V and Marques G E 2004 Influence of quantum dot shape on the Landé g-factor determination *Phys. Rev.* B **69** 201310-14
[15] Laheld U E H and Einevoll G T 1997 Excitons in CdSe quantum dots *Phys. Rev.* B **55** 5184-88

[16] Safwan S A, Hekmat M H and El-Meshad N A 2007 Exciton state in a quantum dot *Fizika* A (Zagreb) **16** 1-10
[17] Bolcatto P G and Proetto C R 1999 Shape and dielectric mismatch effects in semiconductor quantum dots *Phys. Rev.* B **59** 12487-12498
[18] Bolcatto P G and Proetto C R 2001 Partially confined excitons in semiconductor nanocrystals with a finite size dielectric interface *J. Phys.: Condens. Matter* **13** 319-325
[19] Marin J L, Riera R and Cruz S A 1998 Confinement of excitons in spherical quantum dots *J. Phys. Condens. Matter* **10** 1349-61
[20] Li Y, Voskoboynikov O, Lee C P and Sze S M 2001 Energy and coordinate dependent effective mass and confined electron states in quantum dots *Solid State Commun.* **120** 79-83.
[21] Zhou A P and Sheng W D 2009 Electron and hole effective masses in self-assembled quantum dots *Eur. Phys. J.* B **68** 233-236
[22] Romestain R and Fishman G 1994 Exitonic wave function, correlation energy, exchange energy, and oscillatorstrength in a cubic quantum dot Phys. Rev. B **49** 1774-1781
[23] Keldysh L V 1979 Coulomb interaction in thing films of semiconductors and semimetals *Pis'ma Zh. Eksp. Teor Fiz.* **29** 716-719
[24] Fonoberov V A, Pokatilov E P and Balandin A A 2002 Exciton states and optical transitions in colloidal CdS quantum dots: Shape and dielectric mismatch effects *Phys. Rev.* B **66** 085310-23
[25] Boichuk V I and Bilynskii I V 1992 Bound energy of the Wannier exciton in similar heterogeneous double structures *Phys. Stat. Sol.* (b) **174** 463-470
[26] Efros Al L, Rosen M, Kuno M, Nirmal M, Norris D J and Bawendi M G 1996 Band-edge exciton in quantum dots of semiconductors with a degenerate valense band: Dark and bright exciton states *Phys.Rev.* B **54** 4843-57

Appendix

The calculations for finite confinement potential are similar to the infinite potential case but obtained functional $E(\alpha)$ has considerably more complicated form. The components of formula (8) are expressed now by the next formula:

$$T_0 = N^{-1}(\delta)\left[J_e(\delta)+J_h(\delta)\right],\quad T_1 = 3N^{-1}(\delta)\tilde{m}\,\delta\left[I_e(\delta)-I_h(\delta)\right],\quad T_2 = -\tilde{m}\,\delta,$$
$$P_c = -2N^{-1}(\delta)(1-\tilde{m}\,\delta)\Pi_3(\delta),\quad P_p = -3N^{-1}(\delta)\Pi_4(\delta),$$

where $\delta = \alpha a_B$ $\eta_{ie} = \frac{x_{ie}}{a_0}$, $\eta = \eta_e - \eta_h$, $\tilde{m} = \frac{\mu_Q}{\mu_M}$, $m_e = m_{eQ}W_{eQ} + m_{eM}W_{eM}$, $m_h = m_{hQ}W_{hQ} + m_{hM}W_{hM}$;

W_{eQ}, W_{eM} (W_{hQ}, W_{hM}) – probability of finding electron (hole) in quantum dot and matrix respectively; $N(\delta)$ - normalization coefficient that for finite confinement potential has a form:

$$N(\delta) = \int_0^\infty d\eta_1 \int_0^\infty d\eta_2 \int_0^\infty d\eta_3 \prod_{i=1}^{3} F_i(\eta)\exp\left(-2\delta\sqrt{\eta_1^2+\eta_2^2+\eta_3^2}\right),$$

where:

$$F_i(\eta)=\begin{cases} I_1(\eta), 0\le \eta\le \tilde b \\ I_2(\eta),\ \eta>\tilde b \end{cases},\ \tilde b=\frac{b}{a_0},\qquad I_1(\eta)=\sum_{n=1}^{3}K_n(\eta),$$

$$K_1(\eta)=\int_{-\infty}^{-\tilde b+\eta} d\eta'\left[\phi_e\left(\frac{1}{2}(\eta'+\eta)\right)\phi_h\left(\frac{1}{2}(\eta'-\eta)\right)\right]^2,$$

$$K_2(\eta)=\int_{-\tilde b+\eta}^{\tilde b-\eta} d\eta'\left[\phi_e\left(\frac{1}{2}(\eta'+\eta)\right)\phi_h\left(\frac{1}{2}(\eta'-\eta)\right)\right]^2,$$

$$K_3(\eta)=\int_{\tilde b-\eta}^{\infty} d\eta'\left[\phi_e\left(\frac{1}{2}(\eta'+\eta)\right)\phi_h\left(\frac{1}{2}(\eta'-\eta)\right)\right]^2,$$

$$I_2(\eta)=\int_{-\infty}^{\infty} d\eta'\left[\phi_e\left(\frac{1}{2}(\eta'+\eta)\right)\phi_h\left(\frac{1}{2}(\eta'-\eta)\right)\right]^2.$$

It should be noticed that similar to the infinite quantum well [25], the kinetic energy in QD case is expressed through quadratic and linear variational parameters.

$$J_{\left\{e\atop h\right\}}(\delta)=\int_0^\infty d\eta_1\int_0^\infty d\eta_2\int_0^\infty d\eta_3 F_0^{\left\{e\atop h\right\}}(\eta_1)F_2(\eta_2)F_3(\eta_3)\exp\left(-2\delta\sqrt{\eta_1^2+\eta_2^2+\eta_3^2}\right),$$

where: $$F_0^h(\eta)=\begin{cases} I_3(\eta), 0\le \eta\le \tilde b \\ I_4(\eta),\ \eta>\tilde b \end{cases},\qquad I_3(\eta)=\sum_{n=4}^{6}K_n(\eta),$$

$$K_4(\eta)=\int_{-\infty}^{-\tilde b+\eta} d\eta'\phi_e^3\left(\frac{1}{2}(\eta'+\eta)\right)\phi_h^2\left(\frac{1}{2}(\eta'-\eta)\right)\Phi_h(\eta,\eta'),$$

$$K_5(\eta)=\int_{-\tilde b+\eta}^{\tilde b-\eta} d\eta'\phi_e^3\left(\frac{1}{2}(\eta'+\eta)\right)\phi_h^2\left(\frac{1}{2}(\eta'-\eta)\right)\Phi_h(\eta,\eta'),$$

$$K_6(\eta)=\int_{\tilde b-\eta}^{\infty} d\eta'\phi_e^3\left(\frac{1}{2}(\eta'+\eta)\right)\phi_h^2\left(\frac{1}{2}(\eta'-\eta)\right)\Phi_h(\eta,\eta'),$$

$$I_4(\eta)=\int_{-\infty}^{\infty} d\eta'\phi_e^3\left(\frac{1}{2}(\eta'+\eta)\right)\phi_h^2\left(\frac{1}{2}(\eta'-\eta)\right)\Phi_h(\eta,\eta')$$

$$\Phi_h(\eta,\eta')\equiv\frac{\partial^2}{\partial\eta'^2}\phi_h\left(\frac{1}{2}(\eta'-\eta)\right).$$

The function $F_0^e(\eta)$ can be calculated similar to $F_0^h(\eta)$ by substituting all the quantities with the hole indexes with the ones with electron indexes.

The quantities $I_{e,h}(\delta)$, which also are contributing to kinetic energy of electron-hole system, can be determined by the formula

$$I_{e,h}(\delta)=\int_0^\infty d\eta_1\int_0^\infty d\eta_2\int_0^\infty d\eta_3 F_4^{e,h}(\eta_1)F_2(\eta_2)F_3(\eta_3)\frac{\exp\left(-2\delta\sqrt{\eta_1^2+\eta_2^2+\eta_3^2}\right)}{\sqrt{\eta_1^2+\eta_2^2+\eta_3^2}}.$$

Functions $F_4^{e,h}(\eta)$ are expressed in a similar way to $F_0^{e,h}(\eta)$ with:

$$\Phi_{e,h}(\eta,\eta')\equiv\eta\frac{\partial}{\partial\eta}\phi_{e,h}\left(\frac{1}{2}(\eta'\pm\eta)\right),$$

The last two terms in the expression (8) are related to the potential energy. First one of them (Π_3) is connected to electron-hole Coulomb interaction and can be obtained with the help of the formula:

$$\Pi_3=\int_0^\infty d\eta_1\int_0^\infty d\eta_2\int_0^\infty d\eta_3\prod_{i=1}^{3}F_i(\eta_i)\frac{\exp\left(-2\delta\sqrt{\eta_1^2+\eta_2^2+\eta_3^2}\right)}{\sqrt{\eta_1^2+\eta_2^2+\eta_3^2}}.$$

The interface polarization is taken into account by the term Π_4:

$$\Pi_4=\int_0^\infty d\eta_1\int_0^\infty d\eta_2\int_0^\infty d\xi_1\int_{-\infty}^\infty d\xi_2\int_{-\infty}^\infty d\xi_3\,\phi_e(\xi_2)\,\phi_h(\xi_3)\exp\left(-2\delta\sqrt{\eta_1^2+\eta_2^2+(\xi_2-\xi_3)^2}\right)\times$$
$$\times J_0\left(\xi_1\sqrt{\eta_1^2+\eta_2^2}\right)\frac{\left(\varepsilon_Q^2-\varepsilon_M^2\right)e^{\bar{b}\xi_1}ch\left[\xi_1(\xi_2+\xi_3)\right]-\left(\varepsilon_Q-\varepsilon_M\right)^2ch\left[\xi_1(\xi_2-\xi_3)\right]}{\left(\varepsilon_Q+\varepsilon_M\right)^2e^{2\bar{b}\xi_1}-\left(\varepsilon_Q-\varepsilon_M\right)^2}.$$

Journal of Physics: Conference Series **289** (2011) 012005 doi:10.1088/1742-6596/289/1/012005

Hybrid nanostructures based on the conducting polymers incorporated to porous silicon matrix

L Monastyrskyi[1], O Aksimentyeva[2]

[1]Ivan Franko National University of Lviv, Department of Electronics
50 Dragomanov str. Lviv, 79005, Ukraine
[2]Ivan Franko National University of Lviv, Department of Chemistry
6/8 Kyryla-Mefodia str., Lviv, 79005, Ukraine
E-mail: aksimen@ukr.net; liu_mon@yahoo.com

Abstract. An influence of conducting polymer on the structure, morphology and luminescent characteristics of porous silicon – conducting polymer hybrid nanostructures has been studied. The incorporation of conducting polymers such as polyphenylacetylene (PPA), polyepoxypropyle carbazole, polyparaphenylene (PPP), polyaniline to porous silicon matrix was provided by magnetron deposition, electropolymerization and dip coating. The hybrid layers were studied by FT-IR spectroscopy, scanning electron and atom force microscopy and examined on the ability to photo- and cathode luminescence (CL). For the hybrid nanostructured layers based on PPA and PPP a significant increasing in the luminescence intensity has been found. According to FT-IR spectroscopy the interaction between polymer functional groups and porous silicon surface takes a place. Probably it leads to decreasing a rate of non-radiation surface recombination and to increasing a radiation component of recombination as well as increasing in the CL total intensity.

1. Introduction

The nanostructures based on inorganic and organic semiconductors including the conducting polymers have attracted a great attention due to the prospect of using such structures in the solar cells, light emitting diodes, sensors, lasers and memory devices [1-5]. Embedding semiconductor nanocrystals in a conjugated polymer matrix has results in significant change in luminescent and photoelectric properties [2]. Belong the semiconductor materials a special place is occupied by the heterogeneous systems based on nanoporous silicon (PS), having a wide application in optoelectronic and sensor devices [6, 7]. Of great interest are the heterostructures created on the PS and surface organic layers, among which the conjugated conducting polymers, particularly, polyaniline (PANI) have been intensively studied [8, 9]. So far there are only few data on the luminescence properties of heterostructures based on other conducting polymers, such as polyphenylacetylene (PPA), polyepoxypropyle carbazole (PEPC), polyparaphenylene (PPP). At the same time, it is known that PPP, PPA, PEPC exhibit the electroluminescence in the sandwich like structures, mostly developed for PPP [10]. In the present work an influence of integrated polymer on the molecular spectra, morphology of surface and luminescence characteristics of PS – conducting polymer (PPP, PPA, PANI, PEPC) nanostructures has been studied.

Journal of Physics: Conference Series **289** (2011) 012005 doi:10.1088/1742-6596/289/1/012005

2. Experimental

The porous silicon layers were obtained by oxidization of the single crystal silicon wafers both n-type (111) and p-type (100) of conductivity with specific resistance of 4.5 and 6.4 Ohm cm respectively and 460 μm of the thickness. The wafer anodization was carried out in a 40% HF solution in ethyl alcohol at a current density 10 $mA \cdot cm^{-2}$ and under illumination by a 150 W tungsten lamp during 10-20 minutes.

For creation of PS-PANI hybrid layer a method of electrochemical polymerization with cyclic potential sweeping in 0.1M solutions of purified aniline in 0.5 M H_2SO_4 was used. A wafer of PS of the area of 1 cm^2 was as an operating electrode, a platinum wire was used as a counter electrode and saturated Ag/AgCl electrode was employed as a reference. The potentiostat PI-50 was a power source. After coating the film as described in [11] the samples were rinsed with deionized water and dried under vacuum at 353 – 373 K.

Magnetron deposition of the conducting polymers on the porous silicon surface was carried out on the URM-3 equipment in the argon atmosphere at residual pressure $5x10^{-3}$ mm Hg on the frequency of 13,56 MHz in the magnetic field 0,1 T at the power of 100 W. The middle rate of condensation was ≈2 nm/min. Formation of PS-PPA and PS-PEPC heterostructure was carried out by dip coating method from polymer solution in chloroform.

The molecular structure of the obtained hybrid layers were studied by FT-IR spectroscopy with AVATAR 320N spectrophotometer in the spectral region of 4000 – 400 cm^{-1}. The morphology of the hybrid layer was studied by scanning electron microscope JEOL JSM – T220A and by atom force microscopy (AFM) with SZM SOLVER PRO-M microscope.

Photoluminescence (PL) was excited by pulse nitrogen laser LGI-21 (337.1 nm), pulse duration τp = 20 ns, at frequency of 60 Hz. The measurements were carried out both at room temperature and of temperature of the liquid nitrogen. Cathode luminescence was excited by pulse electron flow with energy Ep=9 keV, with pulse frequency f = 50 Hz and τ p = 3 μs. The current density in the electron beam j = 150 A/m^2. The temperature of CL measurements was 80 K.

3. Results and Discussion

To study the character of the surface topology of the obtained heterostructures the two methods – scanning electron microscopy (SEM) and atom-force microscopy (AFM) has been employed. The surface of PS has a shape of vertical nanorods created due to the etching of separate places of a single silicon surface (Fig.1,a). As a result of profile analysis of AFM-images it is defined that the pore depth achieves 180 - 200 nm at total thickness of porous layer about 5 μм.

The PANI films obtained on the surface of PS by electropolymerization method also as dip-coated PPA film in general repeat the relief of the surface forming a continuous polymer layer (Fig.1, b). However, the morphology of the films is difficult interpret due to the irregularity of PS surface. As has been shown by the profile analysis of PS-polymer heterostructure a thickness of PS-polymer layer achieves 40–45 nm. In the process of magnetron deposition the polymer is formed as permanent nanofilm on the porous silicon surface (Figure 2, a) and also may deposit inside silicon pores which is clearly demonstrated for PS-PPP heterostructure (Figure 2, b).

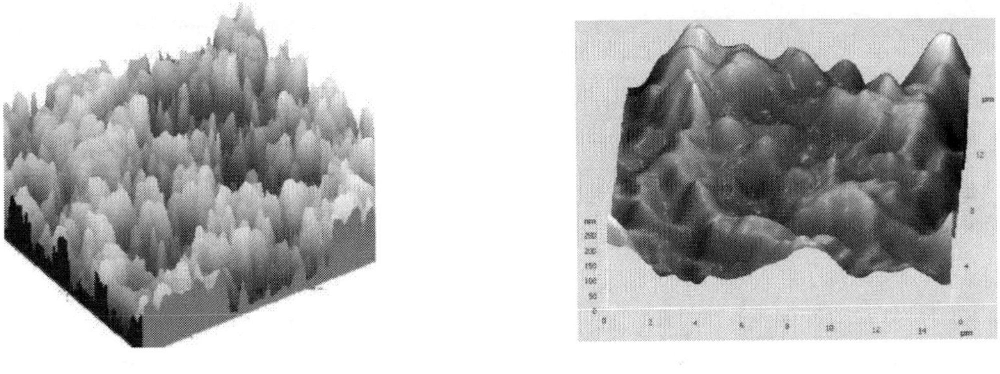

(a) (b)

Figure 1. AFM images of the porous silicon surface (a) and porous silicon – polyaniline hybrid structure (b)

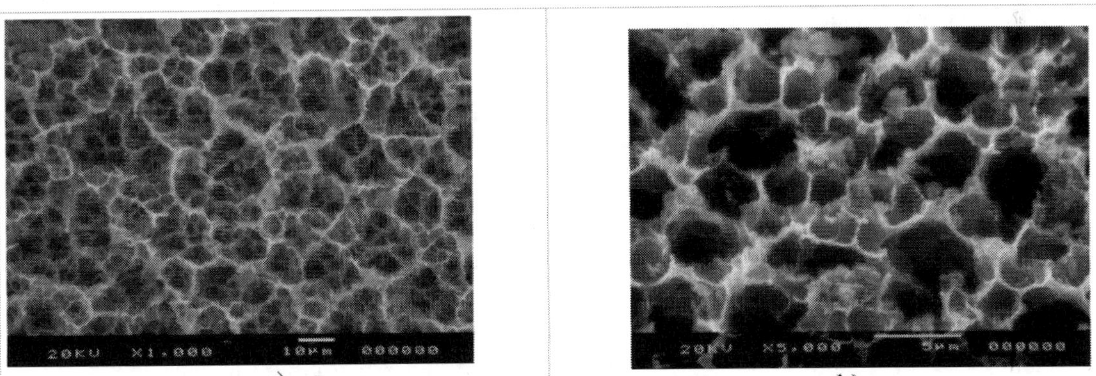

a) b)

FIGURE 2. SEM microphotos of the porous silicon - conducting polymer nanostructures obtained by magnetron deposition: a) PS-PANI (x 1000); b) PS-PPP (x 5000)

The photoluminescence spectra of porous silicon studied in the spectral range of 1.5-3.1 eV are in good agreement with the ones described in literature [6-9]. It has been found that in hybrid PS-polymer nanostructures the "orange - red" photoemission is preserved, but it decreasing by a factor of 2 – 3 takes a place (Figure 3). A weak shift of PL peak may be due to non-uniform transmittance of the conducting polymers (the transmittance window is close to the emission peak of PS).

An appearance of luminescence of the porous silicon in the visible range of spectrum may be explained by some ideas. From the one hand the quantum size effects making possible the annihilation of the excitons which are spatially limited (trapped) in the intermediate area [6]. On the other hand emission may be caused by the complexes of Si:O:H, H:Si:H, siloxen $Si_6H_3O_6$ and its modifications [7] formed during the etching process. PS emission may be considered as a process of monomolecular recombination, limited by tunneling across the potential barrier, or as bimolecular process of recombination of the electron and the hole being free or bound to the exciton [12]. At the same time, it is known that such polymers as PPP, PPA and PEPC may be active layer in the electroluminescence devices [10].

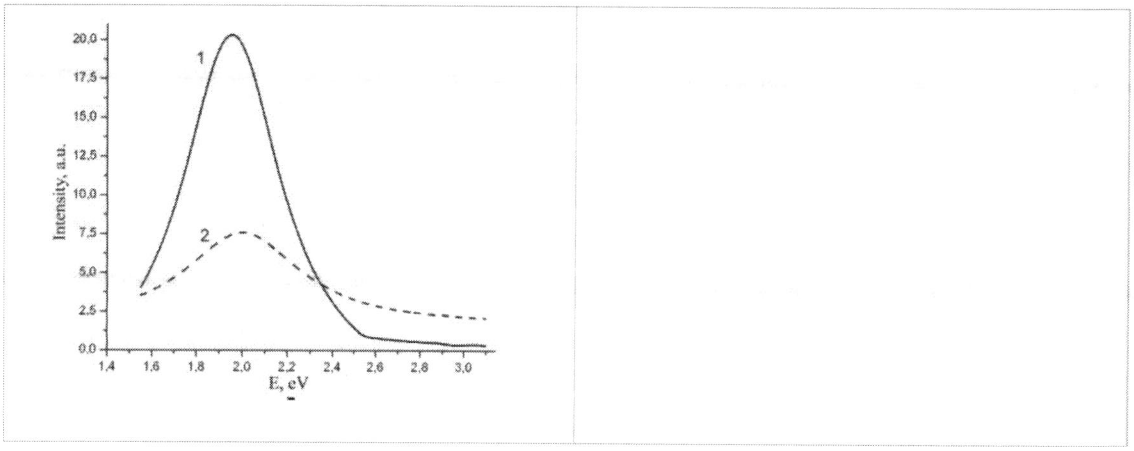

FIGURE 3. Photoluminescence spectra of PS (1); PS-PANI (2); PS-PPP hybrid structure at 293 K.

As a test on the ability to electroluminescence may be study of cathode luminescence exited by electron beam [13] without special metal contacts.

It known that in silicon single crystal any CL (as well as EL) does not develop even at very high excitation energies and at low temperatures. At the same time, the existence of PS nanostructured layer makes it possible to excite the light emission under the action of the cathode beam. We observed the cathode luminescence of the porous silicon with a band maximum near 2.12 eV which is in good agreement with the data described in [14].

The cathode luminescence is observed at electron excitation of heterostructures based on PS and conducting polymers, fabricated by different methods (electropolymerization, magnetron deposition and dip-coating). Furthermore, the shape of the spectrum and the character of its change in comparison with CL spectrum of initial PS (at T = 80 K) depends on the electron and molecular structure of polymer.

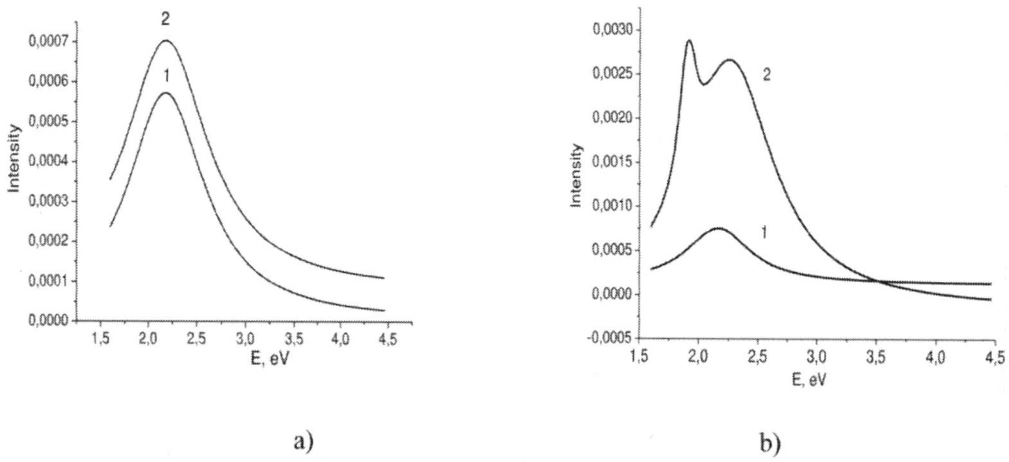

FIGURE 4. CL spectra of (a) PS (1), PS-PEPC (2); (b) PS (1) and PS-PPP (2) hybrid structure

It is found that PANI and PEPC polymers integrated with PS cause a slow increasing in CL intensity (Figure 4, a). At the same time an intensity of CL significantly increases at the presence of PPP and PPA in hybrid layers as compared with CL in porous silicon matrix without polymer (Figure 4, b).

According to the FT- IR spectroscopy (typical spectra presented in Figure 5) the surface of PS, obtained by electrochemical etching the silicon wafer in HF alcohol solution, contents the different functional groups with characteristic absorption bands at 2900-3620 cm^{-1} (-OH), 1100 cm^{-1}, 1240 cm^{-1} (Si-O-Si), 1111 cm^{-1} (C-O), 1072 cm^{-1} (CH-OH), 620 cm^{-1} ($Si-H_2$) and others (Fig.5,a)

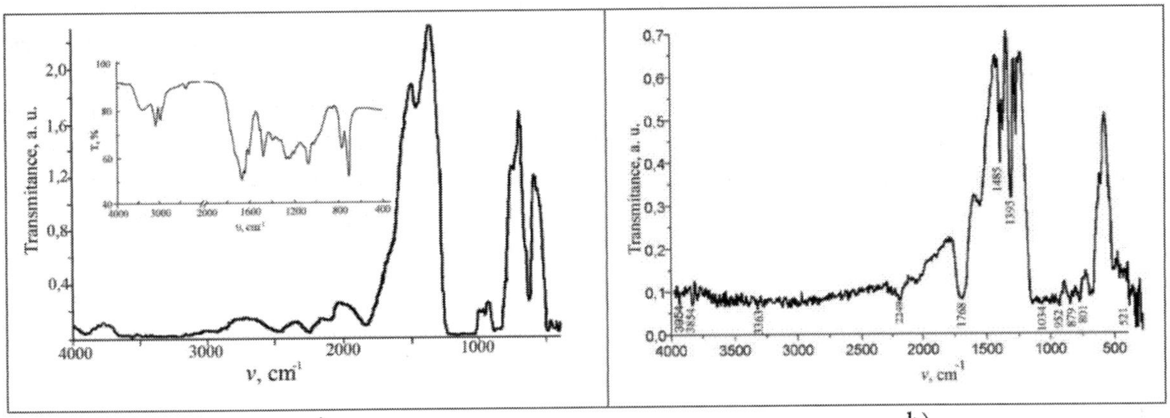

a) b)

FICURE 5. FT-IR spectra (a) of porous silicon (insert – PPA on KBr) and (b) of PS-PPA hybrid structure

In result of polymer adsorption an activation of several bonds (for example -N-H for polyaniline or -C=C-H- for PPA) and creation the new passivative complexes with surface silicon bonds may proceed. The evidences to such interaction are the shifts of the Si-O-Si absorption band from 1100 and 1240 cm^{-1} to 1050 and 1250 cm^{-1}, also as appearance of the new bands, do not observed in porous silicon and polymer spectrum (Fig.5,a, insert), with maximums at 1188 (C-C), 1280 (Ar-0H) and 1770 (C=O) cm^{-1} (Fig. 5, b). These bands may be assigned to formation of the weak covalent bonds between polymer and semiconductor due to polymer chemosorption on the active centres (Si-O-Si) of the porous silicon surface, probably with formation of light emissive complexes [7]. It is assumed that a rate of non-radiation surface recombination decreases at coating the PS surface with polymer due to the passivative action of the polymer. Apparently, this causes the increase in the radiation component of recombination as well as the increase in CL intensity. If recombination proceeds along the radiating channels of the nanocrystal grains, the PS being coating by a conducting polymer could significantly effects the recombination mechanism. From the other hand increasing in CL intensity may be caused by the interband radiation recombination in the conducting polymer [10, 15] deposited in nanopores of silicon.

5. Conclusion

The fabricated hybrid semiconductor – polymer nanostructures based on porous silicon and conducting polymer demonstrate a visible luminescence and the capability to light emission under electron beam excitation. The effect of the hybrid structure on the character of emission processes is defined by the type of conducting polymer integrated into PS. According to FT-IR spectroscopy the some interaction between polymer functional groups and porous silicon surface takes a place. We have assumed, it is leading to decreasing a rate of non-radiating surface recombination and to increasing the radiation component of recombination as well as increasing in the CL total intensity. An increasing the luminescence intensity may be also caused by the interband radiation recombination in the conducting polymer deposited in confinement of the silicon nanopores.

An "anomalous" increasing in quantum efficiency of the cathode luminescence at absorption of polymers by the porous silicon surface can be interpreted by the effects which are inverse those leading to decrease of the emission. In particular, during the process of polymer deposition it is possible an activation of certain polymer bonds and creation the new passivative complexes with dangling silicon bonds. It is clear that such aggregation of silicon atoms gives rise to increasing the quantum efficiency of emission that is observed experimentally. The degree of influence of the polymer's cover has depends on the character of pores filling by polymer and the magnitude of dipole moment of the adsorbed polymer in the hybrid structures. At larger dipole moment there is occurs trapping of an electron or hole on the energy levels, i.e. the rate of Shockley-Read non-radiation recombination via these levels increases.

5. References

[1]. Kang,Y., Park, N.-G., Kim, D. 2005 *Appl. Phys. Lett.* **86** 113101
[2]. Bakueva L., Musikhin S., Sargent E.H. Shik A. 2003 *Surface Science* **532-535** 1051
[3]. Aksimentyeva O.I., Poliovyi D.O., Fechan A.V., Yevchuk O.M. 2008 *MCLC* **497** 362
[4]. Baratto, C., Faglia, G., Sberveglieri, G., et al. 2002 *Sensors* **2** 121
[5]. Aksimentyeva O., Monastyrskyi L., Savchyn V., Stakhira P.Y. 2007 *MCLC* **467** 73
[6]. Svechnikov S.V., Kaganovich E.B., Manoilov E.C. 1998 *Sem.Phys.Quant.Elect.Optoel.* **1** 13
[7]. Beliakov L.V., Makarova T.L., Saharov V.I., Serenkov I.T. 1998 *Phys.Tech. Sem.* **32** 1122
[8]. Jayachandran M., Paramasivam M., Murali K.R., Trivedi D.C. 2001 *Mater.Phys.Mech.* **4** 143
[9]. Halliday, D.P., Eggleston, J.M., Adams, P.N. 1997 *Synth.Metals* **85** 1245
[10]. Lee C.H., Kang G.W., Jeon J.W., et al. 2000 *Thin Sol. Films* **363** 306
[11]. Stakhira, P., Aksimentyeva, O., Cherpak, V. 2005 *Phys.Chem.Sol.State* **6** (2) 235
[12] Blonskyy. I., Brodyn, M., Vakhnin, A., et al. 2001 *Phys.Lett.A.* **279** 391
[13]. Pankove J.I. 1971 *Optical processes in semiconductors* (Englewood Cliffs: New Jersey, USA)
[14]. Yong, Z., Jin Yi, Jin, Yong, N., 1994 Luminescence **60-61** 404
[15]. Heeger, A.J. 2002 *Synth.Metals* **123** 23

Journal of Physics: Conference Series **289** (2011) 012006 doi:10.1088/1742-6596/289/1/012006

Structure changes in $Fe_{73.1}Si_{15.5}B_{7.4}Nb_{3.0}Cu_{1.0}$ amorphous alloy with temperature

S Mudry, Yu Kulyk

Physics of Metals Department, Ivan Franko National University of Lviv, Kyrylo and Mephodii Str. 8, 79005 Lviv, Ukraine

E-mail: kulyk2007@rambler.ru.

Abstract. Structure of $Fe_{73.1}Si_{15.5}B_{7.4}Nb_3Cu_1$ amorphous alloy has been studied within 293–748K temperature range by means of high temperature X–ray diffraction method. Main structure parameters have been calculated from experimental data and then analysed. It was found a prepeak occurence before principal maximum. Formation of nanocrystals with Fe_3Si like chemical short range order in amorphous matrix, is suggested to be responsoble for that prepeak. It is shown also that heating of alloy is accompanied by increasing of inhomogeneities in short range order structure.

1. Introduction

It is well known that amorphous alloys, having a short range order structure in atomic arrangement, posses by some temperature interval where this structure persists with heating. It is of interest to study the structure behavior within this temperature range and upon phase transition from amorphous state to crystalline one. To present time the information on high temperature structure studies of amorphous alloys is scarce. Most of works consider quenched alloys, before annealed at different temperatures. It is clear that quenched alloys have a somewhat another structure than same alloys at elevated temperatures. The temperatures changes of structure are especially important for such amorphous alloys, in whose matrix growing nanocrystals are distributed [1-4]. Typical material, representing this group of alloys is $Fe_{73.1}Si_{15.5}B_{7.4}Nb_{3.0}Cu_{1.0}$ amorphous alloy whose structure at different temperatures has been studied in this work.

2. Experimental

Amorphous ribbons $Fe_{73.1}Si_{15.5}B_{7.4}Nb_{3.0}Cu_{1.0}$ of 30μm thickness were prepared by means of rapid quenching of molten alloy. Composition of samples was cheeked with using of X–ray microanalysis. X–ray diffractometer DRON–3, equipped with high temperature vacuum camera was used for diffraction studies. LiF single crystal, installed into initial beam allowed us to obtain monochromatized Co Kα- radiation. Diffraction patterns were recorded at isothermal conditions within temperature range 293-823 K. Upon accounting the corrections for absorption and polarization, the intensity curves $I(s)$ ($s = \frac{4\pi}{\lambda}\sin\theta$, 2θ- scattering angle; λ- wave length) were analyzed and used for calculation of structure factors and pair correlation functions, from which the main structure parameters have been directly obtained.

Journal of Physics: Conference Series **289** (2011) 012006 doi:10.1088/1742-6596/289/1/012006

3. Results and discussion

Intensity curves for $Fe_{73.1}Si_{15.5}B_{7.4}Nb_{3.0}Cu_{1.0}$ alloy both in amorphous and amorphous–crystalline state are shown in Fig.1. The main structure parameters, obtained from these functions are listed in table 1. As is seen, such the parameters as principal peak position (s_o) and most probable distance to nearest neighbors (R) are slightly changeable with temperature. At the same time the non-monotonic temperature change of structure factor principal peak height $a(s_o)$ occurs. Namely, this parameter shows the tendency to increasing within temperature range 293-563 K that is suggested to be caused by local atomic ordering. On other hand, the topologic disordering of structure with temperature is accompanied by decrease of $a(s_o)$, whereas at heating to pre-crystallization temperatures this parameter increases up to 4,08, that is typical for pre-crystallization processes at topologic and chemical ordering.

The important feature of intensity curves is the occurrence of prepeak, whose position changes from. 18.5 up to 19.5 nm^{-1} within temperature range, where amorphous phase is stable Appearance of new maxima in diffractograms at T=718 K and corresponding to crystalline phase, indicates the formation of nanocrystals in amorphous matrix on the base of α-Fe(Si) solid solution, with DO_3 (Fm3m) ordering for Fe_3Si–phase. As follows from obtained experimental data, the range of thermal stability for nanocrystalline structure is from 728 up to 823 K, (Fig.1).

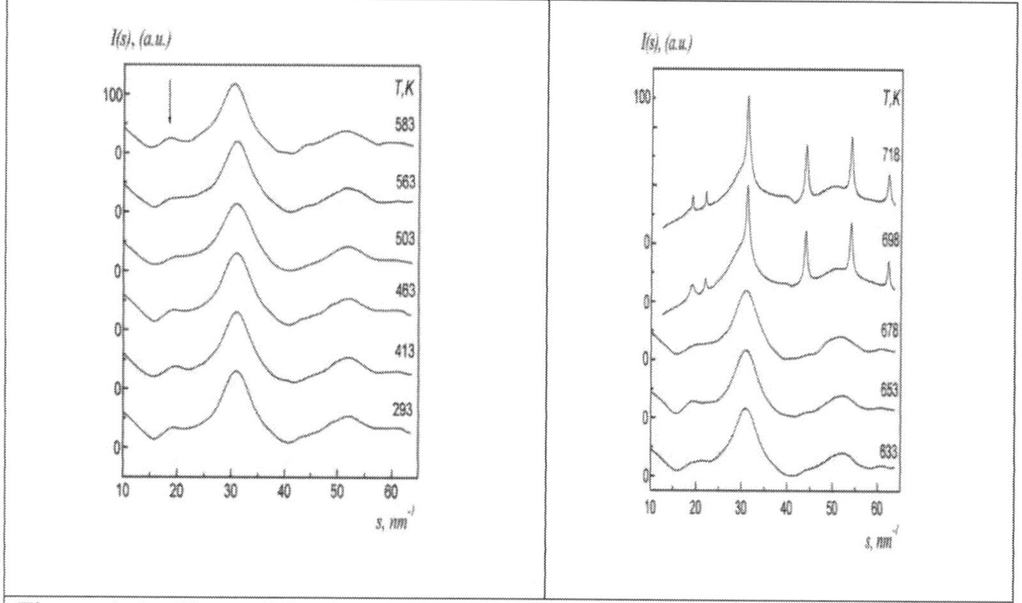

Figure 1a,b. X-ray diffraction pattern of $Fe_{73.1}Si_{15.5}B_{7.4}Nb_{3.0}Cu_{1.0}$ amorphous alloy at different temperatures.

It should be noted that positions of (110) and (200) superstructure reflexes in diffraction patterns of partially crystallized samples are close to prepeak position in intensity curve for amorphous phase (Fig.2b). It is known [5-9] that prepeak appearance is an evidence of correlation in atomic distribution of medium scale due to formation of intermetallic–based nanoclusters. The correlation interatomic distance R_p. can be calculated with using of Erenfest formula $s_pR_p = 7.73$ (s_p- prepeak position). Experimental data, obtained here, allowed us to point out that within 293-653 K, temperature range the correlation distance varies within interval R_p≈*4.0-4.2 Å*. Analysis of crystalline structure for Fe_3Si chemical compound shown that obtained values of R_p are the same as interatomic distances *R*(Si-Si)=*4.01 Å* and *R*(Fe-Fe)=*4.01 Å* in second coordination sphere. Therefore appearance of prepeak in intensity curves for $Fe_{73.1}Si_{15.5}B_{7.4}Nb_{3.0}Cu_{1.0}$ amorphous alloy is caused by correlation of Fe-Fe and Si-Si atomic pairs at distances ~*4 Å* and confirms the formation of nanoclusters with Fe_3Si

Journal of Physics: Conference Series **289** (2011) 012006 doi:10.1088/1742-6596/289/1/012006

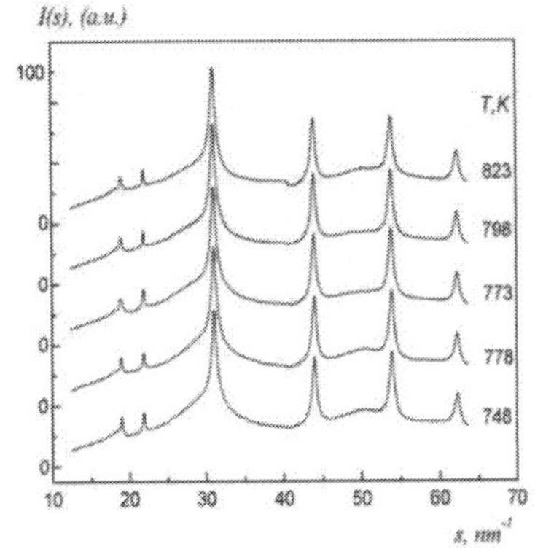

Figure 1c. X-ray diffraction pattern of $Fe_{73.1}Si_{15.5}B_{7.4}Nb_{3.0}Cu_{1.0}$ amorphous alloy at different temperatures.

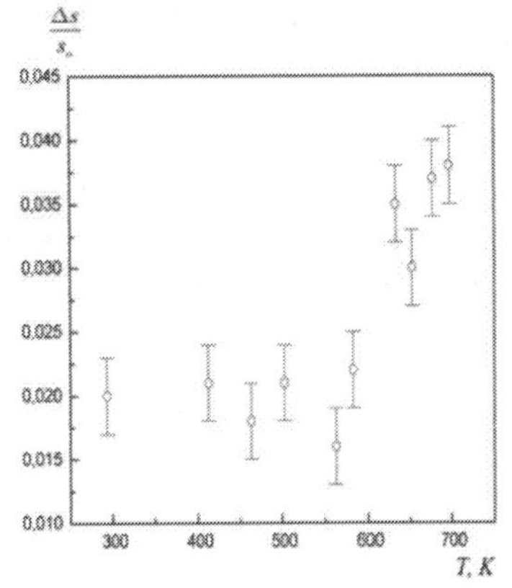

Figure 2. The temperature change of

parameter for $Fe_{73.1}Si_{15.5}B_{7.4}Nb_{3.0}Cu_{1.0}$ amorphous alloy.

like chemical ordering. The size of such nanoclusters, determined from prepeak half-width, was found to be ~1-2 nm.

Table 1. Structure parameters of $Fe_{73.1}Si_{15.5}B_{7.4}Nb_{3.0}Cu_{1.0}$ amorphous alloy at different temperatures.

T, K	s_o, nm^{-1}	s_c, nm^{-1}	$\Delta s/s_o$	$a(s_o)$	r_1, *nm*
293	31.10	30.50	0.020	3.50	0.261
413	31.10	30.45	0.021	3.42	0.261
463	31.05	30.50	0.018	3.67	0.260
503	31.05	30.40	0.021	3.58	0.261
563	31.00	30.50	0.016	3.68	0.261
583	31.00	30.30	0.022	3.54	0.261
633	31.00	29.90	0.035	3.36	0.261
653	31.00	30.00	0.030	3.32	0.260
678	31.05	29.90	0.037	3.59	0.262
698	31.05	29.85	0.038	4.08	0.263

In order to determine the temperature change of structure inhomogeneities in amorphous alloy under investigation we have analyzed the angular distribution of scattered intensity in the region of principal maximum. For that purpose, along with determination of peak position as top point (s_o), the centroid of maximum was determined too according to formula:

$$S_c = \frac{\int_{S_1}^{S_2} I(S)ds}{(S_2 - S_1)}$$

where (s_1,s_2)- limits of integral (s_1=9.8 nm^{-1}, s_2=45.0 nm^{-1}), $I(s)$- scattered intensity distribution. Difference between both values (table 1) is evidence of asymmetric profile of principal diffraction maximum. Most probably that inhomogeneous short range order structure with Fe_3Si clusters is just the main reason of such scattered intensity distribution. One might could consider that $\frac{Ds}{s_o}$ ($Ds = s_o - s_c$) parameter is the measure of inhomogeneity in structure of amorphous alloy.

Temperature dependence of such parameter reveals two individual regions of its change (Fig. 2). This parameter is in fact unchangeable within temperature range (293-583 K), whereas it increases significantly within pre-crystallization interval of temperatures (633-698 K). We suggest that increase of inhomogeneity is caused by increasing both of size and number of Fe_3Si nanoclusters resulting the Nb - atoms enrichment of amorphous matrix. Shift of centroid diffraction maximum (s_c) to low s - values confirms such behavior. This conclusion is in agreement with results, obtained in [8], where the significant structure changes in $Fe_{73.5}Si_{13.5}B_{9.0}Nb_{3.0}Cu_{1.0}$, amorphous alloy, annealed at T=623-673 K were revealed by means of small angle X–ray diffraction. The increasing of inhomogeneous regions of electron density fraction in amorphous alloy at low temperature annealing was suggested.

4.Conclusions

1.It is shown that prepeak in intensity curve of $Fe_{73.5}Si_{13.5}B_{9.0}Nb_{3.0}Cu_{1.0}$, amorphous alloy is related with correlation in distribution of Si-Si and Fe-Fe atomic pairs in Fe_3Si like clusters.
2.Increase of inhomogeneities fraction in short range order structure at near crystallization temperatures is supposed to be caused by increasing both of size and number of Fe_3Si clusters as well as Nb-atoms enrichment of amorphous matrix.
3. At initial crystallization the formation of nanocrystals on the base of α-Fe(Si) ordered solid solution occurs. The structure of such nanocrystals is unchangeable within 713-823 K temperature range.

References

[1] Hono K, Ping D H, Ohnuma M and Onodera H 1999 Cu clustering and Si partioning in the carly crystallization stage of an $Fe_{73.5}Si_{13.5}B_{9.0}Nb_{3.0}Cu_{1.0}$ amorphous alloys. *Acta Mater*. **47**. 997-1006.

[2] Chen W Z and Ryder P L 1995. X-ray and differential scanning calorimetry study of the crystallization of amorphous $Fe_{73.5}$ Cu_1Nb_3 $Si_{13.5}B_9$ alloy**.** *Mat. Sci. Eng*. **34** 204-209.

[3] Mudry S, Bednarska L, Kulyk Yu, Kovbuz AND M, Herstyk O. 2004 Temperatures changes of structure in $Al_{87}Ni_8Y_5$ amorphous alloy. *Archives of Materials Science* **25** 373-379.

[4] Mattern N, Danzig A and Muller M 1995 Effect of Cu and Nb on crystallization and magnetic properties of amorphous $Fe_{77.5}Si_{15.5}B_7$ alloys. *Mat. Sci. Eng*. **194** 77-85.

[5] Ruppersberg H, Lee D and Wagner C N J 1980 Observation of chemical short range order in an amorphous $Ni_{40}Ti_{60}$ alloy. *J. Phys*. F: *Metal Phys.***10** 1645-52.

[6] Sakata M, Cowlam N and Davies H A 1981 Chemical short-range order in liquid and amorphous $Cu_{66}Ti_{34}$ alloys. *J. Phys*. F: *Metal Phys*. **11** 157-162.

[7] Jergel M and Mrafko P 1984. The temperature dependence of the prepeak in the diffraction pattern of the amorphous $Ti_{61}Cu_{16}Ni_{23}$ and $Ti_{62.5}Cu_{12}Ni_{23}Si_{2.5}$ alloys. *Phys. Stat. Sol*. **83A** 113-121.

[8] Xu. R, Verkerk P, Howell W S at al. 1993 Nanometre superstructure in liquid alkali-thallium alloys *J. Phys.:Condens. Mater* **5** 50 9254-9260.

[9] Zhang Lin, Wu Youshi, Bian Xnifang at al. Origin of the prepeak in the structure factors of liquid and amorphous Al-Fe-Ce alloys *J.Phys.:Condens. Mater* **11** 50 7959-69

[10] Maslov V V, Il'inski A G, Nosenko V K, Brovko A P and Evlash I K 2000 Roentgenografic study of structural transformation in a thermotreated amorphous Finemet-type alloys. *Metallofizika i noveishie teknologii* **22** 3 43-55.

Journal of Physics: Conference Series **289** (2011) 012007 doi:10.1088/1742-6596/289/1/012007

On the structural-optical correlations in radiation-modified chalcogenide glasses

T. Kavetskyy[1-3], I. Kaban[4], O. Shpotyuk[2,3], W. Hoyer[4], V. Tsmots[1]

[1]Solid State Microelectronics Laboratory, Drohobych Ivan Franko State Pedagogical University, 24 I. Franko Str., Drohobych, 82100, Ukraine
[2]Institute of Materials, Scientific Research Company "Carat", 202 Stryjska Str., Lviv, 79031, Ukraine
[3]Institute of Physics, Jan Dlugosz University of Czestochowa, 13/15 Al. Armii Krajowej, Czestochowa, 42-201, Poland
[4]Institute of Physics, Chemnitz University of Technology, 09107 Chemnitz, Germany

e-mail: kavetskyy@yahoo.com

Abstract. In this work, we report our recent results on the γ-irradiation-induced structural transformations in the Ge-Sb-S glasses as observed from the structural studies using high-energy synchrotron x-ray diffraction and extended x-ray absorption fine structure spectroscopy in comparison with the optical measurements using VIS/IR spectroscopy techniques. The structural-optical correlations in the radiation-induced effects are established. The structural changes upon irradiation are explained in the frames of the concept of coordination topological defects formation.

1. Introduction

Chalcogenide glasses are known as perspective optoelectronic materials with unique physical and chemical properties, induced phenomena, etc. [1]. Their attractive application in optoelectronics is tightly related to the externally induced structural defects.

The concept of coordination topological defects (CTDs) was applied to interpret both photo- and radiation-induced phenomena in amorphous chalcogenides [2]. Infrared (IR) Fast Fourier Transform spectroscopy [3] is a sensitive method for CTDs' detection due to direct relation between intensity of vibrational mode and concentration of oscillating centres/chemical bonds. Recently, it has been shown [4] that Raman scattering in depolarized configuration is also an informative tool for CTD identification. However, application of vibrational methods is limited because they provide information on the statistics of chemical bonds only and help to construct possible topological CTD-related schemes, but no information on the local atomic structure in the vicinity of CTD can be obtained. Besides, the sensitivity of IR and Raman spectroscopy is rather low if oscillating complexes are present in the glass matrix in a small amount, while the role of these complexes in the externally induced defect formation processes could be significant.

The goal of the present work is to apply direct structural techniques such as high-energy synchrotron x-ray diffraction (XRD) and extended x-ray absorption fine structure (EXAFS) spectroscopy in combination with optical VIS/IR measurements in the visible (VIS) and infrared (IR) spectral ranges for the study of radiation-modified chalcogenide glasses on the example of Ge-Sb-S ternary system.

2. Experimental

The bulk $Ge_xSb_{40-x}S_{60}$ samples of chalcogenide glasses with x = 5, 25, 27 and 35 were prepared by a standard melt-quenching method. Radiation treatment of the samples with accumulated dose of 2.41 and 7.72 MGy was performed at the normal conditions of stationary radiation field, created in a closed cylindrical cavity by a number of concentrically established ^{60}Co radioisotope capsules as shown schematically in Fig. 1. No special measures were taken to prevent uncontrolled thermal annealing of the samples, but maximum temperature in the irradiating camera did not exceed 320-330 K during prolonged γ-irradiation (more than 30 days), providing absorbed dose power $P < 5$ Gy/s. Principal advantages of γ-irradiation as a structure modification factor for amorphous chalcogenides among other types of ionizing irradiation are discussed elsewhere [5].

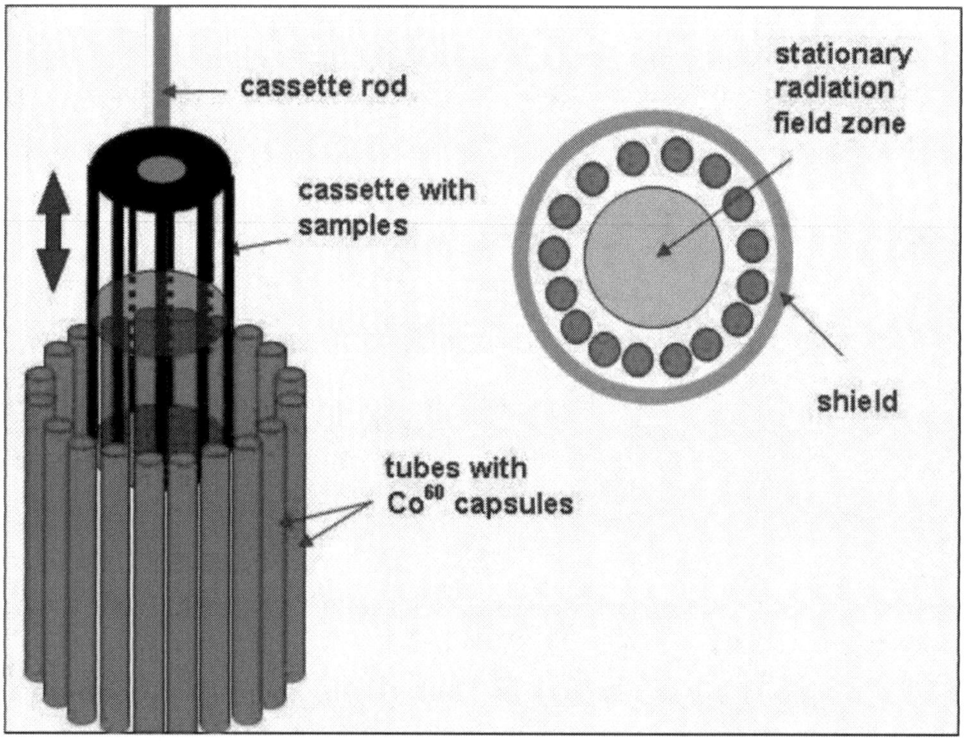

Figure 1. Setup for radiation treatment of the chalcogenide glasses.

High-energy XRD experiments were performed at the synchrotron experimental station BW5 at HASYLAB, DESY in Hamburg (Germany). EXAFS measurements at Ge K-edge were carried out at the synchrotron beam line X1 (HASYLAB) in transmission mode using Si (111) double-crystal monochromator. The measuring time was k-weighted during collection of the signal. The experimental data were treated using VIPER program [6] and FEFF8.4 code [7]. VIS optical transmission measurements were performed with "Specord M40" spectrophotometer. Far IR optical reflectance measurements were conducted with IR spectrometer "KSDI-82".

3. Results and discussion

Fig. 2 shows parts of the pair distribution functions $g(r)$ in the range of 2.0-3.0 Å for $Ge_{25}Sb_{15}S_{60}$ and $Ge_{35}Sb_5S_{60}$ glasses in the unmodified and γ-irradiation-modified states [8]. The main peak at 2.26 Å corresponds to the Ge–S first neighbor correlations; the shoulder for $Ge_{25}Sb_{15}S_{60}$ and peak for $Ge_{35}Sb_5S_{60}$ at 2.65 Å corresponds to the Ge–Sb first neighbor correlations. The radiation-induced structural changes in the Ge–S correlations are more pronounced for composition with 25 at.% of Ge, while the changes in Ge–Sb correlations are better detectable for the alloy with 35 at.% of Ge.

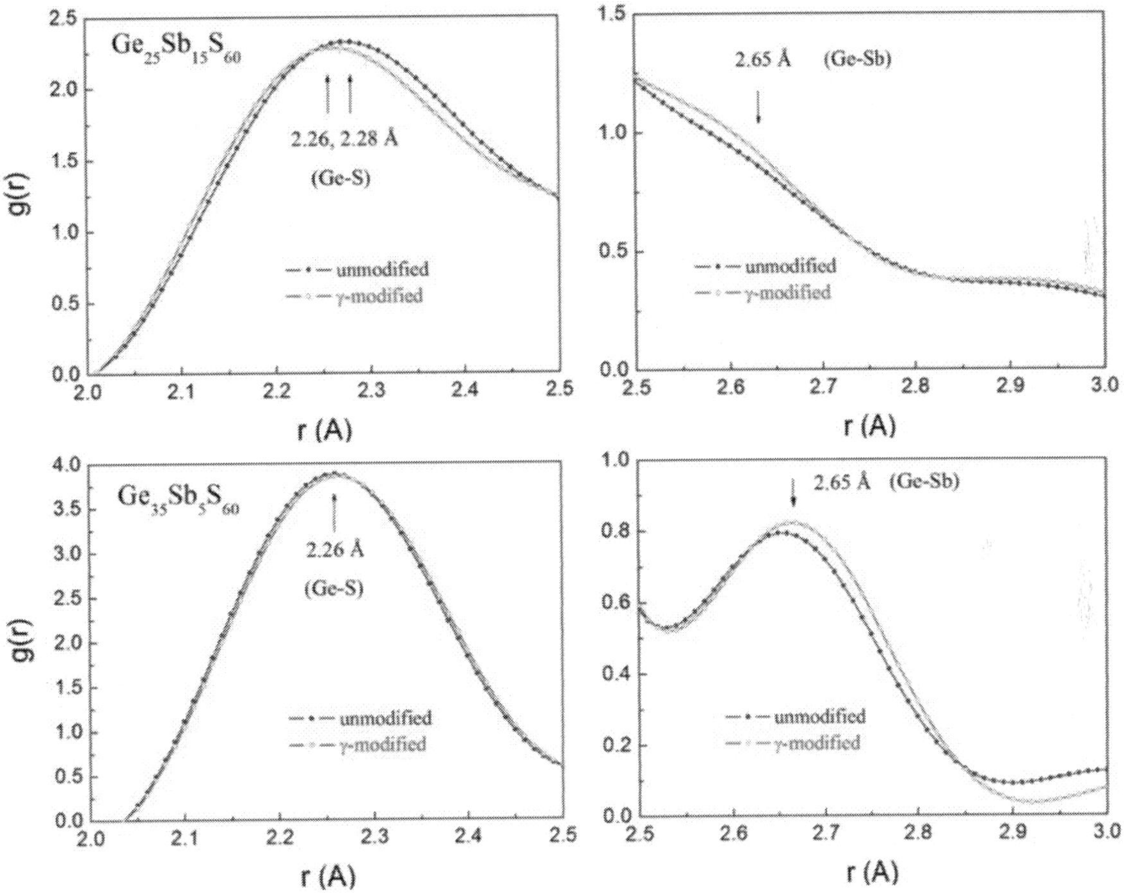

Figure 2. Parts of the pair distribution functions $g(r)$ in the range of 2.0-3.0 Å for $Ge_{25}Sb_{15}S_{60}$ (top) and $Ge_{35}Sb_5S_{60}$ (bottom) glasses in unmodified and γ-irradiation-modified states.

IR optical reflectance measurements [9] support the changes observed on $g(r)$'s, but only for the Ge–S correlations (Fig. 3). Indeed, only the strong peak at 370 cm^{-1} corresponding to the bond stretching modes in GeS_4 tetrahedra (vibrations of Ge–S bonds) [10] is modified upon irradiation. The peak at 290 cm^{-1} (well resolved for the glass with 25 at. % Ge) corresponds to the bond stretching modes in SbS_3 pyramids (vibrations of Sb–S bonds) [11]. A shoulder at 330 cm^{-1} (better seen at higher Ge content) is assigned to the bond stretching modes in SbS_3 pyramids [12]. The Sb–S correlations are found to be stable to radiation impact. The Ge–Sb vibrations are not resolved in the IR spectral region studied probably due to a small amount of these groups in the glass matrix of Ge–Sb–S ternaries as it was predicted earlier by Feltz [13] and observed experimentally in the XRD [8] and EXAFS [14] studies.

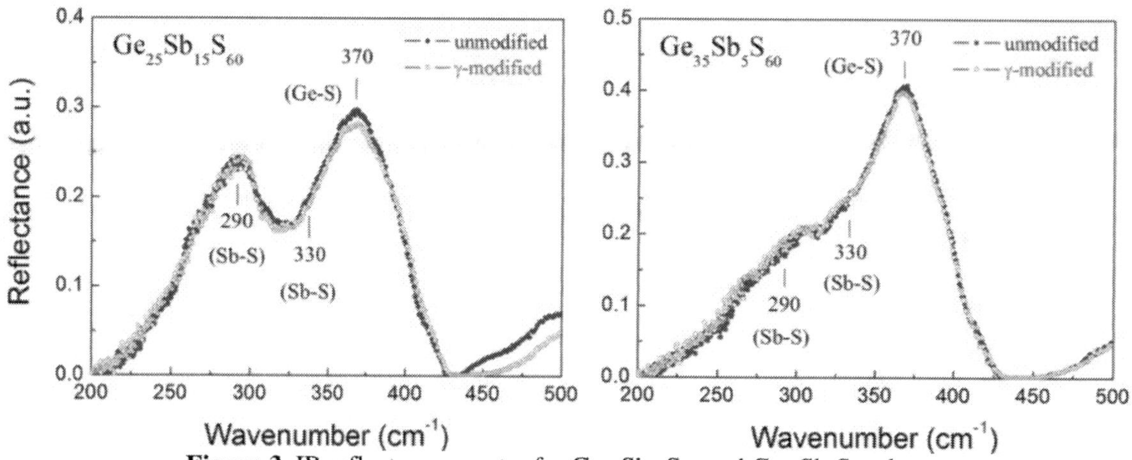

Figure 3. IR reflectance spectra for $Ge_{25}Sb_{15}S_{60}$ and $Ge_{35}Sb_5S_{60}$ glasses in unmodified and γ-irradiation modified states.

In the frames of the concept of coordination topological defects [2], the radiation-induced defective structural transformations detected in the XRD and IR studies of the $Ge_{25}Sb_{15}S_{60}$ and $Ge_{35}Sb_5S_{60}$ glasses can be explained by the formation of CTDs as a result of chemical bond distortion and switching upon irradiation. The changes in Ge–S correlations can be an indicative of the distortion of Ge–S bonds with switching and creation of negatively charged under-coordinated Ge_3^- and positively charged over-coordinated S_3^+ defects. Schematically this mechanism is illustrated in Fig. 4 (left) [8]. The radiation-induced non-defective structural transformations detected only from XRD study are connected with formation of Ge–Sb bonds with normal atomic coordination (Ge_4^0, Sb_3^0, S_2^0) in their vicinity with intermediate configuration accompanied by appearance of unstable Ge_3^- and S_3^+ CTDs as shown in Fig. 4 (right) [8]. The nature of Sb atoms (property of Sb to annihilate a wrong coordination defect in its vicinity) plays a key role for occurrence of the non-defective mechanism.

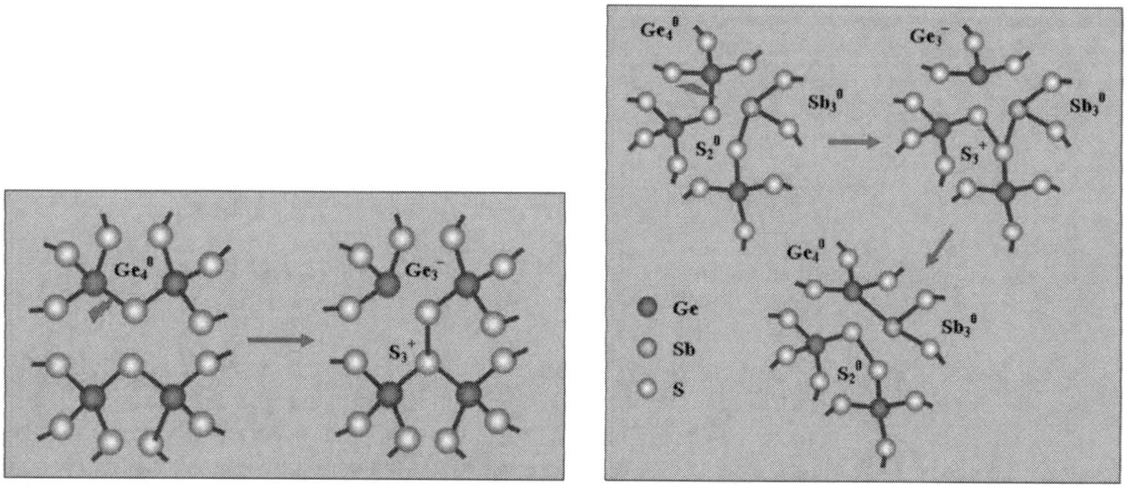

Figure 4. Radiation-induced defective (left; formation of Ge_3^- and S_3^+ CTDs) and non-defective (right; formation of Ge–Sb bonds with normal atomic coordination (Ge_4^0, Sb_3^0, S_2^0) in their vicinity with intermediate configuration accompanied by appearance of unstable Ge_3^- and S_3^+ CTDs) structural transformations in Ge–S sub-system of Ge–Sb–S glasses.

It is understandable that the main radiation-induced structural transformations (defective and non-defective) take place in Ge–S sub-system, whereas no changes in Sb–S correlations demonstrate radiation stability of Sb–S sub-system.

The EXAFS $\chi(k)k^3$ Ge K-edge spectra and their Fourier transforms for $Ge_{27}Sb_{13}S_{60}$ glass in the unmodified and γ-radiation modified states (Fig. 5) give an evidence for the conclusion on the changes in the Ge–S sub-system. Indeed, the first nearest neighbor distance r = 2.24 Å detected in the first coordination shell is attributed to the Ge–S bonding [15]. Upon irradiation, the coordination number of Ge atoms N_{GeS} decreases by 1.3 % and Debye-Waller factor σ^2_{GeS} or MSRD increases by 4.4 % that support the appearance of CTDs or atomic pairs with wrong coordination and static disorder.

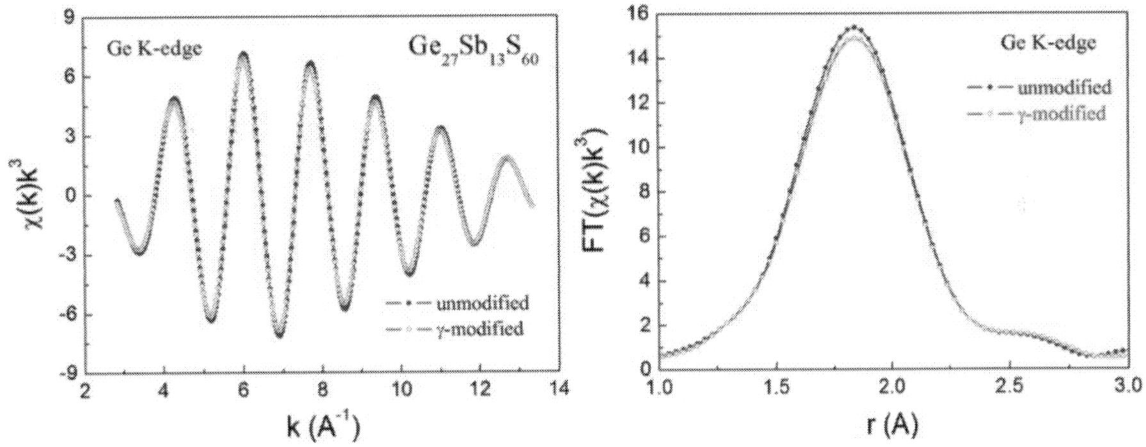

Figure 5. EXAFS $\chi(k)k^3$ Ge K-edge spectra and their Fourier transforms (without phase shift correction) for $Ge_{27}Sb_{13}S_{60}$ glass in unmodified and γ-irradiation-modified states.

The radiation stability or non-sensitivity of Sb–S sub-system is evidenced also from optical transmission measurements as illustrated in Fig. 6. It is clearly seen that the red shift of optical transmittance (or radiation-induced darkening effect) is detected for the Ge-rich composition with 27 at.% Ge and no effect is observed for the Sb-rich alloy with 35 at.% Sb.

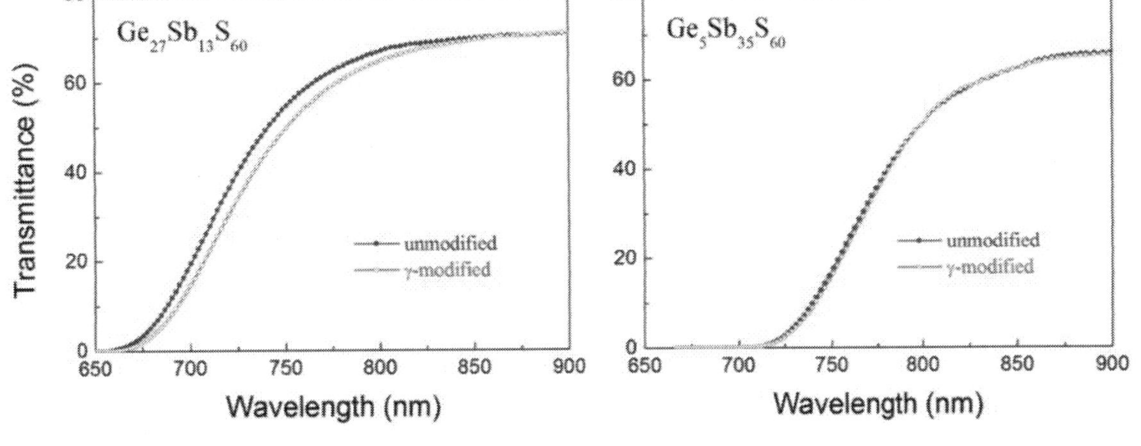

Figure 6. Optical transmission spectra for $Ge_{27}Sb_{13}S_{60}$ and $Ge_5Sb_{35}S_{60}$ glasses in unmodified and γ-irradiation modified states.

4. Conclusion

We report our recent results on the γ-irradiation-induced structural transformations in Ge-Sb-S glasses as observed from the structural studies using high-energy synchrotron XRD and EXAFS in comparison with the optical investigations using VIS/IR spectroscopy techniques. The structural-optical correlations in the radiation-induced effects are established. In contrast to the IR optical measurements, XRD study reveals the role of heteronuclear metal-metal chemical bonding (Ge-Sb), which cannot be negligible in the radiation-induced structural transformations in the glass system investigated. It is concluded that both defective and non-defective mechanisms of the radiation-induced structural changes might be responsible for the effects observed in the XRD and IR experiments. The radiation-induced structural changes can be plausibly explained with the coordination topological defects formation concept.

Acknowledgments

This work is partially supported by the Ministry of Education and Science of Ukraine and Fundamental Researches State Fund of the Ministry of Education and Science of Ukraine (Project #F25.2/138). Two of us (T.K. and I.K.) acknowledge DESY for support of the experiments performed at HASYLAB (Hamburg, Germany). T.K. also acknowledges International Visegrad Fund for Scholarship.

References

[1] Shimakawa K, Kolobov A and Elliott S R 1995 Photoinduced effects and metasltability in amorphous semiconductors and insulators *Adv. Phys.* **44** 475-588
[2] Shpotyuk O I 1996 Photo- and radiation-induced coordination defects in amorphous chalcogenides *Proc. SPIE* **2968** 246-255
[3] Golovchak R Ya and Shpotyuk O I 2005 Radiation-induced bond switching in mixed Ge-As sulphide glasses *Philos. Mag.* **85** 2847-2860
[4] Kavetskyy T, Vakiv M and Shpotyuk O 2007 Charged defects in chalcogenide vitreous semconductors studied with combined Raman scattering and PALS metods *Radiation Measurements* **42** 712-714
[5] Kavetskyy T, Shpotyuk O, Kaban I and Hoyer W 2007 Atomic- and void-species nanostructures in chalcogenide glasses modified by high energy gamma-irradiation J. Optoelectron. *Adv. Mater.* **9** 3247-3252
[6] Klementev K V 2001 Extraction of the fine structure from x-ray absorption spectra *J. Phys. D: Appl. Phys.* **34** 209-217
[7] Ankudinov A L, Ravel B, Rehr J J and Conradson S D 1998 Real-space multiple-scattering calculation and interpretation of x-ray-absorption near-edge structure *Phys. Rev.* B **58** 7565-7576
[8] Kavetskyy T, Shpotyuk O, Kaban I and Hoyer W 2008 Radiation-modified structure of $Ge_{25}Sb_{15}S_{60}$ and $Ge_{35}Sb_5S_{60}$ glasses *J. Chem. Phys.* **128** 244514-1-244514-8
[9] Kavetskyy T, Shpotyuk O, Balitska V, Dovbeshko G, Blonskyy I, Kaban I, Hoyer W, Iovu M and Andriesh A 2008 Vibrational and structural properties of unmodified and radiation-modified chalcogenide glasses for advanced optical applications *Proc. SPIE* **7142** 71420B-1-71420B-8
[10] Vahalova R, Tichy L, Vlcek M and Ticha H 2000 Far infrared spectra and bonding arrangement in some Ge-Sb-S glasses *Phys. Stat. Solidi* A **199** 199-209
[11] Ticha H, Tichy L, Rysava N and Triska A 1985 Some physical properties of the glassy $(GeS_2)_x(Sb_2S_3)_{1-x}$ system *J. Non-Cryst. Solids* **74** 37-46
[12] El Idrissi Raghni M A, Lippens P E, Olivier-Fourcade J and Jumas J C 1995 Local structure of glasses in the As_2S_3-Sb_2S_3 system *J. Non-Cryst. Solids* **192-193** 191-194
[13] Feltz A 1986 Amorphous and Vitreous Inorganic Solids (Moscow: Mir)
[14] Cervinka L, Smotlacha O, Bergerova J and Tichy L 1991 The structure of the glassy Ge-Sb-S system and its connection with the MRO structures of GeS_2 and Sb_2S_3 *J. Non-Cryst. Solids* **137-138** 123-126
[15] Sen S, Ponader C W and Aitken B G 2001 Atomic structure and chemical order in Ge-As sulphide glasses: a combined Ge and As K-edge EXAFS study *J. Non-Cryst. Solids* **293-295** 204-210

Journal of Physics: Conference Series **289** (2011) 012008 doi:10.1088/1742-6596/289/1/012008

Post- irradiation structural relaxation in quasibinary arsenic/antimony trisulphide glasses

Ya Shpotiuk[1,2], **V Baltiska**[1,3]

[1] Scientific Research Company "Carat" , Institute of Materials, 202 Stryjska Str., Lviv, Ukraine
[2]Ivan Franko National University of Lviv, Faculty of Electronics, 50 Drogomanova Str., Lviv, 79005 Ukraine
[3]Lviv State University of Vital Function Safety, 35 Kleparivska Str, Lviv, 79023 Ukraine

E-mail: shpotyuk@novas.lviv.ua

Abstract. Radiation – induced dynamic effects were studied in chalcogenide glasses of quasibinary $(As_2S_3)_{1-x}(Sb_2S_3)_x$ (0<x<0.3) system using optical spectrophotometry technique in the region of fundamental optical absorption edge. It was established that dynamical component of post – irradiation decaying gradually grows with Sb content in the studied glasses, the kinetics of the observed changes begin non-exponential tending towards monomolecular one. The observed changes were explained by compositional dependence of free volume in the studied system.

1. Introduction

Radiation induced effects in chalcogenide glassy semiconductors (ChGS) are in sphere of tight interests because of perspectives for their wide application in IR optics and photonics [1,2]. The first research on radiation-physical properties of ChGS was performed yet in the middle of the 60-s by I.A. Domoryad [3,4]. It was established, in part, that high energy γ-irradiation caused low-energetic shift in fundamental optical absorption edge of ChGSas as well as a number of other changes in their mechanical properties. The observed radiation-induced darkening effect was unstable [5,6]. It tended monotonically towards some residual value, showing a so-called dynamic component in contrast to static one, which kept mostly constant after irradiation. In Ge-based ChGS, this post-irradiation instability was shown to be quite prolonged up to 2-3 months. However, a detailed systematic study of these effects in other ChGS, especially in arsenic sulphides, has not been performed yet.

2. Experimental

In this work, we shall present the results on time degradation in radiation-induced optical darkening of quasibinary $(As_2S_3)_{1-x}(Sb_2S_3)_x$ ChGs. The studied ChGS samples of four based compositions (x=0,0.1,0.2,0.3) were prepared by conventional melt quenching. The ingot was cut into discs of ~1 mm thickness and polished to high optical quality. To remove mechanical strains appeared after quick cooling, the samples were annealed during 1 hour at the temperature of 10-20 K below their glass transition temperature. Radiation treatment of the samples was conducted by γ-quanta with average energy of 1.25MeV and absorbed dose of 1.66 MGy in normal conditions of stationary radiation field created in closed cylindrical cavity surrounded by ^{60}Co sources. No special measures were taken to prevent uncontrolled thermal annealing of the samples, but maximum temperature in the irradaiating camera did not exceed 320-330K during prolonged γ-irradiation (more then 10 days). The spectral dependences of optical transmission τ of the samples were measured in 200-900 nm range in different time periods after irradiation (from 1 to 40 days) using "Specord M-40" device, the maximum error of measurements being no more 0.5%.

3. Results and discussion

The typical spectral dependence of radiation-induced optical darkening $\Delta\tau$ ($\Delta\tau=\tau_{in}-\tau_{irr}$,where τ_{in} is optical transmittance of initial non-irradiated sample, τ_{irr} is optical transmittance of irradiated sample measured in 1,2,3,5 and 40 day after irradiation, respectively) for vitreous As_2S_3 is shown in Figure 1.

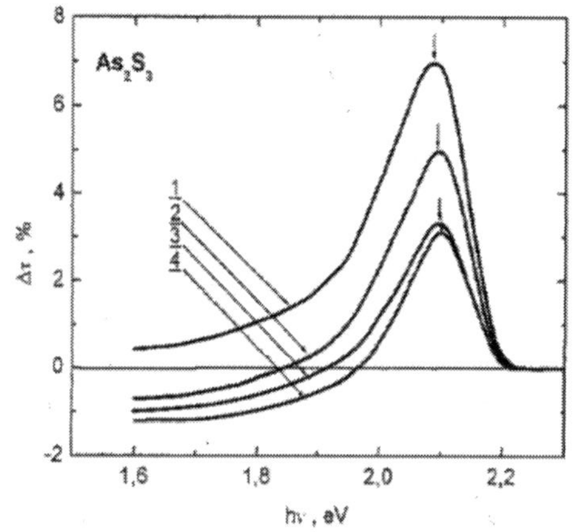

Figure 1. Spectral dependences of radiation induced optical darkening in vitreous As_2S_3 measured in the 1-st (1), 3-rd (2), 5-th (3) and 40-th day (4) after irradiation.

It is obvious that spectral dependence of $\Delta\tau$ has non-monotonic character with well expressed maximum $\Delta\tau_{max}$ at photon energies near 2.0 eV. With increase in time after irradiation, the low-energetic wing of this dependence shifts towards more negative $\Delta\tau$ values attaining a more symmetric shape (see Figure 1).

The similar spectral dependences were obtained for ChGS samples of the studied quasibinary $(As_2S_3)_{1-x}(Sb_2S_3)_x$ system (x=0.1,0.2 and 0.3). The only difference was decrease in the total value of measured radiation –induced effect with Sb content. Simultaneously, the spectral dependences of Sb-contained ChGS demonstrate shift towards negative $\Delta\tau$ values and increase in the photon energies corresponding to maximum $\Delta\tau_{max}$ values.

Thus, the character of dynamic post-irradiation effects in the studied $(As_2S_3)_{1-x}(Sb_2S_3)_x$ ChGS is determined mainly by their chemical composition (the positive $\Delta\tau$ value was only in vitreous As_2S_3 while with Sb_2S_3 content this parameter attained the negative values).

Compositional trends in static and dynamic components of radiarion – induced darkening effect in the studied ChGS at wavelength of maximal changes are well illustrated in Figure 2. The observed effect of post-irradiation decaying sharply enhanced with Sb content so that in ChGS with x=0.3 this effect was not detectable vanishing to small under-margin value.

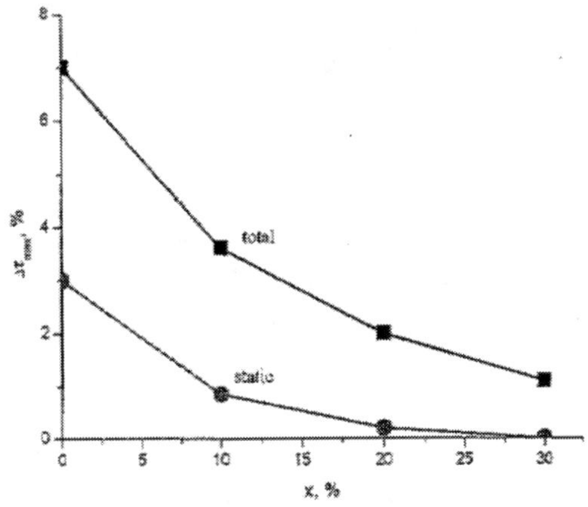

Figure 2. Compositional dependences of total and static components of radiation-induced optical darkening in $(As_2S_3)_{1-x}(Sb_2S_3)_x$ ChGS at the wavelength of maximal changes

The observed behavior of radiation – induced darkening effect in the studied ChGS samples correlates well with compositional dependence of free volume in this quasibinary system (see Fig.3). This feature allows us to put forward the hypothesis on decisive role of atomic compactness in the post-irradiation relaxation efficiency in ChGS. Dynamic tendency in radiation-induced changes significantly grows with Sb_2S_3 content in accordance to the observed rough decrease in molar volume.

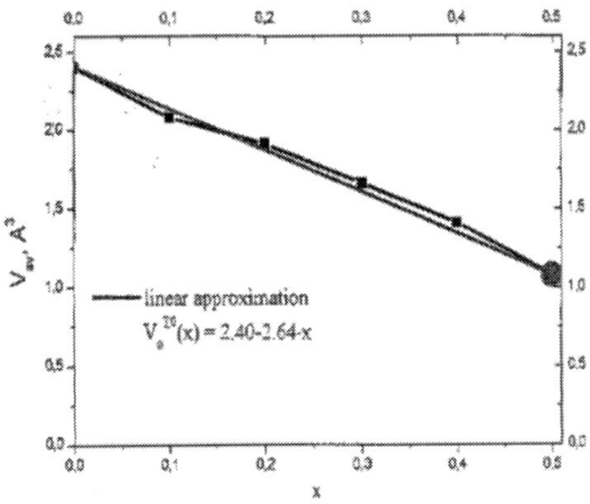

Figure 3 Compositional dependence of free volume per atom in $(As_2S_3)_{1-x}(Sb_2S_3)_x$ ChGS (bold circus denote values calculated as a comparison between atomic densities of isocompositional glasses and crystals, other points correspond to values estimated as input from As_2S_3 and Sb_2S_3 components

In order to quantitatively describe kinetics of the above post-irradiation instability effects, the possible mathematical variants of decay processes in ChGS were taken into account. In general, the rate of decaying in the control physical parameter $\chi(t)$ can be described due to known differential equqtion presented below [7,8].

$$\frac{dn}{dt} = -\lambda n^{\alpha} t^{\beta}, \qquad (1)$$

where α and β are power indexes (α>0, 0<β<1) and λ is constant dependent on ChGS composition.

As it was pointed out in [7,8], there are five typical relaxation functions (RF), which can be simultaneously considered as possible partial solutions of Eq.(10 in dependence on numerical values of α and β parameters (Table 1).

Table 1. Degradation equations and corresponding RF for post-irradiation instability kinetics in ChGS

Degradation equation	RF type	RF parameters
$\frac{d\chi}{dt} = -\lambda\chi$	MonomolecularRF 1 $\chi = \chi_O e^{-\frac{t}{\tau}}$	$\chi_O = e^C, \tau = \frac{1}{\lambda}$, $c - const,\ \lambda \neq 0$
$\frac{d\chi}{dt} = -\lambda\chi^2$	Bimolecular RF 2 $\chi = \frac{\chi_O}{1 + \frac{t}{\tau}}$	$\chi_O = e^C,\ \tau = \frac{1}{\lambda}$, $c - const,\ \lambda \neq 0$
$\frac{d\chi}{dt} = -\lambda\chi^{\alpha}$	Party-polymerized RF 3 $\chi = \frac{\chi_O}{\left(1 + \frac{t}{\tau}\right)^{\kappa}}$	$\chi_O = c^{\frac{1}{\alpha - 1}},\ \tau = \frac{c}{\lambda(\alpha - 1)}$, $\kappa = \frac{1}{\alpha - 1}$, $c - const,\ \alpha \neq 1,\ \lambda \neq 0$
$\frac{d\chi}{dt} = -\lambda\chi t^{\beta}$	Nonexponential RF 4 $\chi = \chi_O exp\left[-\left(\frac{t}{\tau}\right)^{\kappa}\right]$	$\tau = \left(\frac{1 + \beta}{\lambda}\right)^{\frac{1}{1 + \beta}}$, $\kappa = 1 + \beta$, $\chi_O = e^C$, $c - const,\ \beta \neq -1,\ \lambda \neq 0$
$\frac{d\chi}{dt} = -\lambda\chi^{\alpha} t^{\beta}$	Generalized RF 5 $\chi = \frac{\chi_O}{\left(1 + \left(\frac{t}{\tau}\right)^{\kappa}\right)^{r}}$	$r = \frac{1}{\alpha - 1}$, $\kappa = 1 + \beta$, $\tau = \left(\frac{c}{\lambda} \cdot \frac{1 + \beta}{\alpha - 1}\right)^{\frac{1}{1+\beta}}$, $\chi_o = c^{\frac{1}{1-\alpha}}$, $c - const,\ \alpha \neq 1,\ \beta \neq -1,\ \lambda \neq 0$

With a purpose of adequate mathematical description of the observed post-irradiation relaxation kinetics, the numerical values of fitting parameters of the above RF were calculated in such a way to minimize the mean-square deviation of the experimentally measured points from the chosen RF.

As a result (see Fig.4 and table 2), it was established that kinetics of the observed decaying was satisfactorily described on the basis of non-exponential RF 4, taken in the form of so-called suppressed-exponential RF with character non-exponential indexes of 1.8,2.2,3.0,3.5 days for $(As_2S_3)_{1-x}(Sb_2S_3)_x$ with x=0, 0.1,0.2 and 0.3, respectively. This kinetics tends towards single- exponential (or, alternatively, monomolecular) one with character time constants of 2.1,1.9, 1.6 and 1.4 days (for the above ChGS taken in the same sequence).

This result contradicts strongly to previus study on Ge-based ChGS [7], where bimolecular relaxation kinetics proper to annihilation of coordination topological defects was dominant. It means that other type of structural – relaxation processes at the level of medium range order is more essential in the glass under consideration. It should be supposed that in case of studied $(As_2S_3)_{1-x}(Sb_2S_3)_x$ ChGS the process of post-irradiation relaxation attains activation character owing to renovation of destroyed chemical bonds, being described by monomolecular RF 1 rather than bimolecular one. It is quite understandable that such process is quicker in time giving small values of character time constants.

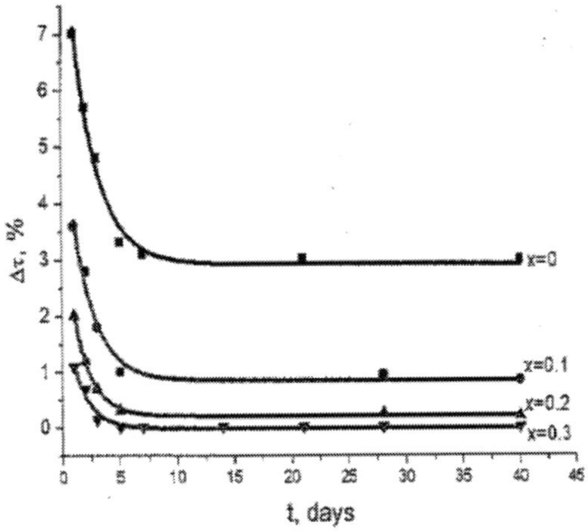

Figure 4. Time dependence of radiation – induced darkening $\Delta\tau_{max}$ in $(As_2S_3)_{1-x}(Sb_2S_3)_x$ ChGS(the bold lines correspond to modeling monomolecular RF)

Table 2. Main fitting parameters of post-irradiation relaxation in $(As_2S_3)_{1-x}(Sb_2S_3)_x$ ChGS

ChGS composition	Time constant in RF 1	Non-eponentiality index in RF 4
As_2S_3	2.1	1.8
$(As_2S_3)_{0.9}(Sb_2S_3)_{0.1}$	1.9	2.2
$(As_2S_3)_{0.8}(Sb_2S_3)_{0.2}$	1.6	3.0
$(As_2S_3)_{0.7}(Sb_2S_3)_{0.3}$	1.4	3.5

4.Conclusion

The process of post-irradiation structural relaxation in quasibinary $(As_2S_3)_{1-x}(Sb_2S_3)_x$ ChGS has an activation character owing to renovation of destroyed covalent chemical bonds, its kinetics being adequately described by suppressed-exponential RF tending towards monomolecular one. The observed compositional features in this decaying correlate well with known free volume tendencies.

Acknowledgements

The authors acknowledge support from Science and Technology Center in Ukraine under regular STCU Project No 3745. In part of kinetic modeling, this work was carried out in the framework of Agreement on Scientific Technological Cooperation between Scientific Research Company "Carat" (Lviv, Ukraine) and Jan Dlugosz University of Czestochowa (Poland), supported by Ukrainian and Polish Ministries of Education and Science.

Journal of Physics: Conference Series **289** (2011) 012008 doi:10.1088/1742-6596/289/1/012008

References

[1] Tanaka K 1996 Photoinduced processes in chalcogenide glasses. *Current Opinion in Solid State and Materials Science* **24** 567-571

[2] Zhang X, Ma H and Lucas J 2003 Application of chalcogenide glass bulks and fibers.*J .Optoelectron. Adv. Mater.* **5** 1327-33

[3] Tekucheva I A and Domoryad I A 1985 Influence of gamma-irradiation on changes in mechanical and optical characteristics of some ChGS, Stekloobraznyje polprovodniki. Austr. Of All-Union Conference, Leningrad 308-309

[4] Domoryad I A and Tekucheva I A 1989 Gamma – induced changes in mechanical and optical properties of some ChGS. Proc.Inter.Conf. "Non-Crystalline Semiconductors -89", Uzhgorod 138-140

[5] Shpotyuk O I, Kovalskiy A P, Skordeva E, Vateva E, Arsova D, Golovchak R Ya and Vakiv M 1999 Effect of gamma-irradiation on the optical properties of $Ge_xAs_{40-x}S_{60}$ glasses. *Physica* B *Condensed Matter* **271** 242-47

[6] Skordeva E, Arsova D,Pamukhieva V, Vateva E, Golovchak R, Kovalskiy A P and Shpotiuk O 2000 γ-induced changes in Ge-As-S glasses. *J. Opto-Electronics and Advanced Materials* **2**(3) 259-66

[7] Balitska V, Filipecki J, Shpotyuk O, Swiatek J and Vakiv M 2001 Dynamic radiation induced effects in chalcogenude glasses. *J.Non-Cryst.Solids* **287** 216-21

[8] Balitska V O, Butkievich B, Shpotiuk O I and Vakiv M M 2002 On the analytical description of ageing kinetics in ceramics manganite – based NTC thermistors. *Microelectronics Reliability* **42** 2003-07

[9] Mazurin O V 1977 Relaxation phenomena in glass. *J. Non-Cryst. Solids* **25** 130-169

[10] Griscom D L, Gingerich M E and Frieble E J 1993 Radiation – induced defects in glasses: origin of power-law dependences of concentration on dose. *Phys.Rev.Lett.* **71** 1019-22

[11] Wolliams G and Watts D C 1970 Non-symmetrical dielectric relaxation behavior arising from a simple empirical decay function. *Faraday Soc.Trans.* **66** 80-85

[12] DeBest J and Gilard P 1963 Variation of the viscosity of glass and relaxation of stresses during stabilization. *Physics and Chemistry of Glasses* **4** 117-28

Structural modification of covalent-bonded networks: on some methodological resolutions for binary chalcogenide glasses

M. Shpotyuk[1,2], Ya. Shpotyuk[1,3], O. Shpotyuk[1]

[1]Lviv Scientific Research Institute of Materials of SRC "Carat", 212, Stryjska str., Lviv, 79031, Ukraine

[2]Lviv Politechnic National University, 12, Bandera str., Lviv, 79013, Ukraine

[3]Ivan Franko National University of Lviv, 1, Universytetska str., Lviv, 79000, Ukraine

shpotyukmy@yahoo.com

Abstract. New methodology to estimate efficiency of externally-induced structural modification in chalcogenide glasses is developed. This approach is grounded on the assumption that externally-induced structural modification is fully associated with destruction-polymerization transformations, which reveal themselves as local misbalances in covalent bond distribution, normal atomic coordination and intrinsic electrical fields. The input of each of these components into the total value of structural modification efficiency was probed for quasibinary $(As_2S_3)_{100-x}(Sb_2S_3)_x$ ChG.

1. Introduction

Chalcogenide glasses (ChG, i.e. vitreous compounds of chemical elements from IV-th and V-th groups of the Periodic table with chalcogens – S, Se or Te) are unique disordered solids widely used in IR optoelectronics and photonics [1-3]. This uniqueness is connected with their extreme sensitivity to the influence of external factors such as temperature, photoabsorbed light or high-energy irradiation. That is why the problem on development of estimation methodology for structural modification (SM) efficiency in ChG seems to be very important, especially in respect to practical implementation of these disordered solids. Unfortunately, the SM model [4-6] developed up to now gives no quite adequate possibilities to resolve this problem.

With this in mind, we have presented a new estimation methodology for externally-induced structural modification in ChG. The experimental static radiation-induced darkening effect in quasi-binary $(As_2S_3)_{100-x}(Sb_2S_3)_x$ ChG was used as a criterion for the developed model.

2. Criterion for structural modification

Our approach is grounded on the assumption that externally-induced structural modification is fully associated with destruction-polymerization transformations [7].

In general, the SM criterion κ for metastable (or, equivalently, permanent) *ex-situ* changes can be estimated conceptually in accordance to the following equation:

$$\kappa = \kappa_1 \cdot (1 - \kappa_2), \qquad (1)$$

where κ_1 is SM criterion for transient *in-situ* changes determined after A.I. Popov [4-6] and κ_2 is structural relaxation (SR) input.

By accepting that Popov's coefficients in [4-6] is restricted by 2.0, it is conveniently to take a half of this parameter for κ_1 value in (1). So the efficiency of radiation-induced structural transformations occurred under irradiation can be find as:

$$\kappa_1^{Popov} = \frac{VEC - Z}{Z(1 - I_C)}, \quad (2$$

where VEC is average valence electron concentration; Z is average coordination number and I_C is average ionicity of chemical bonds.

$$VEC = \frac{\sum_i N_i x_i}{\sum_i x_i}, \quad (3)$$

where N_i and x_i is number of group and atomic fraction of i-th element, respectively.

$$Z = \frac{\sum_c cN_c}{\sum_c N_c}, \quad (4)$$

where N_c is number of atoms with c coordination.

$$I_C = \sum_m S_m N_m, \quad (5)$$

where S_m is ionicity of the bond m determined by electronegativity difference of the atoms

$$S_m = f(\Delta x) \cong 0.25 \cdot \Delta x^2 \quad (6)$$

and N_m is fraction of m bonds in material taken due to chemically-ordered covalent-bonded network [8].

SR input κ_2 describes relaxation effects in ChG after irradiation. It is evident that $\kappa_2 = 0$, if no relaxation in the irradiated ChG samples ($\kappa = \kappa_1$), while $\kappa_2 = 1$, if all changes relaxed after irradiation ($\kappa = 0$). Three different effects have an influence on the SR input κ_2 in ChG:

$$\kappa_2 = f\left(\kappa_2^{\Delta E} \cdot \kappa_2^{FV} \cdot \kappa_2^{M}\right). \quad (7)$$

The first component $\kappa_2^{\Delta E} = f(\Delta E)$ describes SR input owing to local misbalance in covalent chemical bonds distribution (re-switched covalent chemical bonds), being a function of energetic balance of corresponding bond-switching reaction:

$$\kappa_2^{\Delta E} = 1 - \frac{|\Delta E|}{E^0}, \quad (8)$$

where ΔE is energetic balance of induced bond switching; E^0 is average energetic linking of covalent network before bond switching (thermally irreversible bond switching occurs, if $\Delta E > 0$, while thermally reversible bond-switching occurs, if $\Delta E < 0$);

The second component $\kappa_2^{FV} = f(FV)$ describes SR input owing to local misbalance in normal (due to *8-N* rule [9]) atomic coordination, being a function of free volume excess in a glass in respect to crystal:

$$\kappa_2^{FV} = 1 - \frac{\Delta\rho}{\rho_{cr}} = \frac{\rho_g}{\rho_{cr}}, \quad (9)$$

where ρ_g and ρ_{cr} are atomic densities of glass and iso-compositional crystal, respectively.

The third component $\kappa_2^{M} = f(M)$ describes SR input owing to local misbalance in electrical charge distribution, being a function of average metallicity of covalent bonds:

$$\kappa_2^M = \alpha_M . \tag{10}$$

Due to W. Harrison [10], the bond metallicity coefficient α_M in (10) can be determined as:

$$\alpha_M \approx \frac{1.11 \cdot V_1}{\sqrt{V_2^2 + V_3^2}} , \tag{12}$$

where $V_1 = \frac{\varepsilon_P - \varepsilon_S}{4}$ is energy of metallic bond (ε_P and ε_S are atomic terms); $V_2 = \frac{2.16 \cdot \hbar^2}{m \cdot d^2}$ is energy of covalent bond ($\frac{\hbar^2}{m}$ = 7.62 eV•Å^2, d – bond length) and $V_3 = \frac{\varepsilon_P^A - \varepsilon_P^C}{2}$ is energy of ionic bond (ε_P^A and ε_P^C are anion and cation atomic terms, respectively, $V_3 = 0$ for homopolar bonds).

3. Experiment

The investigated samples of $(As_2S_3)_{100-x}(Sb_2S_3)_x$ ChG (x = 0, 10, 20 and 30) were obtained by conventional melt quenching (regime of switched-off furnace) in the evacuated quartz ampoules from a mixture of high purity (99.999 %) precursors. Amorphous state of prepared ChG was controlled visually by a character conch-like fracture and data of X-ray diffraction. Before the experiment samples were sliced and optical quality polished up to the 0.7-1.0 mm thickness.

The gamma-irradiation procedure was carried out in the closed cylindrical cavity of concentrically established Co^{60} sources (mean energy – 1.25 MeV) with a few Gy/s power at a temperature not exceeding 35 °C. The total duration of irradiation was nearly 2 months, the accumulated dose being as high as 1.66 MGy.

SPECORD M-40 spectrometer was used for the optical characterization of samples. Optical spectra of investigated ChG were recoded just after irradiation and 40 days later. Radiation-induced darkening was determined by normalized value of optical transmission change in the range of fundamental optical absorption edge (at the wavelength of maximum changes).

4. Results and discussion

$(As_2S_3)_{100-x}(Sb_2S_3)_x$ ChG system is characterized by absence of any structural anomalous in the concentration range 0<x<30. It means that all physical properties of these glasses are described mainly by the linear dependences. Some of them are shown in Table 1. Average energetic linking of covalent network E^0 is calculated as in [11]. Values of atomic densities of glass ρ_g and iso-compositional crystal ρ_{cr} are taken from [12].

Table 1. Some physical properties of $(As_2S_3)_{100-x}(Sb_2S_3)_x$ ChG.

Composition	X	E^0, eV CO CBN	ρ_g, g/cm^3	ρ_{cr}, g/cm^3
As_2S_3	0	2.120	3.19	3.49
$(As_2S_3)_{90}(Sb_2S_3)_{10}$	10	2.126	3.31	3.57
$(As_2S_3)_{80}(Sb_2S_3)_{20}$	20	2.132	3.41	3.66
$(As_2S_3)_{70}(Sb_2S_3)_{30}$	30	2.138	3.52	3.75

Experimentally-obtained compositional dependence of total and static radiation-induced darkening effect in $(As_2S_3)_{100-x}(Sb_2S_3)_x$ ChG is shown on the Fig. 1. One can observe that post-irradiation darkening decay increases with Sb content. Moreover, the radiation-induced darkening disappears completely in $(As_2S_3)_{70}(Sb_2S_3)_{30}$ ChG.

Journal of Physics: Conference Series **289** (2011) 012009 doi:10.1088/1742-6596/289/1/012009

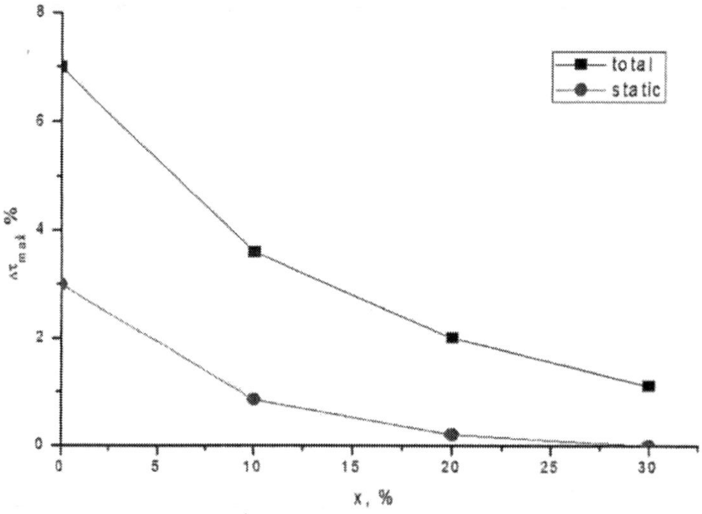

Fig. 1. Compositional dependence
of total and static radiation-induced darkening effect in $(As_2S_3)_{100-x}(Sb_2S_3)_x$ ChG.

The calculated values of SM criteria for $(As_2S_3)_{100-x}(Sb_2S_3)_x$ ChG are presented in Table 2. The energetic $\kappa_2^{\Delta E}$ criterion is calculated for two possible cases: $(As_4^+;S_1^-)$ coordination topological defect (CTD) formation due to (As–S) → (As–As) bond switching and $(Sb_4^+;S_1^-)$ CTD formation due to (Sb–S) → (Sb–Sb) bond switching in addition to the first one. In according to the above stated SM estimation principles, the compositional dependence of experimental static radiation-induced darkening effect in quasibinary $(As_2S_3)_{100-x}(Sb_2S_3)_x$ ChG can be explained only in the case of SM criteria κ_2 increasing with Sb content. In compliance with this, the $\kappa_2^{\Delta E}$ (As) and κ_2^{FV} parameters correlate well with experimental results. In contrast, the κ_2^{M} SM criterion changes negligibly, while $\kappa_2^{\Delta E}$ (As, Sb) have a tendency to decreasing with x.

On the basis of the above consideration, one can state that the input of $\kappa_2^{\Delta E}$ (As) SM criterion is greatest. At the same time, the tendencies in $\kappa_2^{\Delta E}$ (As, Sb) testify in a favour of absence of CTD based on Sb atoms.

Table 2. Numerical SM criteria for $(As_2S_3)_{100-x}(Sb_2S_3)_x$ ChG.

Composition	X	κ_1	$\kappa_2^{\Delta E}$ (As)[1]	$\kappa_2^{\Delta E}$ (As, Sb)[2]	κ_2^{FV}	κ_2^{M}
As_2S_3	0	0.6944	0.656	0.656	0.914	0.8190
$(As_2S_3)_{90}(Sb_2S_3)_{10}$	10	0.6966	0.691	0.650	0.926	0.8187
$(As_2S_3)_{80}(Sb_2S_3)_{20}$	20	0.6988	0.726	0.644	0.931	0.8184
$(As_2S_3)_{70}(Sb_2S_3)_{30}$	30	0.7011	0.761	0.639	0.939	0.8181

(As)[1] – $(As_4^+;S_1^-)$ CTD formation due to (As–S) → (As–As) bond switching
(As,Sb)[2] – $(Sb_4^+;S_1^-)$ CTD formation due to (Sb–S) → (Sb–Sb) bond switching in addition to (As)[1]

5. Conclusions

Compositional dependence of experimental static radiation-induced darkening effect in quasibinary $(As_2S_3)_{100-x}(Sb_2S_3)_x$ ChG correlates well with $\kappa_2^{\Delta E}$ (As) and κ_2^{FV} parameters. At the same time, $\kappa_2^{\Delta E}$ (As, Sb) and κ_2^{M} criterions are contrary to this tendency. Local misbalance in bond distribution and normal atomic coordination have an essential influence on structural relaxations in ChG with Sb content. In addition, the bond switching criterion give evidence for only As-based bond switching, but not Sb-based.

6. Acknowlefgements

This work was supported by the Science and Technology Center in Ukraine under Project No 3745.

References

[1] Bureau B, Zhang X H, Smektala F, Adam J-L, Troles J, Ma H, Boussard-Pledel C, Lucas J, Lucas P, Le Coq D, Riley M R and Simmons J H 2004 Recent advances in chalcogenide glasses J. Non-Cryst. Solids **345&346** 276-283.

[2] Ohta T, Birukawa M, Yamada N and Hirao K 2002 Optical recording; phase-change and magneto-optical recording J. Magn. Mater. **242-245** 108-115.

[3] Sanghera J S and Aggarwal I D 1999 Active and passive chalcogenide glass optical fibers for IR applications: a review J. Non-Cryst. Solids **256-257** 6-16.

[4] Popov A I, Michalev N I and Shemetova V K 1983 Structural modification of some glassy chalcogenides Phil. Mag. B **47** (1) 73-81.

[5] Popov A I, Domoryad I A and Michalev N I 1988 Structural modification of arsenic chalcogenide glasses under γ-irradiation Phys. Stat. Sol. (a) **106** 333-337.

[6] Popov A 2004 Atomic structure and structural modification of glass (Semiconducting Chalcogenide Glass 1: Glass Formation, Structure, and Simulated Transformations in Chalcogenide Glassess. – Semiconductors and Semimetals, vol 78 / Ed. R. Fairman and B. Ushkov) (Elsevier Academic Press: Amsterdam, Boston, London, New York, Oxford, Paris, San Diego, San Francisco, Singapore, Sydney, Tokyo, , vol 78, p 51-95).

[7] Shpotyuk O I 2004 Radiation-induced effects in chalcogenide vitreous semiconductors (Chapter 6) In.: Semiconductors and Semimetals, vol. 78 / Ed. R. Fairman, B. Ushkov (Elssevier Academic Press: Amsterdam, Boston, London, New York, Oxford, Paris, San Diego, San Francisco, Singapore, Sydney, Tokyo) 215-260.

[8] Zachariasen W H 1932 The atomic arrangement in glass J. Am. Cheram. Soc. **54** 3841-3851.

[9] Urusov V S 1987 Theoretical crystallochemistry (Moskow: Moskow University Press).

[10] Harrison W A 1980 Electronic structure and the properties of solids. The physics of the chemical bond (San Francisco: W.H. Freeman and Company).

[11] Pauling L 1967 The nature of the chemical bond (3rd ed) (Oxford and IBH New Dehli).

[12] Gutenev M S and Ivanova N I 1987 Distribution of structural groups in the quasi-binary glass systems $(AX_n)_x(BX_m)_{1-x}$ (where A, B=Ge, As, Sb; X=S, Se) due to dielectrometry and magnetochemistry data Phys. and Chem. Glass. **13** (3) 454-458.

Journal of Physics: Conference Series **289** (2011) 012010 doi:10.1088/1742-6596/289/1/012010

Structural studies of spinel manganite ceramics with positron annihilation lifetime spectroscopy

H Klym[1,2], A Ingram[3], O Shpotyuk[1], J Filipecki[4] and I Hadzaman[1]

[1]Institute of Materials of SRC "Carat", 202 Stryjska str., Lviv, 79031, Ukraine
[2]Lviv Polytechnic National University, 12 Bandera str., Lviv, 79013, Ukraine
[3]Opole University of Technology, 75 Ozimska str., Opole, 45370, Poland
[4]Institute of Physics of Jan Dlugosz University,
13/15 Armii Krajowei, 42201, Czestochowa, Poland

E-mail: shpotyuk@novas.lviv.ua; klymha@yahoo.com

Abstract. The new transition-metal manganite $Cu_{0.1}Ni_{0.8}Co_{0.2}Mn_{1.9}O_4$ ceramics for temperature sensors with improved functional reliability are first proposed. It is established that the amount of additional NiO phase in these ceramics extracted during sintering play a decisive role. This effect is well revealed only in ceramics having a character fine-grain microstructure, while the monolithization of ceramics caused by great amount of transferred thermal energy reveals an opposite influence. The process of monolitization from the position of evolution of grain-pore structure was studied in these ceramics using positron annihilation lifetime spectroscopy.

1. Introduction

The spinel-type $Cu_{0.1}Ni_{0.8}Co_{0.2}Mn_{1.9}O_4$ ceramics are widely used as one of the most perspective materials for application as negative temperature coefficient thermistors, precise temperature sensors, in-rush current limiters, etc. [1, 2]. That is why the problem of their functional stability and reliability is very important.

Previous investigations have shown that the quantity of the additional phase and its distribution in bulk and on the surface of ceramic samples are essentially influenced by temperature-time sintering regimes [3]. The reduction of NiO-phase content in ceramics unexpectedly essentially inhibited the development of thermal ageing, where the observed relative resistance changes didn't exceed 2-3 %.

As a rule, to eliminate the degradation effects in ceramics, the method of chemical modification by metallic additives at the initial technological stages of ceramics preparation is usually used [4, 5]. These metallic additives, being located in inter-granular regions closely to grain boundaries, diminish thermally-activated ageing phenomena owing by stabilizing the cationic distribution within individual ceramics grains. As a result, the chemically-modified ceramics show higher stability in comparison with non-modified ones.

Because of significant complications in the structure of these spinel-type ceramics revealed at the levels of individual grains, grain boundaries and pores [1], the further progress in this field is dependent to a great extent on the development of new characterization techniques, which can be used in addition to traditional ones. This concerns, in part, the positron annihilation lifetime (PAL) spectroscopy, the method only recently applied to fine-grained powders and ceramics [6, 7]. In general, the PAL being it sensitive to low electron density reflects a so-called void-species distribution within structural network of solids [6]. But meaningful interpretation of PAL data is well reliable only for some types of crystals. It was shown that in the case of ceramics these data were determined mainly by crystallographical features of individual grains, while structural disturbances due to intergranular contacts within ceramics were a subject for additional complications [6, 8-10].

The aim of this work is grain-pore structural studies of high-reliable transition-metal manganite $Cu_{0.1}Ni_{0.8}Co_{0.2}Mn_{1.9}O_4$ ceramics using PAL spectroscopy in addition to traditional methods of structural characterization.

2. Experimental

The precise amounts of high purity and previously tested carbonate salts were weighted and wet mixed. This mixture was thermally decomposed in the air at 700±5°C for 4 h. Then the obtained powders were milled, blended with organic binder and pressed into the disks of approximately 10 mm in diameter and 1 mm in thickness. The prepared four batches of blanks were sintered in the air in respect to the time-temperature regimes (see Fig.1).

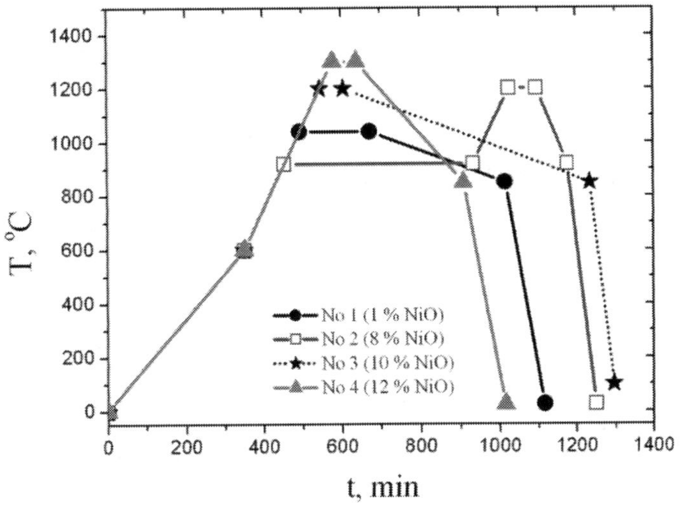

Figure 1. Time-temperature sintering regimes for $Cu_{0.1}Ni_{0.8}Co_{0.2}Mn_{1.9}O_4$ ceramics

It should be noted the sintering route of ceramics was performed to ensure all necessary conditions for inhibition effect in degradation [11], the content of additional NiO phase with NaCl-type structure having decisive role in the final ceramics structure. In fact, we deal with Ni-deficient ceramics in respect to stoichiometric $Cu_{0.1}Ni_{0.8}Co_{0.2}Mn_{1.9}O_4$ composition taken as start one in disproportionality calculations. Four batches of $Cu_{0.1}Ni_{0.8}Co_{0.2}Mn_{1.9}O_4$ ceramics with 1-12 % of NiO phase were prepared owing to different amounts of thermal energy transferred during the sintering (batch No 1 – 1 % NiO, batch No 2 – 8 % NiO, batch No 3 – 10 % NiO, batch No 4 – 12 % NiO). The latter was numerically determined as square value restricted by temperature-time curve above straight line corresponding to 920 ^{0}C, which is the temperature of monophase $Cu_{0.1}Ni_{0.8}Co_{0.2}Mn_{1.9}O_4$ ceramics formation [11].

In respect to the results of microstructure characterization with X-ray diffractometry method, the lattice constant of the main spinel was slightly grown from 8.38 до 8.41 Å without significant changes in the parameter of additional NiO phase (at the level of 4.17-4.19 Å) despite increase in the content of this phase from 1 to 12 %.

The microstructure of the sintered ceramics was probed using electron microscope JSM-6700F, cross-sections morphology of the samples being tested near surface (0-70 μm depth) and chip centers.

The PAL measurements were performed with an ORTEC spectrometer using ^{22}Na source placed between two sandwiched ceramic samples [12-14]. The obtained data were treated with LT computer program [15], the best results corresponding to two-component fitting procedures. The numerical values of trapping parameters (positron lifetime in defect-free bulk τ_b, average positron lifetime $\tau_{av.}$ and positron trapping rate of defect κ_d) were calculated using short and long positron-trapping lifetimes τ_1 and τ_2, as well as component intensities I_1 and I_2 ($I_1 + I_2 = 1$) [7, 11]:

$$\kappa_d = \frac{I_2}{I_1}\left(\frac{1}{\tau_b} - \frac{1}{\tau_2}\right), \quad (1)$$

$$\tau_b = \frac{I_1 + I_2}{\frac{I_1}{\tau_1} + \frac{I_2}{\tau_2}}, \quad (2)$$

$$\tau_{av.} = \tau_1 I_1 + \tau_2 I_2. \quad (3)$$

In addition, the difference $(\tau_2 - \tau_b)$ can be accepted as a size measure of extended defects where positrons are trapped, the τ_2/τ_b ratio represents the nature of these defects [7, 11-14].

3. Results and Discussion

To explain the above phenomena, microstructure of the prepared ceramics was studied. As it follows from Fig. 2, the prepared ceramics samples differ significantly by their grain-pore microstructure.

a – batch No 1 (1 % NiO)

c – batch No 3 (10 % NiO)

b – batch No 2 (8 % NiO)

d – batch No 4 (12 % NiO)

Figure 2. Microstructure of $Cu_{0.1}Ni_{0.8}Co_{0.2}Mn_{1.9}O_4$ ceramics

The samples of batch No 1 (Fig. 2a) are characterized by fine 1-3 μm grains. The numerous intergranual pores are small enough in these samples, their sizes not exceeding 1-2 μm. White film, which can be attributed to additional NiO phase extractions, weakly appears in these ceramics mainly near intergrain boundaries, sometimes it partially fills of pores. The samples of batch No 2 (Fig. 2a) are characterized by larger grains with sizes near 5-7 μm, some of them achieving 10 μm. The white NiO film appears in these ceramics only in the regions of grain boundaries. The grain structure of the samples of batch No 3 (Fig. 2c) gradually changes. The corresponding chip structure of these ceramics is more monolithic, it being characterized only by separate pores with 1-3 μm in sizes. White NiO film appears as bright layer of 10 μm thickness on the grain surface of these samples. In contrast, the grain structure of the samples of batch No 4 (Fig. 2d) attains fully monolithic shape. Only some individual pores of relatively large sizes (near 3-5 μm) are observed in these ceramics, the NiO appearing as uniform layer on the whole ceramics surface. The observed additional NiO phase is non-uniformly distributed within ceramics bulk, being more clearly pronounced near grain boundaries. These phase extractions serve as specific trapping centers for positrons penetrating ceramics.

By accepting two-state positron trapping model [13, 14], for spinel-type ceramic materials the first component of spectra was connected with main spinel structure and the second one – with extended defects located near grain boundaries in the vicinity of additional extracted phases. The intensity I_1 corresponds to the amounts of the main spinel phase, while the I_2 intensity – to the amount of addition NiO phase near grain boundaries.

The lifetime of the first and second components for $Cu_{0.1}Ni_{0.8}Co_{0.2}Mn_{1.9}O_4$ ceramics is typically for spinel-structured materials and equal ~ 0.37 and 0.20 ns, accordingly (see Table 1). The lower τ_1 value in the batch No 2 (0.17 ns) well correlated with Ni content in different crystalographical positions. Since the amount of grain/pores in the sample of batch No 2 was greater, the process of positron trapping in these ceramics was more intensive (the positron trapping rate of defects increased from 0.48 to 0.62 ns^{-1}).

Table 1. PAL characteristics of $Cu_{0.4}Co_{0.4}Ni_{0.4}Mn_{1.8}O_4$ ceramics mathematically treatment within two-component fitting procedure

Sample batch	Fitting parameters				Component input		Positron trapping modes				
	τ_1, ns	I_1, a.u.	τ_2, ns	I_2, a.u.	$\tau_{av.}^{1}$, ns	$\tau_{av.}^{2}$, ns	$\tau_{av.}$, ns	τ_b, ns	κ_d, ns^{-1}	$\tau_2 - \tau_b$, ns	τ_2/τ_b
No 1	0.19	0.82	0.38	0.18	0.16	0.07	0.23	0.21	0.48	0.17	1.8
No 2	0.17	0.79	0.36	0.21	0.14	0.07	0.21	0.19	0.62	0.17	1.9
No 3	0.20	0.86	0.37	0.14	0.17	0.05	0.22	0.21	0.34	0.16	1.7
No 4	0.21	0.84	0.37	0.16	0.18	0.06	0.23	0.22	0.33	0.15	1.7

The intensity of the second component for samples of batch No 1 is 18 %, although amount of additional NiO phase in this batch is small (1 %). This additional phase is localized near grain boundaries and partly fills pores. In samples of batch No 2 the intensity I_2 increase to 21 % (see Table 1) correspondingly to amount of NiO phase. In batch No 3 ceramics, the grain-pore structure was not developed because of occurred monolithization process accompanied by surface extraction of additional NiO phase. The same was character for batch No 4 ceramics samples. These transformations were in good agreement with positron trapping parameters.

Nevertheless, there were no significant changes in $\tau_{av.}$, τ_b, τ_2/τ_b and (τ_2- τ_b). In all cases, the same type of positron trapping centre is formed. The character size of these extended positron traps near grain boundaries estimated due to (τ_2 - τ_b) difference is close to single-double atomic vacancies [6].

Conclusions

Thus, it should be concluded that the results of PAL measurements confirming the interphase mass-transfer processes occurred in mixed transition-metal manganite ceramics caused by monolitization processes during technological modification and owing to optimal content of additional NiO phase. Obtained results are in well agreement with microstructural X-ray diffractometry and electron microscopy data, confirming the structural changes in these ceramics caused by their technological modification.

References

[1] Sheftel I T 1973 *Thermoresistors* (Nauka, Moscow) p 415

[2] Bekker J A, Green C B and Pearson G L 1946 Properties and uses of thermistors – thermally sensitive resistors. *Trans. Amer. Inst. Elect. Engrs.* **65** 711-725

[3] Balitska V, Hadzaman I, Klym H and Shpotyuk O 2008 Sintering effects in degradation of mixed transition-metal manganite ceramics. Select. Proc. XIV Intern. Sem. on Phys. and Chem of Solids 206-211

[4] Metz R, Brieu M, Legros R and A. Rousset 1990 Intergranular phases in electroceramics. *Colloque de Physique* **51** 1003-1008

[5] Castelan P, Bui A, Loubiere A, Rousset A and Legros R 1992 Ageing study of NTC thermistors by thermopower measurements. *Sensors and Actuators* A **33** 119-122

[6] Krause-Rehberg R and Leipner H S 1999 *Positron Annihilation in Semiconductors. Defect Studies* (Springer-Verlag, Berlin-Heidelberg-New York) p 378

[7] Shpotyuk O and Filipecki J 2003 Free volume in vitreous chalcogenide semiconductors: possibilities of positron annihilation lifetime study (Wydawnictwo Wyzszej Szkoly Pedagogicznej w Czestochowie) p 144

[8] Nambissan P M G, Upadhyay C and Verma H S 2003 Positron lifetime spectroscopic studied of nanocrystalline $ZnFe_2O_4$. *J. Appl. Phys.* **93(10)** 6320-6326

[9] Ghosh S, Nambissan P M G and Bhattacharya R 2004 Positron annihilation and Mössbauer spectroscopic studies of In^{3+} substitution effects in bulk and nanocrystaline $MgMn_{0.1}Fe_{1.9-x}O_4$. *Phys. Lett. A* **325** 301-308

[10] He J, Lin L, Lu T and Wang P 2002 Effects of electron- and/or gamma-irradiation upon the optical behavior of transparent $MgAl_2O_4$ ceramics: Different color centers induced by electron-beam and γ-ray. *Nucl. Instrum. Methods Phys. Res.* B **191** 596-599

[11] Bodak O, Akselrud L, Demchenko P, Kotur B, Mrooz O, Hadzaman I, Shpotyuk O,Aldinger L, Seifert H, Volkov S and Pekhnyo V 2002 Crystal structure and electrical properties of $Cu_{0.8}Ni_{0.1}Co_{0.2}Mn_{1.9}O_4$ electroceramics. *J. Alloys and Compounds* **347** 14-23

[12] Shpotyuk O, Ingram A, Klym H, Vakiv M, Hadzaman I, and Filipecki J 2005 PAL spectroscopy in application to humidity-sensitive $MgAl_2O_4$ ceramics. *J. Europ. Ceram. Soc.* **25**, 2981-2984

[13] Klym H, Ingram A, Shpotyuk O, Filipecki J and Hadzaman I 2007 Extended positron-trapping defects in insulating $MgAl_2O_4$ spinel-type ceramics. *Phys. Stat. Sol.* (c) **4(3)** 715-718

[14] Balitska V, Filipecki J, Ingram A and Shpotyuk O 2007 Defect characterization methodology in sintered functional spinels with PALS technique *Phys. Stat. Sol.* (c) **4(3)** 1317-1320

[15] Kansy J 2000 Positronium trapping in free volume of polymers. *Radiation Physics and Chemistry* **58** 427-431

Journal of Physics: Conference Series **289** (2011) 012011 doi:10.1088/1742-6596/289/1/012011

Multifunctional thick-film structures based on spinel ceramics for environment sensors

M Vakiv[1], I. Hadzaman[1], H Klym[1,2], O Shpotyuk[1] and M Brunner[3]

[1]Institute of Materials of SRC "Carat", 202 Stryjska str., Lviv, 79031, Ukraine
[2]Lviv Polytechnic National University, 12 Bandera str., Lviv, 79013, Ukraine
[3]Fachhochschule Köln / University of Applied Sciences,
2 Betzdorfer str., Köln, 50679, Germany

E-mail: shpotyuk@novas.lviv.ua, klymha@yahoo.com

Abstract. Temperature sensitive thick films based on spinel-type $NiMn_2O_4$-$CuMn_2O_4$-$MnCo_2O_4$ manganites with p- and p^+-types of electrical conductivity and their multilayer p^+-p structures were studied. These thick-film elements possess good electrophysical characteristics before and after long-term ageing test at 170 °C. It is shown that degradation processes connected with diffusion of metallic Ag into film grain boundaries occur in one-layer p-and p^+-conductive films. Some part of the p^+-p structures were of high stability, the relative electrical drift being no more than 1 %.

1. Introduction

Spinel ceramics based on mixed transition-metal manganites and/or magnesium aluminates are known to be widely used for temperature measurement, in-rush current limiting, liquid and gas sensing, flow rate monitoring and indication, etc. [1-5]. But their sensing functionality is sufficiently restricted because of bulk performance allowing, as a rule, no more than one kind of application.

At the present time, a number of important problems connected with hybrid microelectronic circuits, multilayer ceramic circuits, temperature sensors, thermal stabilizers, etc. requires such resolution, when not bulk (e.g. sintered as typical bulk ceramics), but only thick-film performance of electrical components (possessing the possibility to group-technology route) is needed [5]. The well-known advantages of screen printing technology revealed in high reproducibility, flexibility, attainment of high reliability by glass coating as well as excellent accuracy, yield and interchangeability by functional trimming are expected to be very attractive now, for new-generation sensing electronics [6]. No less important is the factor of miniaturization for developed thick-film elements and systems, realized in a variety of their possible geometrical configurations. Thus, the development of high-reliable nanostructured thick films and their multilayers based on spinel-type compounds for multifunctional environment sensors operating as simultaneous negative temperature coefficient thermistors and integrated temperature-humidity sensors are very important task [6-8].

To fabricate the integrated temperature-humidity thick-film sensors, only two principal approaches have been utilized, they being grounded on temperature dependence of electrical resistance for humidity-sensitive thick films and/or on humidity dependence of electrical resistance for temperature-sensitive thick films.

The first approach was typically applied to perovsite-type thick films like to $BaTiO_3$ [9]. Within second approach grounded on spinel-type ceramics of mixed Mn-Co-Ni system with RuO_2 additives, it was shown that temperature-sensitive elements in thick-film performance attain additionally good humidity sensitivity [10]. Despite improved long-term stability and temperature-sensitive properties with character material B constant value at the level of 3000 K, such thick-film elements possess only small humidity sensitivity. This disadvantage occurred because of relatively poor intrinsic pore topology proper to semiconducting mixed transition-metal manganited in contrast to dielectric aluminates with the same spinel-type structure.

Thick-film performance of mixed spinel-type manganites restricted by $NiMn_2O_4$-$CuMn_2O_4$-$MnCo_2O_4$ concentration triangle has a number of essential advantages, non-available for other ceramic composites. Within the above system, can be prepare the fine-grained semiconductor materials possessing p^+-conductive $Cu_{0.1}Ni_{0.1}Mn_{1.2}Co_{1.6}O_4$ and p-conductive $Cu_{0.1}Ni_{0.8}Mn_{1.9}Co_{0.2}O_4$. So, a real possibility to prepare multilayer thick-film spinel-type structures for principally new device application, such as thermoelectric transformers in a power supply, high-accurate temperature sensors and compensators exploring current-voltage dependence, temperature difference detecting elements utilizing thermoelectromotive force, etc. seems to be a quite realistic one. In addition, the prepared multilayer thick-film structures involving semiconductor $NiMn_2O_4$-$CuMn_2O_4$-$MnCo_2O_4$ and dielectric $MgAl_2O_4$ spinels can be potentially used as simultaneous thermistors and integrated temperature-humidity sensors with extremely rich range of exploitation properties.

The aim of this work is development of high-reliable temperature and humidity sensitive thick films and multilayered structures based on spinel-type ceramics for multifunctional application in integrated temperature/humidity sensors.

2. Experimental

Bulk temperature sensitive ceramics were prepared by a conventional ceramics processing route using reagent grade cooper carbonate hydroxide and nickel (cobalt) carbonate hydroxide hydrates [11]. Chemical composition of these ceramics and the main points in their sintering schedules are presented in Table 1.

Table 1. Characteristics of temperature sensitive bulk ceramics

Chemical composition	Sintering temperature/duration	Phase composition	B, K
$Cu_{0.1}Ni_{0.1}Co_{1.6}Mn_{1.2}O_4$	1040 °C/ 4 h	spinel	3540
$Cu_{0.1}Ni_{0.8}Co_{0.2}Mn_{1.9}O_4$	920 °C/8 h + 1200 °C/1 h +920 °C/24 h	spinel + NiO (11.5 %)	3378

The bulk $MgAl_2O_4$ ceramics were prepared via conventional sintering route as was described in more details elsewhere [12]. The pellets were sintered in a special regime with maximal temperature 1300 °C during 5 h.

Temperature sensitive $Cu_{0.1}Ni_{0.1}Co_{1.6}Mn_{1.2}O_4$/$Cu_{0.1}Ni_{0.8}Co_{0.2}Mn_{1.9}O_4$-based and humidity sensitive $MgAl_2O_4$-based pastes were prepared by mixing powders of basic ceramics (sintered bulk ceramics were preliminary destroyed, wet-milled and dried) with ecological glass powders (without PbO), inorganic binder Bi_2O_3 and organic vehicle (Table 2).

Table 2. Composition of temperature/humidity sensitive pastes

Constituents	Content, % mass	
	$Cu_{0.1}Ni_{0.1}Co_{1.6}Mn_{1.2}O_4$/ $Cu_{0.1}Ni_{0.8}Co_{0.2}Mn_{1.9}O_4$-based paste	$MgAl_2O_4$-based paste
Basic ceramics	72.8	58
Bi_2O_3	2.9	4
Ecological glass	2.9	8
Organic vehicle	21.4	30

The prepared pastes were printed on alumina substrates (Rubalit 708 S) with Ag-Pt electrodes using a manual screen-printing device equipped with a steel screen. Then thick films were fired in furnace PEO-601-084.

To prepare multifunctional temperature/humidity sensitive elements we used typical design performance in respect to the scheme shown in Fig. 1. In the case under consideration, the main advantages proper to bulk transition-metal manganite ceramics (wide range of electrical resistance with high temperature sensitivity) and humidity-sensitive $MgAl_2O_4$ ceramics were transformed into thick-film multilayers resulting in a principally new and more stretched functionality. The spinel-type $Cu_{0.1}Ni_{0.1}Mn_{1.2}Co_{1.6}O_4$ compound with p^+-type of electrical conductivity, $Cu_{0.1}Ni_{0.8}Mn_{1.9}Co_{0.2}O_4$ compound with p-type of electrical conductivity and dielectric magnesium aluminate d-$MgAl_2O_4$ were designed as overall integrated p^+-p and p-d structures shown in the topological scheme (Fig. 1).

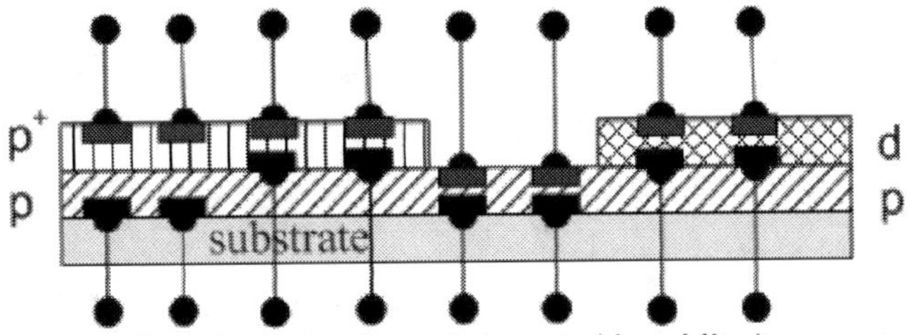

Figure 1. Topological scheme integrated humidity/temperature sensing functionality in the developed spinel-type thick films

The topology of the obtained thick films was investigated using 3D-profilograph Rodenstock RM600. The electrical resistance of temperature-sensitive thick films was measured using temperature chambers HPS 222. The temperature constant B for these thick films was calculated according to the equation:

$$B = 2.3026 \cdot \log\left(\frac{R_1}{R_2}\right) \cdot \frac{T_1 \cdot T_2}{T_2 - T_1}. \qquad (1)$$

where R_1 and R_2 were corresponding resistance at $T_1 = 25$ °C and $T_2 = 85$ °C. accordingly. The current-voltage I-V characteristics were measured at 50, 25 and 5 °C (±0.1 °C) using a precise digital multimeter. The humidity-sensitivity of thick-film elements based on $MgAl_2O_4$ ceramics was evaluated on dependence of electrical resistance from relative humidity (RH). The measurements were performed at 20 °C and 1000 Hz frequency in direction of RH increase and in reverse one. The long-term ageing test at 170 °C for p-, p^+-type thick films and p+-p structures was carried for study of their reliability. The relative change of electrical resistance ($\Delta R/R_o$) was used as a controlled parameter (R_o– initial value of electric resistance, ΔR – absolute change of electric resistance caused by a ageing test).

3. Results and Discussion

In respect to the obtained 3D-profilogramph data, the thickness of temperature sensitive p^+-conductive thick films based on $Cu_{0.1}Ni_{0.1}Co_{1.6}Mn_{1.2}O_4$ ceramics and p-conductive thick films based on $Cu_{0.1}Ni_{0.8}Co_{0.2}Mn_{1.9}O_4$ ceramics were near 47-49 μm and 54-63 μm, accordingly. The thickness of thick films based on humidity sensitive dielectric $MgAl_2O_4$ ceramics was ~ 90 μm. The topology of multilayered p^+-d and p^+-p thick-film structures are shown in Fig. 2.

The temperature sensitive p^+/p-conductive thick films and their p^+-p structures based on spinel-type $NiMn_2O_4$-$CuMn_2O_4$-$MnCo_2O_4$ ceramics posses good linear electrophysical characteristics in the region from 298 to 358 K in semi-logarithmic scale (Fig. 3). The values of B constants were 3589, 3630 and 3615 K for p-, p^+-conductive thick films and p^+-p structure, respectively.

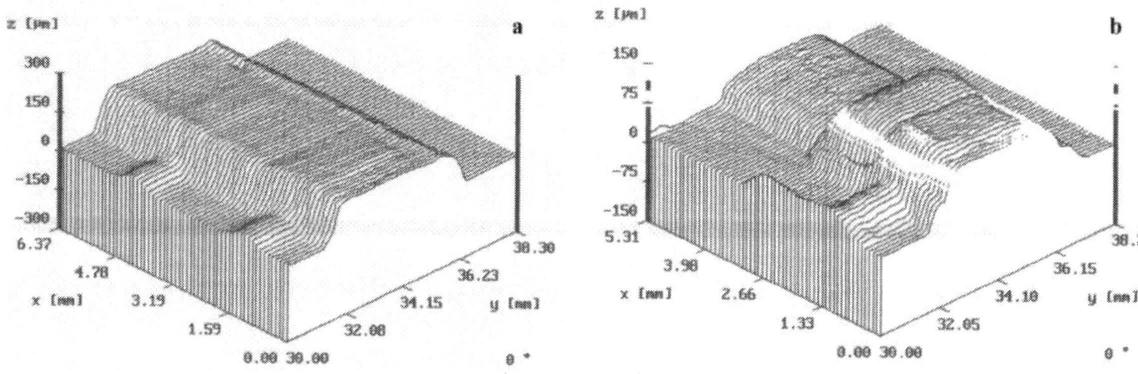

Figure 2. Topology of p^+-d (a) and p^+-p (b) thick-film structures

As was shown early [12], the bulk humidity-sensitive ceramics characterizes of hystereses in desorption cycles. It's connected with peculiarities of pore-grain structure and quantity of addition phases localized near grain boundaries. These failings were succeeded in thick films based on $MgAl_2O_4$ ceramics using the optimum amount of Bi_2O_3, organic solvent, organic copula and pine oil. So, the studied d-type $MgAl_2O_4$ thick films possess linear dependence of electrical resistance from relative humidity without hysteresis in the range of RH ~ 40-99 % (see Fig. 4).

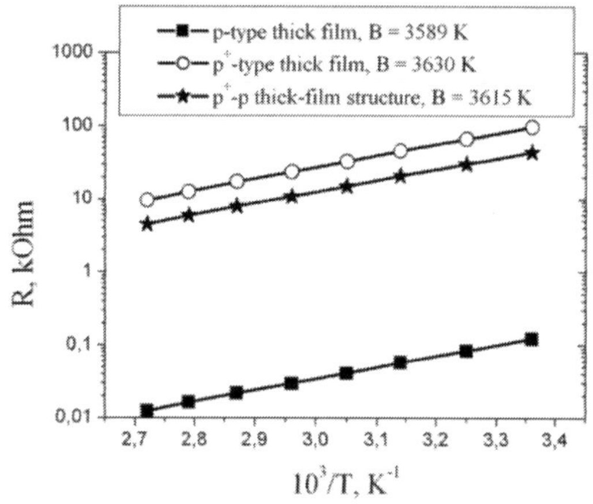

Figure 3. R/T characteristics for temperature-sensitive p- and p^+-conductivity thick films and p^+-p thick-film structure

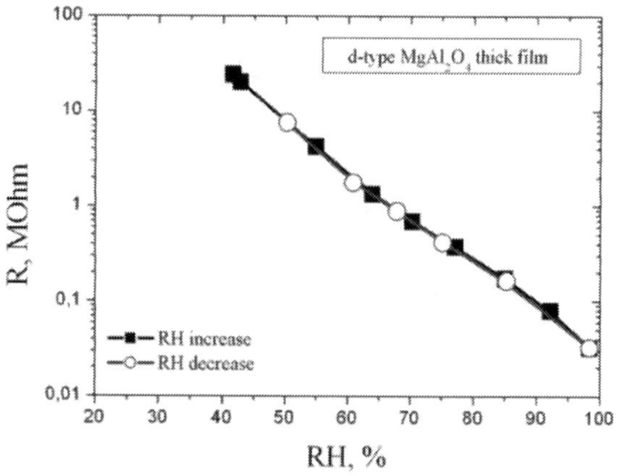

Figure 4. Typical work characteristics for humidity- sensitive thick films

It is shown that electrical resistance of p-and p^+-conductive thick films incidentally decreases in the process of degradation test (see Fig. 5). This effect is supposed to be connected with thermally-induced compression of thick films and diffusion of metallic Ag into the grain boundaries. The value of $\Delta R/R_o$ reaches -(2-5) %. However, the p+-p thick-film structures shows high reliability after long-term ageing test at 170 °C. The relative electrical drift is no more than 1 %.

The values of temperature constant B are decreases on 20-50 K after degradation test both p/p^+-conductive thick films and p^+-p structures, while the activation energy of electrical conductivity was not changed significantly, being at the level of 0.31 eV. Typical current-voltage characteristics for p-conductive thick films are presented in Fig. 6.

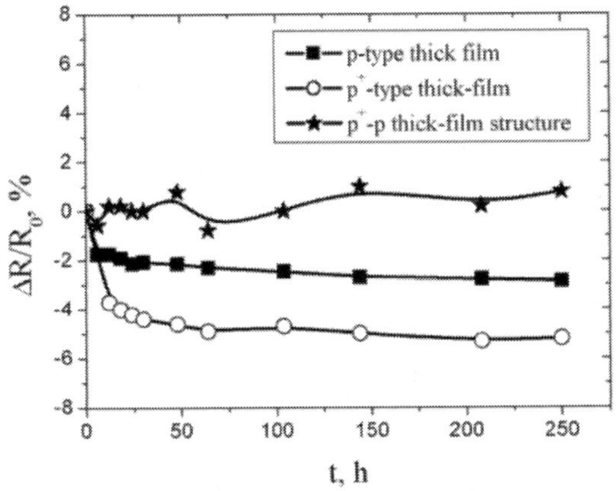

Figure 5. Thermally-induced (170 °C) relative resistance drift ($\Delta R/R_0$) in thick-film elements

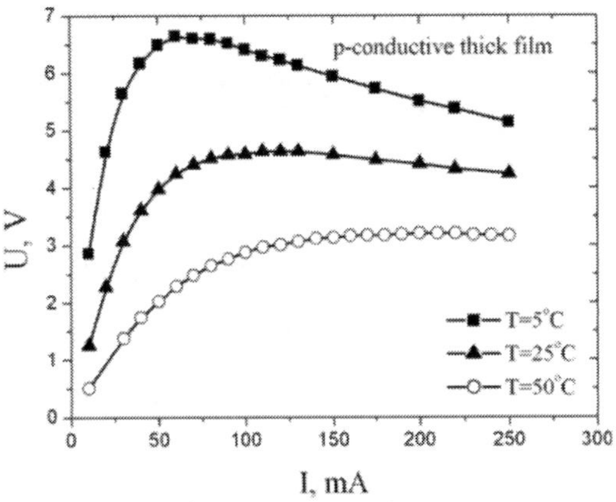

Figure 6. I-V characteristics for p-conductive thick films

4. Conclusions

The separate temperature and humidity sensitive thick-film elements based on spinel-type $NiMn_2O_4$-$CuMn_2O_4$-$MnCo_2O_4$ manganites with p^+/p-type of electrical conductivity, p^+-p structures and dielectric magnesium aluminate $MgAl_2O_4$ were prepared using ecological glass constituents. These thick films can be used to produce multifunctional high-reliable integrated temperature/humidity sensors for effective environment monitoring and control.

Acknowledgements
The authors acknowledge research support from Science and Technology Center in Ukraine under regular STCU Project No 4277.

References

[1] Sheftel I T 1973 Thermoresistors (Nauka. Moscow) p 415
[2] Zaharov V I and Olesk A O 1989 Materials and technology for NTC thick-film thermistors manufacturing. *Elektronnaja Tehnika: Ser. Radiodetali i Komponenty* **63** 30-34
[3] Zaharov V I and Olesk A O 1983 Film thermistors. *Zarubeznaja Elektronnaja Tehnika* **5** 43-74
[4] Zhong J and Bau H H 2001 Thick-film thermistors printed on LTCC tapes. *Am. Ceram. Soc. Bull.* **80** 39-42
[5] Feingold A H, Wahlers R L, Amstutz P, Huang C, Stein S J and Mazzochette J 2000 New microwave applications for thick-film thermistors. *Microwave Journal* **1** 90-98
[6] Qu W 2000 Development of multi-functional sensors in thick-film and thin-film technology. *Meas. Sci. Technol.* **11** 1111-1115
[7] White N W and Turner J D 1997 Thick-film sensors: past, present and future. *Meas. Sci. Technol.* **8** 1-4
[8] Dziedzic A 1997 Thick-film resistive temperature sensors. *Meas. Sci. Technol.* **8** 78-81
[9] Holc J 1995 Temperature characteristics of electrical properties of (Ba,Sr)TiO_3 thick-film humidity sensors. *Sensors and Actuators* **B 26/27** 99-102
[10] Huang J 2003 Preparation and characteristic of the thermistor materials in the thick-film integrated temperature-humidity sensor. *Mat. Sci. Eng.* **B 99** 523-526
[11] Vynnyk I, Hadzaman I, Klym H, Mrooz O and Shpotyuk O 2006 Peculiarities of thermodegradational effects in thick film of mixed transition-metal oximanganites. *Technology and Design in Electronics* **2** 60-62
[12] Shpotyuk O, Ingram A, Klym H, Vakiv M, Hadzaman I and Filipecki J 2005 PAL spectroscopy in application to humidity-sensitive $MgAl_2O_4$ ceramics *J. Europ. Ceram. Soc.* **25**, 2981-2984

Raman spectroscopic study of trivalent lanthanides in concentrated aqueous solutions

M Tsvirko[1,2], **A Svetashev**[2]

[1]Institute of Chemistry and Environmental Protection, Jan Dlugosz University, Armii Krajowej 13/15, 42-200 Częstochowa, Poland
[2]National Ozone Monitoring Research&Education Centre, Belarusian State University, Kurchatov str. 7, 220064 Minsk, Belarus

E-mail: m.tsvirko@ajd.czest.pl

Absrtact. The Raman spectra in the region of the O-H stretching vibration band are used to investigate the effect of lanthanide (III) ions on liquid water structure in concentrated aqueous solutions of $CeCl_3$ and $GdCl_3$ (C~0.5÷3 M). The differential spectra, derived from the consequent Raman spectra, have reflected the substantial discrepancy between the spectra of lanthanide (III) salt solutions and pure water as well as the spectra of standard LiCl (C~1.5÷9M) solutions. The new sharp bands (ν_{max} ~ 3180÷3190 cm^{-1}) and (ν_{max}~3360 cm^{-1}), observed in the differential spectra, are assumed to two types of aqua-ions: $Ln^{3+}[H_2O]_mH_2O_n$ and $Ln^{3+}[H_2O]_mCl^-_p$ having a difference in the 2nd coordination sphere. The discrepancy, observed in the spectra of the equmolar $CeCl_3$ and $GdCl_3$ aqueous solutions, is interpreted as a result of the different hydration numbers and ion radiuses of Ce^{3+}_{aq} and Gd^{3+}_{aq}.

1.Introduction

Despite the fact that the influence of inorganic as well as organic solutes on the liquid water structure has been intensively investigated by various methods, including X-ray, NMR and IR/Raman spectroscopy, some aspects of the phenomenon still remain uncertain [1-8]. In this connection studying the Raman spectra of aqueous solutions is considered to be rather informative owing to high sensitivity of the O-H stretching band (ν~2800–3800 cm^{-1}) to the local environment of water molecules. The O-H stretching band shape were established to reflect interactions between water molecules and solutes, polymers, biological molecules, etc., however, the main problem remains to select and/or identify every bonding effect, especially when the band intensity is employed as a standard in spectrofluometry or lidar measurements [8].

In absolute majority of previous works pure water (or dilute aqueous inorganic salt solution) is generally regarded as a compound of differently sized hydrogen bonded water clusters, which bonds may be broken with the temperature increasing. Some soluted ions were established demonstrating a similar perturbing effect but its distinction from that of temperature has been rarely shown clearly yet [1, 4-8]. The influence of halogen anions upon the water Raman spectra was appeared to be much stronger than that of cations and increased in the sequence of $Cl^- < Br^- < I^-$. At the same time the role of metal ions in water bonding was not adequately investigated and the similar sequences for single- or doubly charged cations were difficult to obtain exactly because of their effects insignificance [1, 4, 8].

In this study the O-H stretching Raman spectra in the range of ν~2500–4000 cm^{-1} were used to determine the effect of lanthanide (III) chlorides on the water microstructure.

2. Experimental

Raman spectra have been measured using a thermostatted cell at 20±0.05°C under continuous argon laser excitation (Spectra Physics, λ=514, 488 nm, P=1W) and registered by the optical system: a double monochromator Aston and an optical multi-channel analyzer OMA-1 (PARS, USA).

Cerium (III), Gadolinium (III) and Lithium chlorides of high analytical purity grade and doubly distilled water were used for the solutions preparing. The lanthanide (III) salts solutions were prepared to be equimolar along the concentration sequence of 0.53, 1.06, 2.12 and 3.18 M, where the highest concentration was close to the solubility limit. The comparative LiCl solutions were three times higher in concentration for having equal anion amounts.

3.Results

The Raman spectra of $CeCl_3$, $GdCl_3$ and LiCl water solutions in the range of ν=2700-3800 cm^{-1} measured under equal excitation conditions are presented at Figures 1-3. At the Figures 1b-3b the differential spectra are shown that were derived from the consequent spectra of lanthanides salt solutions and pure water. Tables 1, 2 contain the main spectral parameters. All intensities of the Raman as well as differential spectra are normalized to the peak height of pure water Raman spectrum. Figure 4 shows the differential spectra obtained subtracting the LiCl and $CeCl_3$ solutions spectra, consequently, from that of $CeCl_3$ and $GdCl_3$ solutions.

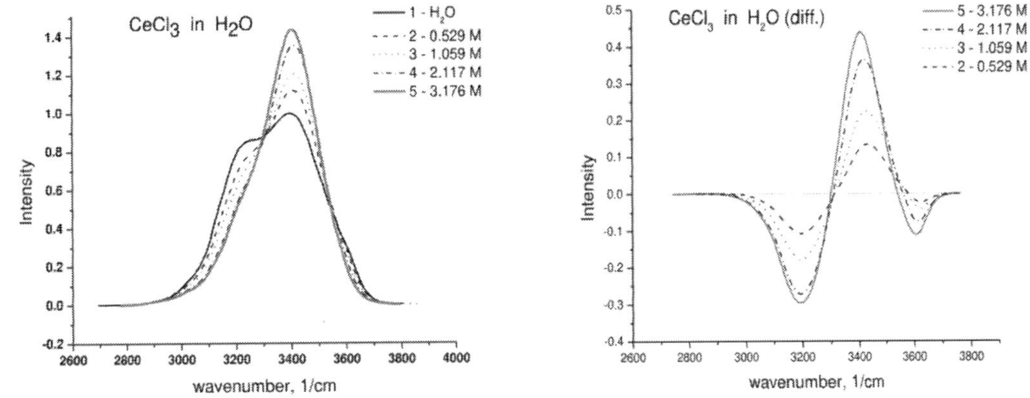

Figure 1. O-H stretching Raman spectra of $CeCl_3$ aqueous solutions (1a) and differential spectra of $CeCl_3$ - pure water (1b): 1 – pure water, 2- 0.529 M, 3- 1.059 M, 4- 2.177 M, 5 – 3.176 M

Our findings manifest the increasing dissolved lanthanide salt concentration leads to monotonous changing of the O–H stretching Raman spectra. The intensity of the low-frequency band shoulder ν ~ 3200 cm^{-1} decreases whereas the maximum intensity occurring near 3405 ± 5 cm^{-1} increases significantly up to 30-40%. Furthermore, in comparison with the pure water spectrum the weak high-frequency band shoulder at ν ~ 3650 cm^{-1} practically disappears so the total band width decreases from Δν~360 cm^{-1} to Δν~240-270 cm^{-1} but the spectra of the salts solutions change in such a way that the spectral band areas remain constant to a high degree of accuracy (±0.1%).

Observed changes in general are similar to spectra transformation registered for LiCl solutions having the equal anion concentrations (see Table 1) but in the case of the lanthanide (III) chlorides the spectra maximums are shifted to low frequencies (Δν~20 cm^{-1}) wile their band widths are distinctly higher.

Much more clear the spectral band modification can be seen from the differential spectra (Figures 1b-3b, Table 2). An important fact is that in all cases under salt concentration increase opposite

Table 1. O-H stretching Raman spectra parameters of concentrated aqueous solutions of Ce(III), Gd(III) and Li(I) chlorides

Solution	C_{solt}, M	C_{anion}, M	ν_{max}, cm^{-1}	$\Delta\nu$, cm^{-1}	I_{max}
Pure H_2O	0	0	33 97.2	395.2	1
LiCl in H_2O	1.588	1.588	3409.3	350.8	1.136
	3.176	3.176	3417.3	314.6	1.239
	6.352	6.352	3425.4	242.0	1.446
	9.528	9.528	3425.4	221.8	1.530
$CeCl_3$ in H_2O	0.529	1.587	3405.2	358.9	1.121
	1.059	3.177	3409.3	326.7	1.206
	2.117	6.351	3409.3	270.2	1.352
	3.176	9.528	3401.2	242.0	1.426
$GdCl_3$ in H_2O	0.530	1.590	3405.2	362.9	1.098
	1.059	3.177	3409.3	334.7	1.172
	2.118	6.354	3405.2	290.4	1.268
	2.987	8.961	3401.2	274.2	1.301

Table 2. The parameters of 'positive' (ν_{max}, I_{max}) and 'negative' (ν_{min}, I_{min}) bands in differential spectra of concentrated aqueous solutions of Ce(III), Gd(III) and Li(I) chlorides and pure water

Solution	C_{solt}, M	C_{anion}, M	ν_{max}, cm^{-1}	I_{max}	ν_{min}, cm^{-1}	I_{min}	N_{isosb}, cm^{-1}
LiCl in H_2O	1.588	1.588	3437.5	0.15	3191.5	0.12	3320.5
	3.176	3.176	3441.5	0.27	3195.5	0.21	3324.6
	6.352	6.352	3437.5	0.48	3195.5	0.34	3324.6
	9.528	9.528	3437.5	0.58	3195.5	0.40	3320.5
$CeCl_3$ in H_2O	0.529	1.587	3429.4	0.13	3191.5	0.11	3316.5
	1.059	3.177	3425.4	0.23	3191.5	0.18	3316.5
	2.117	6.351	3421.4	0.37	3191.5	0.27	3308.5
	3.176	9.528	3405.2	0.44	3191.5	0.30	3300.4
$GdCl_3$ in H_2O	0.530	1.590	3437.5	0.11	3195.5	0.10	3312.5
	1.059	3.177	3437.5	0.19	3195.5	0.17	3308.5
	2.118	6.354	3417.3	0.27	3195.5	0.25	3300.4
	2.987	8.961	3405.2	0.30	3195.5	0.26	3292.3

changes occur simultaneously at three comparatively narrow spectral intervals: intensity increase in the range of 3330-3550 cm^{-1} and intensity decrease in the regions of 2950-3330, 3550-3700 cm^{-1} but there was observed a substantial behaviour discrepancy between the water solutions spectra of lanthanide (III) chlorides and that of LiCl.

In the differential spectra of LiCl solutions: the maximum ($\nu \sim 3437$ cm^{-1}), the high- and low-frequency minimums ($\nu \sim 3600$ and 3195 cm^{-1}) as well as the isosbestical point ($\nu \sim 3324$ cm^{-1}) do not change their positions via concentration changing. In contrast, for the lanthanide (III) chlorides solutions the differential spectra minimum position ($\nu \sim 3192$ - 3195 cm^{-1}) remains unchanged whereas the spectra maximums for higher concentrations shift to low frequencies ($\Delta\nu \sim 25$-30 cm^{-1}) and the isosbestical points are not pronounced.

The areas of the differential spectra positive bands ($\nu \sim 3437$ cm^{-1}) as a function of the anion concentration are presented at Figure 5 showing at the concentration rise the dependencies tend to a plateau but none satiation limit has been achieved.

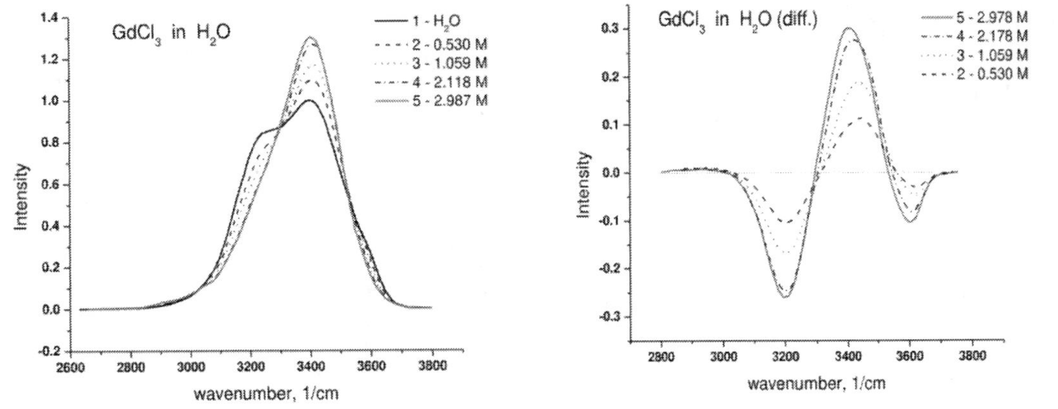

Figure 2. O-H stretching Raman spectra of $GdCl_3$ aqueous solutions (2a) and differential spectra of $GdCl_3$ - pure water (2b) 1 – pure water, 2- 0.530 M, 3- 1.059 M, 4- 2.188 M, 5 – 2.987 M

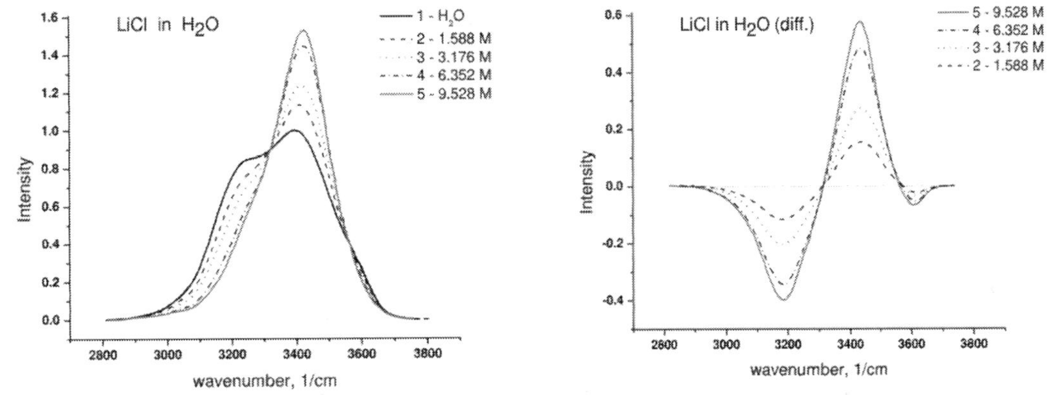

Figure 3. O-H stretching Raman spectra of LiCl aqueous solutions (3a) and differential spectra of LiCl - pure water (3b) 1 – pure water, 2- 1.588 M, 3- 3.176 M, 4- 6.352 M, 5 – 9.527 M

Thus, the differential spectra evidence that at equal Cl^- concentrations the O-H stretching Raman spectra of lanthanide (III) chloride solutions and their behavior under the concentration growth differ from that of LiCl and that the discrepancy between $CeCl_3$ and $GdCl_3$ solutions spectra are also observed.

4. Discussion

The halogenides of alkali metals, alkaline metals and the first-row transition metals are known to act upon the water O-H stretching Raman spectra in the similar way [1, 4, 6-8].

As proposed, the effect of halogen anions on the water structure consists in breaking water-water hydrogen bonds and on the other hand in forming the hydrogen bonds between anions and water molecules that results in the specific changing of the O-H stretching band contour [1, 8, 9] .

The metal ions (cations) effects are not perfectly understood yet, though some differences were observed between the O-H stretching Raman spectra of the aqueous solutions with monovalent (Na^+, K^+, Li^+) and divalent (Mg^{2+}, Ca^{2+}, Mn^{3+}) or trivalent (Al^{3+}, Fe^{3+}) cations. The strength of metal ion effects was found depending on the ionic charge, the outmost electronic structure and ionic size, the ionic charge to be most important, nevertheless, no metal ion solvated forms or their specific impacts have been registered [1, 4, 8].

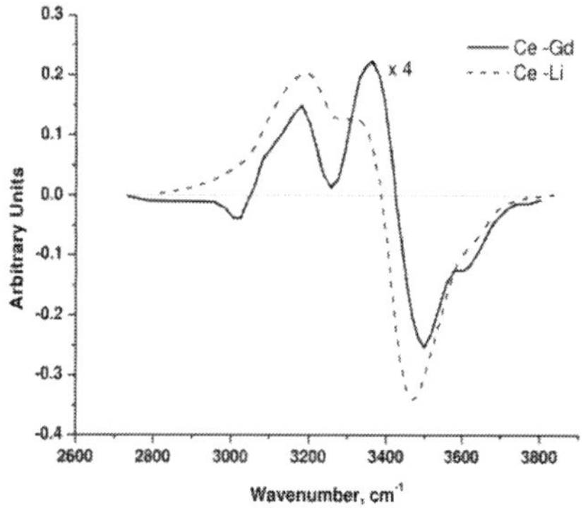

Figure 4. Differential spectra, obtained by the Raman spectra subtraction in solution pairs: (a) – $CeCl_3$ (C=2.177 M) and LiCl (C=6.352 M); (b) - $GdCl_3$ (C=1.694 M) and $CeCl_3$ (C=1.694 M)

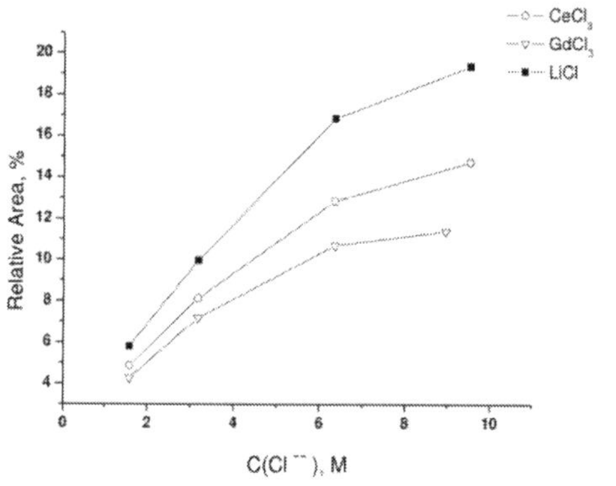

Figure 5. The area of the differential spectra 'positive' band as a function of the anion concentration.

We have also to assume that in the case of LiCl solutions Cl^- anions break the hydrogen-bonded (tetrahedral etc.) water-water $(H_2O)_r$ structures and make the hydrogen-bonded anion-water $(H_2O)_sCl^-$ complexes wile the Li^+ ions do not sufficiently affect the O–H stretching Raman spectra (the sub-symbols 'r' and 's' represent numbers of water molecules associated into the cluster). Thus, the increasing with Cl^- concentration rise 'positive' band at ν~3430 cm^{-1} in the LiCl-H_2O differential spectra (see Figure 3b) should be assigned to the anion-water complexes $(H_2O)_sCl^-$, whereas the 'negative' bands low-frequency at ν~3195 cm^{-1} and high-frequency at ν~3650 cm^{-1} belong probably to the decreasing both anion-free, consequently, hydrogen-bonded water-water structures $(H_2O)_r$ and non-bonded free single water molecules, H_2O. As long as the satiation plateau of the differential spectrum changing were not achieved even at the highest concentration of 9.53 M (see Figure 5), we can not assume the total disappearing of the hydrogen-bonded anion-free water-water forms and their conversion into the anion-bonded structures.

Now, if we have recourse to the Gaussian analysis, which is usually applied to analyze the O–H stretching Raman band for more information of water structure and where the band contour is deconvoluted into 5, 4, 3, or 2 Gaussian components [1], we should assume that the number of components change from 5 or 4 to 2 as the anion concentration increase from 0÷0.5 to 9÷10 M. In that way for the highest LiCl concentration of 9.53 M we have got two Gaussian components: 1) ν_{max} = 3420±10 cm^{-1}, $\Delta\nu$=170±10 cm^{-1}, 80% of the total spectra area; and 2) ν_{max} = 3240±10 cm^{-1}, $\Delta\nu$=140±10 cm^{-1}, 18%, which conform well to the positive and main negative differential spectral bands, consequently. As may be seen, this model neglects the impact of Li^+ cations, which may probably reduce to a slight maximum shift and broadening of the hydrogen-bonded water-water components spectrum.

Undoubtedly, the breaking of water-water hydrogen bonds as well as making of anion-water hydrogen-bonded complexes $(H_2O_n)Cl^-$ take also place in aqueous solutions of lanthanide (III) chlorides but our data obtained for the $CeCl_3$ and $GdCl_3$ solutions show a sufficient distinction in comparing with LiCl. The differences between the O-H stretching Raman spectra of $CeCl_3$ and $GdCl_3$ solutions are also observable but much slighter. These results denote the formation of some new hydrated forms influencing the Raman spectra of lanthanide (III) chlorides in aqueous solutions.

The difference between the Raman spectra of $CeCl_3$ and LiCl aqueous solutions, having the equal Cl^- concentration of 6.35 M, in the case of $CeCl_3$ solution demonstrates the intensity deficiency in the spectral area belonging to anion-water $(H_2O_n)Cl^-$ complexes (ν ~ 3465 cm^{-1}) and the intensity excess in the spectral interval of $\Delta\nu = 3250 \div 3440$ cm^{-1}, where two overlapping 'positive' bands appear with maximums at 3192 and 3324 cm^{-1} (see Figure 4a).

From the Figure 4b, where the differential spectrum is shown of the $CeCl_3$ and $GdCl_3$ equmolar aqueous solutions (C=1.694 M), one can see in the case of $CeCl_3$ the concentration decrease in the anion-water $(H_2O)_sCl^-$ complexes (the 'positive' band ν_{max} ~ 3498 cm^{-1}) and the increase in anion-free water-water $(H_2O)_r$ clusters (ν_{max} ~ 3179 cm^{-1}) as well as the appearing of a new sharp separate spectral band at ν_{max}~3360 cm^{-1}.

To interpret these spectral data we should attract the results obtained by various methods on the coordination of trivalent lanthanide ions (Ln^{3+}) in solutions [2, 3, 10, 11].

According to the majority of previous results, in dilute aqueous solutions any Ln^{3+} ion forms a surroundings, which is called as 'the 1st coordination sphere', with the hydration number changing from nine for the light ions (La^{3+} - Nd^{3+}) to eight for the heavy ions (Tb^{3+} - Lu^{3+}). In the Eu-Tb regions of the lanthanide (III) series the intermediate (between 9 and 8) hydrated numbers are proposed [10].

Our early investigations of the Ce^{3+} ion coordination in aqueous solutions and neat inorganic acids [12, 13], using the methods of UV-absorption and luminescence spectroscopy, deduced that: 1) in dilute aqueous solutions a temperature dependent equilibrium exists of the hydrated $Ce^{3+}{}_{aq}$ forms with different numbers of water molecules in the 1st coordination sphere, $Ce^{3+}[H_2O]_9 \leftrightarrow Ce^{3+}[H_2O]_8$; by our estimation at room temperature the concentration percentage of the $Ce^{3+}[H_2O]_8$ form does not exceed 1.2% in $CeCl_3$ solution (C~10^{-3} M); 2) at concentrations C < 6 M inorganic monovalent inorganic anions such as ClO_4^- or Cl^- do not enter the 1st coordination sphere; 3) in neat inorganic acids, such as HCl (C~11 M), anion enters the 1st coordination sphere that drastically changes the UV spectra and luminescence characteristics of the $Ce^{3+}{}_{aq}$.

Some authors have registered an outer-sphere Ln^{3+}- Cl^- complexes at lower concentrations (~5 M) [2] but the inner-sphere Ln^{3+}- Cl^- complexation have been fixed for La^{3+}, Ce^{3+}, Nd^{3+} and Eu^{3+} in 14 M aqueous solution of LiCl only [3].

We carried out an additional investigation having shown no changes in the $Ce^{3+}{}_{aq}$ spectral or luminescence characteristics (UV absorption and luminescence spectra, quantum yield and decay time of luminescence) up to Cl^- concentrations ~6.6 M. The measurements were made for $CeCl_3$ + $CaCl_2$ aqueous solutions at room temperature.

Our data correspond with the results, obtained in [14] for aqueous solutions of the dysprosium (III) chloride using the UV-vis/near-IR, time-resolved laser-induced fluorescence spectroscopy experiments and molecular-dynamic calculations conducted for dilute to concentrated solutions (ca. 3.1 mol kg^{-1}). The authors concluded that the 1st coordination sphere of $Dy^{3+}{}_{aq}$ was filled with 8 water molecules, while there were 17÷18 molecules in the 2nd sphere; when the solution concentration was increased to 3.1 mol kg^{-1} the average number in the second sphere decreased from 17.5 to 10 and the average number of Cl^- ions increased from 0.5 to 6.3; up to concentrations C~3 M the chloride ion did not form the inner-sphere complexes with dysprosium (III) [14].

Taking into account all these results, we can interpret our data in the following way:

1) Apart from the hydrogen-bonded anion-free water-water $(H_2O)_r$ and anion-water $(H_2O)_sCl^-$ forms in the solutions of lanthanide (III) chlorides the $Ln^{3+}[H_2O]_mH_2O_n$ and $Ln^{3+}[H_2O]_mCl^-_p$ forms arise, where $[H_2O]_m$ represents the 1st coordination sphere of lanthanide (III) ion; these spheres include 9 water molecules in the case of $Ce^{3+}{}_{aq}$ and between 9 and 8 for $Gd^{3+}{}_{aq}$; n~10÷18; p~1÷3; the appearing of inner-sphere Ln^{3+}- Cl^- complexes at used concentration range seemed to be not likely.

2) The spectral band (ν_{max} ~ 3180÷3190 cm^{-1}), appearing at the differential spectra between $CeCl_3$-LiCl and $CeCl_3$-$GdCl_3$ solutions (Figure 4), corresponds most likely to the new anion-free $Ln^{3+}[H_2O]_mH_2O_n$ forms. This band is slightly low-shifted regarding to the band (ν_{max} ~ 3195 cm^{-1}) assumed to the 'traditional' anion-free water-water $(H_2O)_r$ clusters.

3) The second type of new bands at the Figure 4 (ν_{max}~3360 cm^{-1}) can be assigned to outer-sphere complexes $Ln^{3+}[H_2O]_mCl^-_p$. The band maximum is also low-shifted, comparing with the band at ν~3430 cm^{-1} which represents the anion-water $(H_2O)_sCl^-$ forms.

4) The differences, observed in the spectra of the equmolar $CeCl_3$ and $GdCl_3$ aqueous solutions (see Figure 4), may be interpreted as a result of the discrepancy in hydration numbers (9 and 8) and ion radiuses of $Ce^{3+}{}_{aq}$ and $Gd^{3+}{}_{aq}$.

4) All Raman O-H stretching spectra differences, reflected by the differential spectra, and their behavior under the solution concentration growth may be interpreted in terms of the equilibrium of four main hydrated forms: $(H_2O)_r$, $Ln^{3+}[H_2O]_mH_2O_n$, $Ln^{3+}[H_2O]_mCl^-{}_p$ and $(H_2O)_sCl^-$, two of which are the kinds of lanthanide (III) aqua-ions having a difference in the 2nd coordination sphere ion.

Acknowlegment
The authors express their gratitude to Professor V. V. Fadeev for the opportunity to use the Raman equipment.

References

[1] Eisenberg D and Kauzmann W 1969 *The Structure and Properties of Water* (Claredonpress, Oxford) (Oxford Classics Series 2005, Reprinted 2006)
[2] Choppin G. R and Peterman D. R 1998 Applications of Lanthanide Luminescence Spectroscopy to Solution Studies of Coordination Chemistry *Coordination Chemistry Reviews* **174** 283-299
[3] Allen P.G and Bucher J.J 1999 Coordination Chemistry of Trivalent Lanthanide and Actinide Ions in Dilute and Concentrated Chloride Solutions *Inorg. Chem.* **39** 595-601
[4] Li R and Jiang Z 2003 Raman Spectra and ^{17}O NMR study effects of $CaCl_2$ and $MgCl_2$ on Water Structure *Journal of Molecular Structure* **645** 69-75
[5] Pastorczak M and Kozanecki M 2008 Raman Resonance Effect in Liquid Water *J. Phys. Chem.* A **112** 10705-07
[6] Wang Z and Pakoulev A 2004 Vibrational Substructure in the OH Stretching Transition of Water and HOD *J. Phys. Chem.* **108** 9054-63
[7] Dillon S.R. and Dougherty R.C. 2002 Raman Studies of the Solution Structure of Univalent Electrolytes in Water *J. Phys. Chem.* A **106** 7647-50
[8] Dolenko T.A and Churina I.V 2000 Valence Band of Liquid Water Raman Scattering: Some Pecularities and Applications in the Diagnostics of Water Media *Journal of Raman Spectroscopy* **31** 863-870
[9] Li R and Jiang Z 2003 Hydrogen Bonded Structure of Water and Aqueous Solutions of Sodium Halides: a Raman Spectroscopic Study *Journal of Molecular Structure* **707** 83-88
[10] Cotton S. A 2005 Establishing Coordination Numbers for the Lanthanides in Simple Complexes *C.R. Chimie* **8** 129-145
[11] Kimura T and Kato Y 1998 Luminescence Study on the Inner-sphere Hydration Number of Lanthanide(III) Ions in Concentrated Aqueous Salt Solutions in Fluid and Frozen States *Journal of Alloys and Compounds* **278** 92-97
[12] Svetashev A G and Tsvirko M P 1984 Luminescence of Ce^{3+} Ion in Aqueous Solutions *Opt. Spectrosc.* **56** 120–125.
[13] Svetashev A G and Tsvirko M P 1984 Temperature Effect on the Spectral and Luminescence Properties of Cerium (III) Salt Solutions *Theor. Exp. Chem.* **20** 696–701.
[14] Ruas A and Guilbaud P 2006 Experimental and Molecular Dynamics Studies of Dysprosium (III) Salt Solutions for a Better Representation of the Microscopic Features Used within the Binding Mean Spherical Approximation Theory *J. Phys. Chem.* A **110** 11770-79

Journal of Physics: Conference Series **289** (2011) 012013 doi:10.1088/1742-6596/289/1/012013

A simple analytical model for electronic conductance in a one dimensional atomic chain across a defect

Antoine Khater [1] and Dominik Szczęśniak [1, 2]

[1]Laboratoire de Physique de l'Etat Condensé UMR 6087, Université du Maine, 72085 Le Mans, France
[2]Institute of Physics, Jan Długosz University of Częstochowa, al. Armii Krajowej 13/15, 42-218 Częstochowa, Poland

Abstract. An analytical model is presented for the electronic conductance in a one dimensional atomic chain across an isolated defect. The model system consists of two semi infinite lead atomic chains with the defect atom making the junction between the two leads. The calculation is based on a linear combination of atomic orbitals in the tight-binding approximation, with a single atomic one s-like orbital chosen in the present case. The matching method is used to derive analytical expressions for the scattering cross sections for the reflection and transmission processes across the defect, in the Landauer-Buttiker representation. These analytical results verify the known limits for an infinite atomic chain with no defects. The model can be applied numerically for one dimensional atomic systems supported by appropriate templates. It is also of interest since it would help establish efficient procedures for ensemble averages over a field of impurity configurations in real physical systems.

1. Introduction

Impurities, whether chemical or structural, are common place inside physical devices, and exist at unpredictable points in the components material. Their presence affects physical quantities to the extent that the calculated predictions of any theory should be evaluated over the field of impurity configurations. The quantitative evaluation of the consequences of impurity effects is hence of crucial importance to nanoelectronics since device properties are strongly influenced by or even built on such effects [1]. Typical examples to cite are electron scattering by doping elements in semiconductor nanowires [2] and field effect transistors, the spin scattering by disorder in magnetic tunnel junctions [3], and transport of spinpolarized current in dilute magnetic semiconductors [4].

The main aim of the present work is to describe a simple theoretical model for the electronic transport in a one-dimensional atomic chain with a single defect [5]. This model is of interest insofar that if one could present a clear and transparent *analytical* model for electronic transport via a single defect, this should help establish efficient procedures for ensemble averages over the field of impurity configurations. A widely used technique for these averages is the coherent potential approximation (CPA) [6] as implemented in the Korringa-Kohn-Rostoker [7] and linear muffin-tin orbital (LMTO) [8] first principles methods. CPA has been applied to *equilibrium* electronic structure and transport calculations [9]. However, since most devices operate under *nonequilibrium* conditions it is also important to develop appropriate techniques for impurity averaging.

This paper is organized in the following manner. In section 2 we present our model, based on the tight-binding approximation [10-11], and on the matching method [12-13]. The transmission and reflection probabilities, in a Landauer-Buttiker representation [14-15], are calculated in section 3. Finally we present our conclusions in section 4.

Journal of Physics: Conference Series **289** (2011) 012013 doi:10.1088/1742-6596/289/1/012013

2. Model

The system under study is presented in figure 1, along with the associated physical quantities. As can be clearly seen, the system consists of two semi-infinite chains of *white* atoms, namely the leads, connected by a single *black* atom, the defect, where lattice spacing is the same everywhere and is equal to a. The model assumption is that each atom has one s-like orbital.

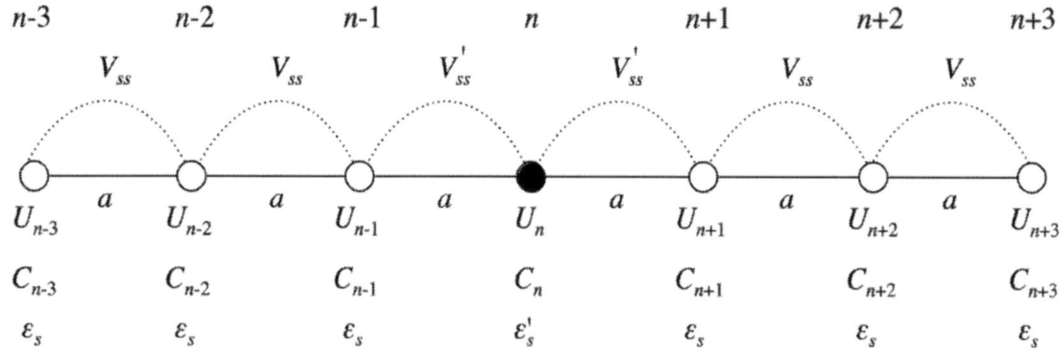

Figure 1. Schematic representation of a one-dimensional chain of atoms along with associated physical quantities. The top row numbers the atoms in the chain. The first bottom row shows the overlap integrals V_{ss} which describe the interactions between nearest neighbor atoms, the lattice spacing a, the potential energy of electrons U_n and the wavefunction site coefficients C_n. The second bottom row shows onsite matrix elements for the electronic 1s orbitals ε_s.

The *defect* has values for the potential energy of its electrons given by U_n, onsite matrix element for orbitals ε'_s and associated overlap integrals V'_{ss}, different than those for the atoms on the lead chains. These latter are given by the same symbols without the apostrophe.

For the nano junction of Fig.1, the motion of electrons may be described by a Schrödinger equation discretized in real space, which is a common method used for solving one-dimensional quantum mechanics problems. Differences between this version and the continuous Schrödinger equation are generally viewed as a weakness of the discrete model from a purely mathematical point of view, where the discrete equation is often presented as the numerical approximation to the continuous one, exact only in the limit of infinitesimal lattice spacing [11].

The tight-binding (TB) or atomic orbital (AO) methods represent a more adequate approach which we adopt in what follows. Since the pioneer work of Slater and Koster [10], the TB method has been extensively applied to semiconductors. Several different tight-binding Hamiltonians have been proposed in the past [16] as many complicated problems can only be managed using this economic yet realistic approach. Consider that the matrix elements of the Hamiltonian that couple nearest neighbor orbitals, are given in this approach by

$$\langle s; j'a \,|\, \hat{H} \,|\, s; ja \rangle = \left(\varepsilon_s + U_j\right)\delta_{j',j} + V_{ss}\left(\delta_{j',j+1} + \delta_{j',j-1}\right), \tag{1}$$

$|\,s;na>$ signifies an s-orbital on the atom located at na, and the δ's are the Dirac deltas. In our case the assumed overlap integrals would be equal to zero unless the two neighbor atoms are close enough so that their orbitals overlap to an appreciable extent. Further, the electronic wave function may be written as

$$|\,\psi >= \sum_j C_j \,|\, s; ja > . \tag{2}$$

Using equation (1) and assuming that j' is the nth site, such that $< s; j'a| = < s; na|$, our Schrödinger equation (1) appears as a tri-diagonal matrix eigenvalue equation with successive rows

$$\left\langle s;na\,|\,\hat{H}-\hat{1}E\,|\,\psi\right\rangle = V_{ss}C_{n-1}+\left(\varepsilon_s+U_n-E\right)C_n+V_{ss}C_{n+1}=0. \tag{3}$$

Equation (3) is a key relation which is the basis for our discussion in the present work. The coefficients C_n can be viewed as the amplitudes of the wave function of the propagating electron on the nth site. In this respect, this equation is equivalent in its form to the equation of motion of a phonon on a 1D atomic chain [12], and to the spin dynamics of a magnon on a magnetically ordered system [13]. Note, furthermore, that the present approach, streamlined here, is effectively comparable to the real-space finite-difference method for electronic conductance calculations [17-18].

3. The scattering cross sections for transmission and reflection

In this section we derive the transmission and reflection coefficients for the system illustrated by Fig.1. To do so we specifically use the *matching method*; this allows one to effectively solve the infinite set of equations of motion [12-13].

Applying equation (3), we may write the Schrödinger's equations for every atom in our system. In this way we obtain an infinite number of coupled equations of motion, of which the first five are

$$\begin{aligned}
&\left\langle s;na\,|\,\hat{H}-\hat{1}E\,|\,\Psi\right\rangle = V'_{ss}C_{n-1}+\left(\varepsilon'_s+U_n-E\right)C_n+V'_{ss}C_{n+1}=0,\\
&\left\langle s;(n+1)a\,|\,\hat{H}-\hat{1}E\,|\,\psi\right\rangle = V'_{ss}C_n+\left(\varepsilon_s+U_{n+1}-E\right)C_{n+1}+V_{ss}C_{n+2}=0,\\
&\left\langle s;(n+2)a\,|\,\hat{H}-\hat{1}E\,|\,\psi\right\rangle = V_{ss}C_{n+1}+\left(\varepsilon_s+U_{n+2}-E\right)C_{n+2}+V_{ss}C_{n+3}=0,\\
&\left\langle s;(n-1)a\,|\,\hat{H}-\hat{1}E\,|\,\psi\right\rangle = V_{ss}C_{n-2}+\left(\varepsilon_s+U_{n-1}-E\right)C_{n-1}+V'_{ss}C_n=0,\\
&\left\langle s;(n-2)a\,|\,\hat{H}-\hat{1}E\,|\,\psi\right\rangle = V_{ss}C_{n-3}+\left(\varepsilon_s+U_{n-2}-E\right)C_{n-2}+V_{ss}C_{n-1}=0.
\end{aligned} \tag{4}$$

We make some parametric simplifications by introducing the new variables: $U_s=U_{n+1}=U_{n+2}=U_{n-1}=U_{n-2}$, $U'_s=U_n$, $\Omega=\left(\varepsilon_s+U_s-E\right)/V_{ss}$, $\Omega'=\left(\varepsilon'_s+U'_s-E\right)/V_{ss}$, $\rho=V'_{ss}/V_{ss}$. Based on these we can rewrite equations (4) in the simpler form

$$\begin{aligned}
&\rho C_n+\Omega C_{n+1}+C_{n+2}=0,\\
&C_{n+1}+\Omega C_{n+2}+C_{n+3}=0,\\
&\rho C_n+\Omega C_{n-1}+C_{n-2}=0,\\
&C_{n-1}+\Omega C_{n-2}+C_{n-3}=0.
\end{aligned} \tag{5}$$

As it was mentioned earlier the set of equations (5) cannot be solved in a classical manner, since the above five equations are coupled to an infinite set of equations on the semi-infinite lead chains. To overcome this problem we use the matching method. Let us write this set of equations in matrix form as

$$A(5\times7)\vec{B}(7)=0, \tag{6}$$

where

$$A(5\times7)=\begin{pmatrix}\Omega' & \rho & \rho & 0 & 0 & 0 & 0(15)\\ \rho & \Omega & 0 & 1 & 0 & 0 & 0(16)\\ 0 & 1 & 0 & \Omega & 1 & 0 & 0(17)\\ \rho & 0 & \Omega & 0 & 0 & 1 & 0(18)\\ 0 & 0 & 1 & 0 & 0 & \Omega & 1\end{pmatrix}=\begin{pmatrix}\Omega' & \rho & \rho & 0 & 0 & 0 & 0(19)\\ \rho & \Omega & 0 & 1 & 0 & 0 & 0(20)\\ \rho & 0 & \Omega & 0 & 0 & 1 & 0(21)\\ 0 & 1 & 0 & \Omega & 1 & 0 & 0(22)\\ 0 & 0 & 1 & 0 & 0 & \Omega & 1\end{pmatrix}, \tag{7}$$

and

$$\vec{B}(7)=\begin{pmatrix} C_n \\ C_{n+1} \\ C_{n-1} \\ C_{n+2} \\ C_{n+3} \\ C_{n-2} \\ C_{n-3} \end{pmatrix}. \tag{8}$$

In the matching method we may write the vector B using a generalized phase factor z [12-13], which relates the phases on successive lattice site, and employ a Hilbert space spanned by two base vectors R and T, for the present one dimensional case. It follows that the vector B may be expressed as

$$\vec{B}(7)=C(7\times 5)\vec{D}(5)+\vec{E}(7), \tag{9}$$

where

$$C(7\times 5)=\begin{pmatrix} 1 & 0 & 0 & 0 & 0 \\ 0 & 1 & 0 & 0 & 0 \\ 0 & 0 & 1 & 0 & 0 \\ 0 & 0 & 0 & z^2 & 0 \\ 0 & 0 & 0 & z^3 & 0 \\ 0 & 0 & 0 & 0 & z^2 \\ 0 & 0 & 0 & 0 & z^3 \end{pmatrix},\ \vec{D}(5)=\begin{pmatrix} C_n \\ C_{n+1} \\ C_{n-1} \\ T \\ R \end{pmatrix},\ \text{and}\ \vec{E}(7)=\begin{pmatrix} 0 \\ 0 \\ 0 \\ 0 \\ 0 \\ z^{-2} \\ z^{-3} \end{pmatrix}. \tag{10}$$

Equation (9) may then be substituted into equation (6) to yield the scattering relation

$$A(5\times 7)\left[C(7\times 5)\vec{D}(5)+\vec{E}(7)\right]=A(5\times 7)C(7\times 5)\vec{D}(5)+A(5\times 7)\vec{E}(7)=0. \tag{11}$$

To solve equation (11), use of the evanescent dynamics on the perfect lead chains is made so that it is recast in a square matrix form as

$$F(5\times 5)\vec{D}(5)=-\vec{G}(5). \tag{12}$$

In equation (12), we have

$$F(5\times 5)=\begin{pmatrix} \Omega' & \rho & \rho & 0 & 0 \\ \rho & \Omega & 0 & z^2 & 0 \\ \rho & 0 & \Omega & 0 & z^2 \\ 0 & 1 & 0 & -z & 0 \\ 0 & 0 & 1 & 0 & -z \end{pmatrix},\ \text{and}\ \vec{G}(5)=\begin{pmatrix} 0(37) \\ 0(38) \\ z^{-2}(39) \\ 0(40) \\ -z^{-1} \end{pmatrix}. \tag{13}$$

Equations (12) and (13) yield analytically the transmission and reflection amplitudes, R and T, as respectively

$$R=-\frac{\rho^2 z+\rho^2 z^{-1}+\Omega'}{\Omega'+2\rho^2 z}, \tag{14}$$

and

$$T = \frac{\rho^2\left(z - z^{-1}\right)}{\Omega' + 2\rho^2 z}. \qquad (15)$$

To verify the validity of these results, the limits when $\Omega' = \Omega$, and $\rho \to 1$, are analyzed. These known limits correspond indeed to the situation where the defect atom becomes *identical* to the atom of the lead chains. We have calculated R and T under these limits, and verified that they do go respectively to the obviously known limits of zero and unity, using equations (14) and (15).

Furthermore, using the analytical results of equations (14) and (15), we obtain in general the scattering cross-sections for the reflection and transmission processes on the one dimensional atomic chain with a single atomic one s-like orbital per site, as follows

$$r = |R|^2,$$
$$t = |T|^2. \qquad (16)$$

For the present case of a single mode of propagation, the electronic conductance across the defect is given by the basic Landauer-Buttiker relation

$$G = \frac{2e^2}{h}\sum_i |T_i|^2 = \frac{2e^2}{h}|T|^2. \qquad (17)$$

Equation (17) may be applied numerically to calculate the ballistic electronic transport for the case of one-dimensional atomic chains with a single defect by adopting the appropriate physical quantities. The case of a silicon defect atom on a carbon atomic chain, for example, needs to generalize the presented approach to sites with a greater number of orbitals which is a straight forward procedure.

4. Conclusions

This work is a contribution to the study of electronic transport in nanostructures. It presents a model calculation in the tight-binding approximation with a single s-state electronic orbital, to calculate the ballistic transport in one-dimensional atomic chains across a single defect. This system is analyzed specifically to illustrate the validity of the approach, insofar that if one could present a clear and transparent *analytical* model for electronic transport via a single defect, this may eventually help establish efficient procedures for ensemble averages over a field of impurity configurations in physical devices.

Analytical relations are derived to express the transmission, T, and reflection, R, amplitudes. The corresponding cross sections for ballistic transmission and reflection processes across the defect are hence derived in the framework of the Landauer-Büttiker approach. The accuracy of these relations is verified in the appropriate limits of an infinite atomic chain with no defects.

The presented model can be applied numerically to calculate the electronic transport in one dimensional atomic systems supported by appropriate templates. It may be generalized to higher dimensions and to more complex systems, and is effectively a precursor model which would allow one to treat more complex nanostructures. The model is also of interest since it should help establish efficient procedures for ensemble averages over a field of impurity configurations in real physical systems, as has been done [19] for the propagation of magnons in two-dimensional disordered magnetic alloys using a modified CPA approach.

Acknowledgments: A Khater would like to thank H. Guo for useful discussions at McGill University.

References

[1] Ke Y, Xia K and H. Guo H 2008 *Phys. Rev. Lett.* **100**, 166805
[2] Markussen T, Rurali R, Jauho A P and Brandbyge M 2007 *Phys. Rev. Lett.* **99**, 076803

[3] Bratkovsky A M 1997 *Phys. Rev.* B **56,** 2344; Tsymbal E Y, Sokolov A, Sabirianov I F and Doudin B 2003 *Phys. Rev. Lett.* **90,** 186602
[4] Ohno H, 1998 *Science* **281,** 951
[5] Szczesniak D 2009 *M.Sc. research project*, Université du Maine, France
[6] Elliott R J, Krumhansl J A and Leath P L 1974 *Rev. Mod. Phys.* **46,** 465 ; Soven P 1967 *Phys Rev.* **156**, 809; Velicky B, Kirkpatrick S and Ehrenreich H 1968 *Phys. Rev.* **175**, 747
[7] Temmerman W M, Gyorffy B L and Stocks G M 1978 *J. Phys.* F **8**, 2461 ; Stocks G M and Winter H 1982 *Z. Phys.* B **46**, 95 ; Stefanou N, Zeller R and Dederichs P H 1987 *Solid State Commun.* **62**, 735
[8] Turek I *et al.*1997 *Electronic Structure of the Disordered Alloys, Surfaces and Interfaces* (Kluwer, Boston); Kudrnovsky´ J and Drchal V, 1990 *Phys. Rev.* B **41**, 7515 ; Kudrnovsky´ J, Drchal V and Masek J 1987 *Phys. Rev.* B **35**, 2487.
[9] Carva K, Turek I, Kudrnovsky´ J and Bengone O 2006 *Phys. Rev. B* **73**, 144421
[10] Slater J C and Koster G F 1954 Phys. Rev. **94**, 1498
[11] Boykin T B and Klimeck G 2004 *Eur. J. Phys.* **25**, 503
[12] Pennec Y and Khater A 1995 *Surf. Sci. Lett.* **348**, L82 ; Fellay A, Gagel F, Maschke K, Virlouvet A, and Khater A 1997 *Phys. Rev.* B **55**, 1707 ; Khater A and Belhadi M 2009 *Surf. Rev. Lett.* **16**, 271
[13] Abou Ghantous M and Khater A 1999 *Eur. Phys. J.* B **12**, 335 ; Bourahla B, Khater A, Tigrine R, Rafil O and Abou Ghantous M 2007 *J. Phys.: Cond. Matt.* **19**, 266208 ;Tigrine R, Khater A, Bourahla B, Abou Ghantous M and Rafil O 2008 *Eur. Phys. J.* B **62**, 59
[14] Landauer R 1957 *IBM J. Res. Dev.* **1**, 223 ;Landauer R 1970 *Philos. Mag.* **21**, 863
[15] Büttiker M 1986 *Phys. Rev. Lett.* **57**, 1761
[16] Vergés J A and Yndurain F 1988 *Phys. Rev.* B **37**, 4333
[17] Khomyakov P A and Brocks G 2004 *Phys. Rev.* B **70**, 195402
[18] Zwierzycki M et al 2008 *Phys. stat. sol.* (b) **245**, 623
[19] Khater A and Abou Ghantous M, to be published

Figures

Fig.1: Schematic representation of a one-dimensional chain of atoms along with associated physical quantities. The top row numbers the atoms in the chain. The first bottom row shows the overlap integrals V_{ss} which describe the interactions between nearest neighbor atoms, the lattice spacing a, the potential energy of electrons U_n and the wavefunction site coefficients C_n. The second bottom row shows onsite matrix elements for the electronic 1s orbitals ε_s.

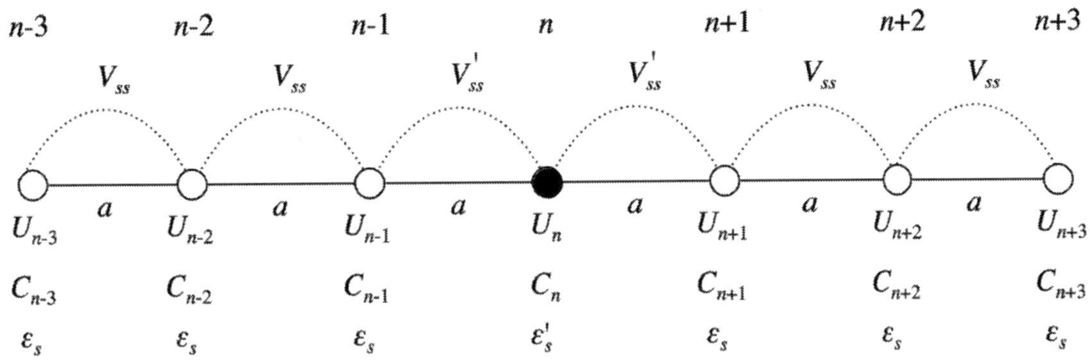

Fig.1

$EuNiO_3$ thin films- growth and characterization

K Bilewska[1], E Wolna[1,2], M Edely[3], P Ruello[3] and J Szade[1]

[1]University of Silesia, Institute of Physics, Uniwersytecka 4, 40-007, Katowice, Poland

[2]Hutchinson Poland Sp. z o.o. Żywiec 1, Leśnianka 73, 34-300 Żywiec, Poland

[3]University of Maine, LPEC, CNRS-UMR 6087, 72085 Le Mans, France

E-mail of corresponding author: k.bilewska@gmail.com

Abstract. We report on the process formation of $EuNiO_3$ thin films on various substrates, studies of optimization conditions for synthesis process and primary investigation. $EuNiO_3$ thin films were deposited by Radio Frequency Magnetron Sputtering in Ar-O atmosphere using a ceramic target with additional nickel pellets. We used Si, $NdGaO_3$ and $SrTiO_3$ substrates. Afterwards samples were annealed in high pressure and temperature in order to achieve a desired oxygen stoichiometry. Thickness of the film was controlled by deposition time and calculated by X-ray Reflectivity. X-ray diffraction measurements revealed that obtained thin films exhibit $EuNiO_3$ phase on the surface still accompanied by Eu_2NiO_4 and NiO phases in the bulk. Therefore annealing seems to be a very surface sensitive process. That inhomogeneity was confirmed in X-ray photoemission depth profiling studies for $EuNiO_3$/Si (60nm) thin film. Surface contamination with adsorbed oxygen has been observed with grazing angle analysis. Ni2p photoemission line analysis revealed a possible mixed-valence character of Ni electronic state.

1. Introduction

Rare earth nickelates are a family of transition metal oxides that received a great deal of attention as it exhibits a wide range of properties. Presumably the most important is that $RNiO_3$ compounds undergo different types of phase transitions: metal-insulator and antiferromagnetic-paramagnetic. The temperatures of transitions appear to be related to the size of the rare earth cation, subsequently the Ni-O bond length and Ni-O-Ni superexchange angle.

$RNiO_3$ compounds crystallize in a perovskite-type structure ABX_3 of $GdFeO_3$ type. The ideal structure is of cubic symmetry with B cations in the cell corner and A cation in the middle. Each B cation is surrounded by six X anions, forming together so-called BX_6 octahedron. In the $RNiO_3$ series the positions of A, B and X atoms are occupied by R, Ni and O ions respectively. This ideal structure is however fairly distorted for most of the nickelates due to small size of the R^{3+} cations, which force the NiO_6 octahedra to tilt in order to optimize the R-O distances. The distortion can be described by so-called tolerance factor or Goldschmidt factor.

Depending on the temperature and R radius size, three regimes can be distinguished where $RNiO_3$ crystal is either a paramagnetic metal, an antiferromagnetic insulator or a paramagnetic insulator. Apart from $LaNiO_3$ all nickelates exhibit at least one transition. When the size of rare earth's radius (or the tolerance factor) decreases, the temperature of the metal-insulator transition (T_{MI}) increases. It is however different when considering the magnetic phase transition. For less distorted perovskites, like $NdNiO_3$ and $PrNiO_3$, the Neel temperature increases for decreasing tolerance (t) factor, similarly to T_{MI}. Yet, when considering medium and small-sized rare earths (R= Sm, Eu, Y, Lu) T_N decreases with diminishing t factor. Such diversity is of top interest and few theories arose in order to explain it from which the most known are Zaanen-Sawatzky-Allen framework and charge disproportionation model.

For many years also Eu mixed valence in various compounds has been investigated. However this effect is not fully understood yet. The phenomenon of mixed valence is related to basic properties of solids like electrical conductivity, magnetic properties, superconductivity, and metal-insulator transition. Eu shows mixed valence in several Eu compounds. For some of them (EuF_3, EuS, EuO_3, $EuCl_3$) which are characterized by trivalent bulk europium state the surface valence transition was found.

$EuNiO_3$ is a rare earth nickelate of monoclinically distorted perovskite structure in room temperature. Bulk compound has a tolerance factor equal 0.892 and has separated phase transitions with T_{MI}=463K and T_N=205K [4].

In this work we report the deposition of $EuNiO_3$ thin films of various thicknesses and on different substrates and we describe primary investigation of their structure.

2. Experimental details

$EuNiO_3$ thin films were deposited by RF Magnetron Sputtering using Argon-Oxygen atmosphere. In order to optimize densification, the target was synthesized by a technique derived from Peccini sol-gel method. The resulting powder was pressed in an isostatic press and sintered in air at 1300°C for 12 hours. The target was analyzed by X-ray diffraction and revealed a mixture of Eu_2NiO_4 and NiO.

The composition of the target and the films was controlled using energy dispersive X-ray analysis in a scanning microscope at 20 kV. The cationic ratio Eu/Ni of target was equal to one but the first thin films exhibited lack of Ni. In order to obtain stoichiometric films three additional nickel pellets of 4mm diameter were added onto the target. Thin films were deposited on Si(100), $NdGaO_3$(100), $NdGaO_3$(110) and $SrTiO_3$(100) substrates.

The residual pressure and working pressure of the deposition process were close for each sample and equal $(4.9\pm0.1)\times10^{-6}$ mbar and $(4.7\pm0.1)\times10^{-2}$ mbar respectively. The films were deposited at the temperature of 600°C. The deposition time was varied in order to obtain films with different thicknesses. The process was followed by annealing at high temperature (800°C) and under high oxygen pressure (180mbar) for 48 hours in order to achieve stoichiometric $EuNiO_3$.

X-ray reflectivity was used in order to measure the film thicknesses. We used a Cu Kα radiation of 1.54Å wavelength.

We have also performed X-ray Photoelectron Spectroscopy measurements at room temperature with Al Kα radiation of 1486.7eV photon energy. In order to study the very top part of the film a grazing geometry was applied. Incident beam of X-rays was irradiating the surface and the photoelectrons were collected under the angle of about 20° with respect to the surface of the film and compared with standard measurements performed with 45° angle between surface and the analyzer. We performed depth profile measurements to investigate the electronic and chemical structure of the sample deeper than few top layers of surface. We used Ar^+ ions of energy of 2 keV for gradual sputtering an area of 2.5mm×2.5mm in ten-minute cycles. After each cycle of sputtering the spectra of Ni2p, O1s, Eu4d and valence band were collected from area of 800μm diameter. Some standard room temperatures XPS spectra were taken as well.

3. Results and discussion

EDS measurements revealed that the best atomic composition (Eu/Ni=1) is obtained when using three additional nickel pellets, so the deposition process was continued with this amount. In figure 1 a linear dependence of the Eu/Ni ratio is showed as a function of surface of nickel pellets.

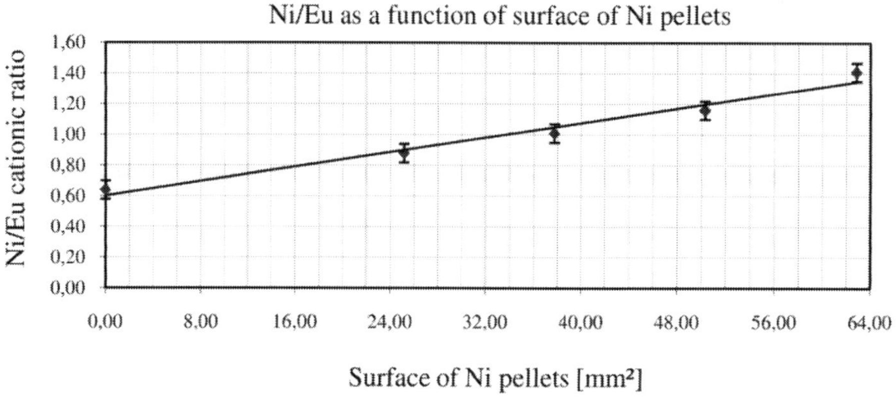

Figure 1. Eu/Ni ratio in the final $EuNiO_3$ films derived from the EDS data as a function of surface of Ni pellets added to the ceramic target. The best composition is achieved for 33 mm^2 of Ni which corresponds to three additional pellets.

Depending on the deposition time samples of various thicknesses were obtained. To measure the thickness of the sample an electron density gradient's sensitive technique, the X-ray Reflectivity, was used. It allowed not only to extract information about the free surface and the interfaces, but also to determine the mass density and the thickness of thin layers along the direction normal to the specimen surface. Thickness t was calculated using the dependence on the difference between two consecutive minima of periodicity displayed in equation (1).

$$t = \frac{\lambda}{2\Delta\theta} = \frac{2\pi}{\Delta q} \text{ where } q = \frac{4\pi \sin\theta}{\lambda} \quad (1)$$

A clear single periodicity identified in the studied samples allowed performing a calculation of the film thickness. Additionally a computation of thickness with MATLAB program was made, and results were in good agreement with experimental data. In figure 2 a spectrum of X-ray reflectivity of sample deposited on silicon is given.

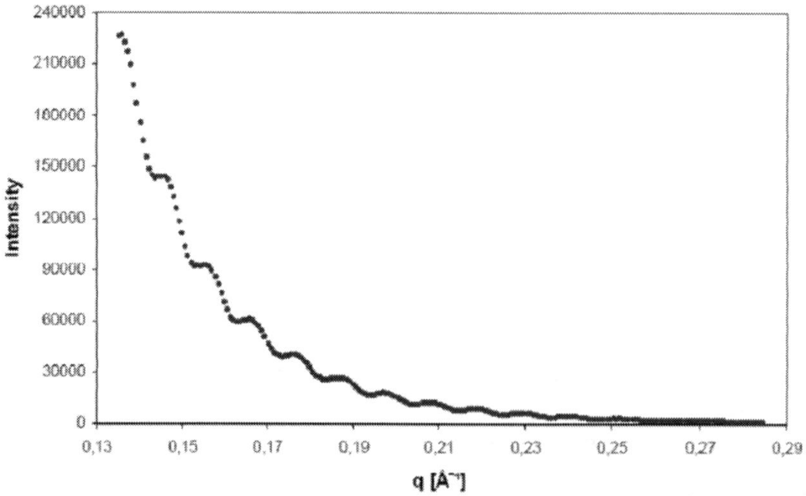

Figure 2. X-ray reflectivity spectrum of $EuNiO_3$/Si(100) of 60nm calculated thickness that corresponds to 10 minutes of deposition.

We have studied selected samples with few techniques described in experimental section. Basic properties of those samples are displayed in table 1.

Table 1. Properties of selected samples. Thickness of the sample was estimated on the base of deposition time.

Sample number	Substrate	Thickness (nm)	Annealing conditions: Temperature (°C)	Annealing conditions: Oxygen pressure (mbar)
ES8	Si(100)	60	800	168
ES15	$SrTiO_3$(100)	120	710	182
ES17	$NdGaO_3$(110)	210	720	183

X-ray diffraction investigation revealed several important features of the obtained samples. Before annealing, all thin films have shown diffraction peaks that can be indexed as a mixture of Eu_2NiO_4 and NiO. The samples were poorly crystallized and strongly oriented although the direct relationship between Si(100) substrate and film is not straightforward. This is probably related to relatively low oxygen pressure of 0.2 bar during film growth. It is in agreement with earlier results received for rare earth nickelate perovskites (e.g. in the work Laffez *et al.* [5]). After annealing we may obtain thin film of the desired composition $EuNiO_3$ or a film with some oxygen deficiency – $EuNiO_{3-\delta}$. Both compounds can be present in various crystalline forms: polycrystalline randomly oriented $EuNiO_{3-\delta}$ thin film, highly textured $EuNiO_{3-\delta}$ thin film or $EuNiO_3$ film with strong preferential orientation. In diffraction patterns we have found that apart from $EuNiO_3$, some NiO and Eu_2NiO_4 are still present in the film. In some cases, products of reaction with the substrate were visible.

The XPS survey spectra have shown that beside the elements which were the standard components of the films some contaminations with carbon, silicon and for one sample with chromium were found. The atomic composition derived from the survey XPS spectra indicated to a value of the Eu/Ni ratio much lower than 1. This fact arises mainly from the surface contamination with carbon. In order to verify the composition through the film a depth profile analysis has been performed.

The most prominent XPS lines from all elements were acquired using high resolution analysis. All XPS spectra were normalized with respect to the maximum value of count rate.

The Eu spectra are known to be sensitive to the valence state due the large (up to 9 eV) shift between the lines attributed to 2+ and 3+ valence states. In all investigated $EuNiO_3$ films both Eu 3d and 4d lines have shown the dominant 3+ valence state. A very slight contribution from the 2+ state at lower binding energies were found. Both 2+ and 3+ valence states appear to be of the same intensity for both standard and grazing geometries of analysis indicating to the lack of surface valence transition.

Ni 2p XPS lines have shown a complex multiplet structure (figure 3). No change was observed when changing the surface sensitivity of the measurement by changing the geometry of the experiment. This again, as for Eu, may be ascribed to uniform distribution of nickel 2p states.

In figure 4 a comparison between spectra obtained in standard and grazing analysis is given for oxygen 1s and valence band. In O1s spectrum a change of ratio between the standard perovskite line around 529eV and line at 531.5eV originating from surface contaminants is observed. The ratio is smaller for grazing analysis geometry which indicates that the $EuNiO_3$/Si(100) surface has more contaminants than can be found deeper in the film [6,7]. In the valence band not only small changes in intensity of line around 2eV are observed, but also a shift towards higher binding energies for grazing analysis geometry. Generally a very low photoelectron intensity was observed at the Fermi level. It is in accordance with the most likely insulating state of the film at the room temperature.

The relative intensity of the feature at around 2eV decreases for the more surface sensitive geometry. It can be thus attributed to the bulk states of the film.

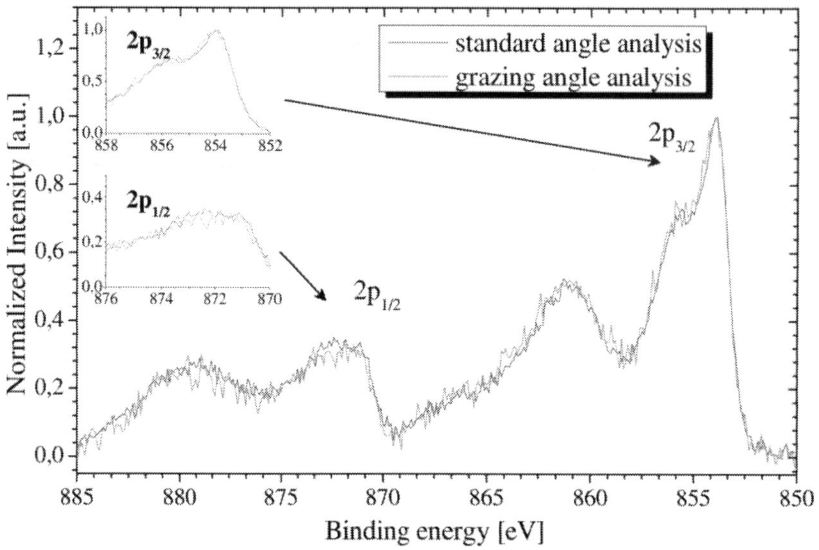

Figure 3. Ni2p spectra for standard and grazing analysis angle for $EuNiO_3/Si$ (60 nm). In the insets the main lines $Ni2p_{3/2}$ and $Ni2p_{1/2}$ are given. No change in the spectrum for various geometries indicates uniform distribution of Ni2p states in the sample.

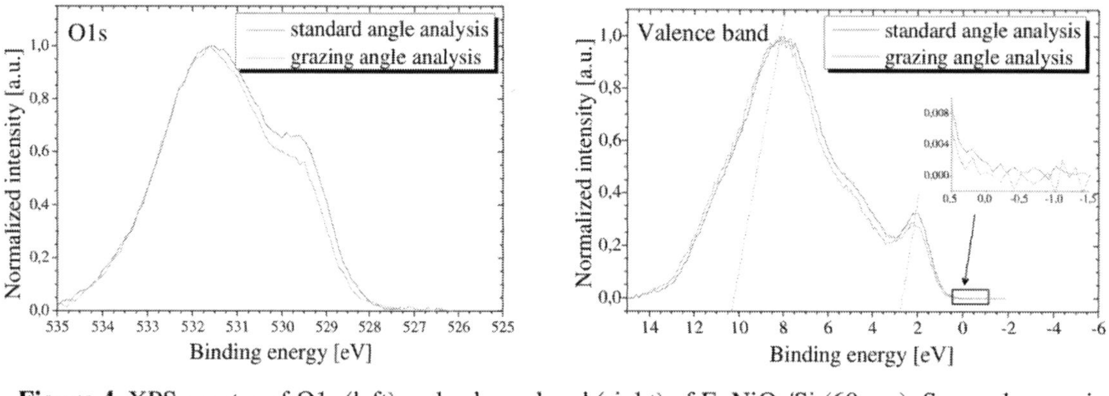

Figure 4. XPS spectra of O1s (left) and valence band (right) of $EuNiO_3/Si$ (60 nm). Some changes in the structure are observable dependent mainly on the concentration of surface adsorbed oxygen.

In figure 5 a depth profile of $EuNiO_3/Si$ thin film of thickness 60nm is presented. It contains information of atomic concentration of each element in the film. One can notice that proper for the perovskites structure Eu:Ni:O atomic ratio equal 1:1:3 is observed only for the first thirty minutes of sputtering. For deeper parts of the film it is closer rather to 2:1:4. The two regions in the figure are roughly separated by the red dashed line. It may indicate that $EuNiO_3$ is followed by a layer of compound similar to Eu_2NiO_4 before reaching the substrate. From optical measurements of $EuNiO_3/Si$ film one can conclude that the Eu_2NiO_4 layer thickness is few hundred nm. The SiO_2 layer starts to be fully visible after about 400 minutes of sputtering. In the measurement we did not reach the pure silicon substrate however some of the silicon substrate was present at every depth of the sample.

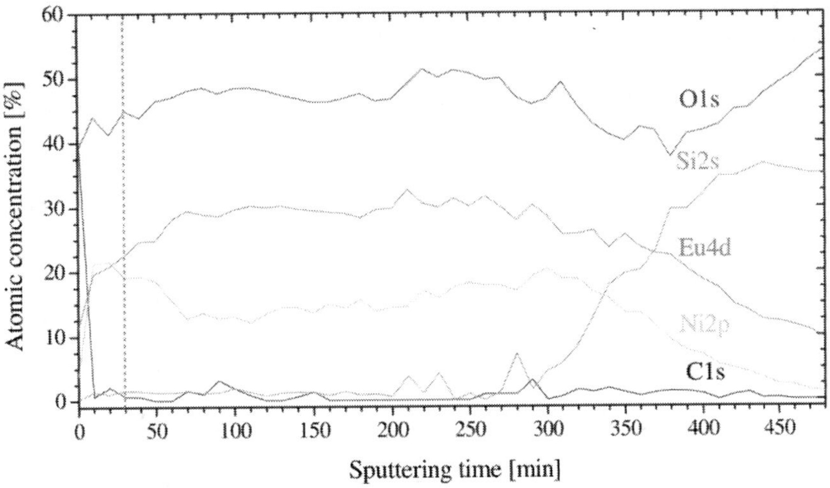

Figure 5. Depth profile of $EuNiO_3/Si$ (60 nm). 2keV Ar^+ ions were used for sputtering.

In the oxygen depth profile first two lines relate to the surface oxygen with noticeable high ratio of contaminants over the perovskite oxygen (figure 6). The perovskite oxygen is present in the film up to 400th minute of measurement where the intensity of 529 eV line diminishes to zero and a state at about 534-544 eV from SiO_2 occurs and at this depth can be partly charged.

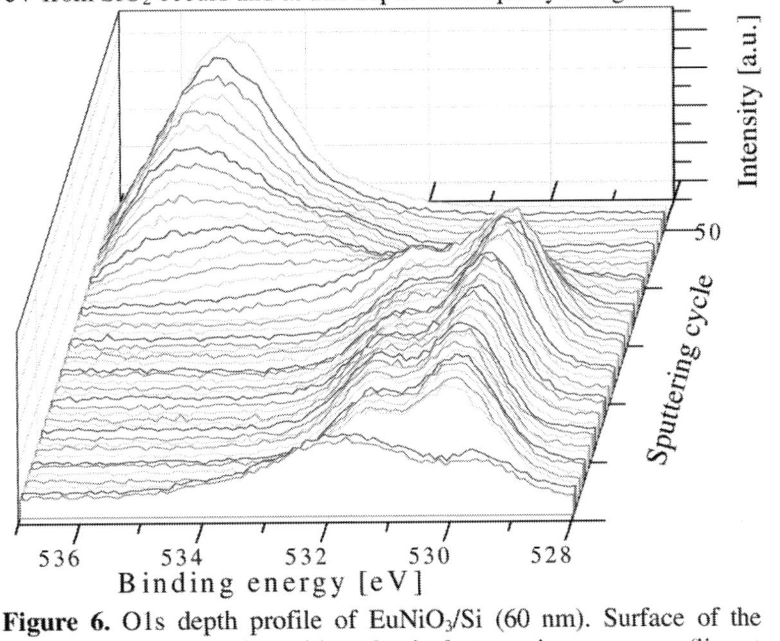

Figure 6. O1s depth profile of $EuNiO_3/Si$ (60 nm). Surface of the sample is clearly dominated by adsorbed contaminant oxygen (line at around 532 eV) whereas deeper in the sample perovskite oxygen is the main component.

Nickel 2p depth profile has been displayed in figure 7. This profile confirms the thesis of existence of Eu_2NiO_4 layer between silica and $EuNiO_3$. The trivalent state at around 855-856eV disappears slowly after fifth cycle (50 minutes of sputtering) leaving only the divalent state around 854 eV. This state can be also partly induced by reduction due to sputtering. Such effect is known for transition element oxides. A shift of lines visible in the region close to SiO_2 layer is probably caused by variation of the neutralization conditions necessary because of charging effect.

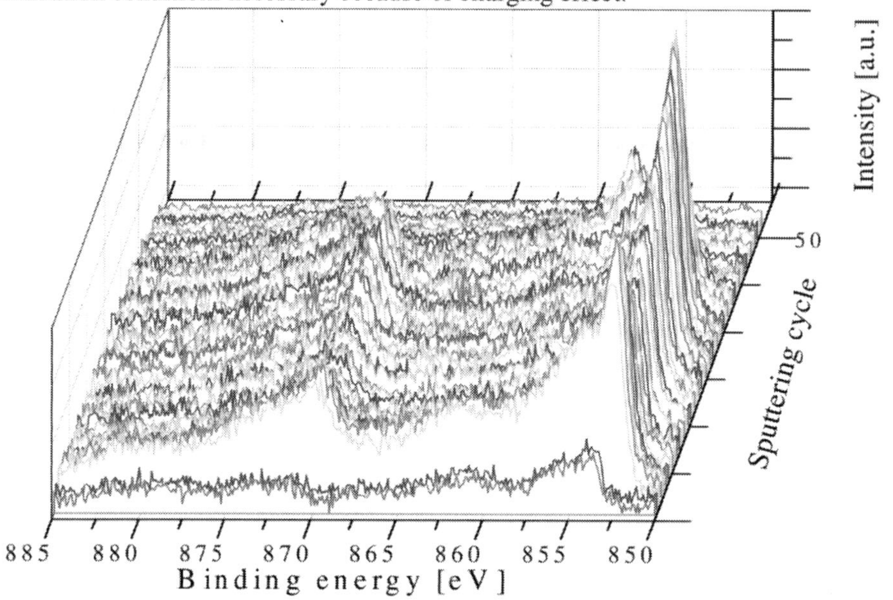

Figure 7. Profile of Ni2p of $EuNiO_3$/Si (60 nm). Ni chemical state is at least partly reduced due to sputtering. At region close to the SiO_2 surface there is probably some charging shifting the lines. The layer of $EuNiO_3$ can be discontinuous (islands on insulating SiO_2) and this is the reason for charging.

In valence band no large changes are visible for various depths of the film (figure 8). The part connected with the film valence band structure includes a peak at around 2eV which corresponds to Eu4f induced 2+ valence state. The structure slightly changes at the film-silica border and a clear shift to the higher binding energies is visible giving rise to a higher energy gap.

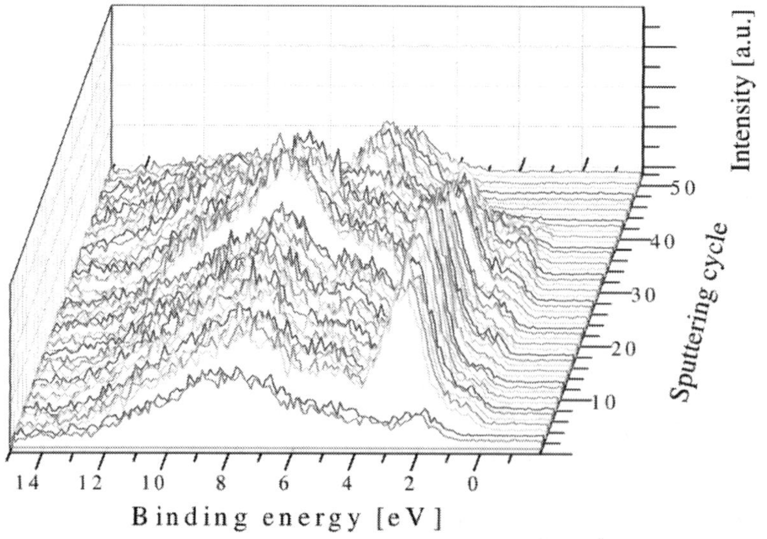

Figure 8. Valence band profile of $EuNiO_3$/Si (60 nm). The peak at about 2eV comes from the induced Eu4f states with valence 2+.

In figure 9 spectra of Ni2p band for $EuNiO_3$ on various substrates are given. The first peak at around 855.4eV can be attributed to the nickel oxidation states Ni^{2+}. The second peak at around 856.7eV corresponds to trivalent Ni. Analysis of the Ni multiplets is based on the fitting results and has to take into account other result obtained for the similar systems [8]. All spectra have characteristic satellites at around 860eV and 878eV. The intensity of these satellites in relation to the main peak of the spin – orbit doublet is known to depend on the valence state of Ni. It is very pronounced for NiO [8]. Each sample shows presence of Ni divalent and trivalent states. However in the case of $EuNiO_3/SrTiO_3$ the trivalent state is almost negligible.

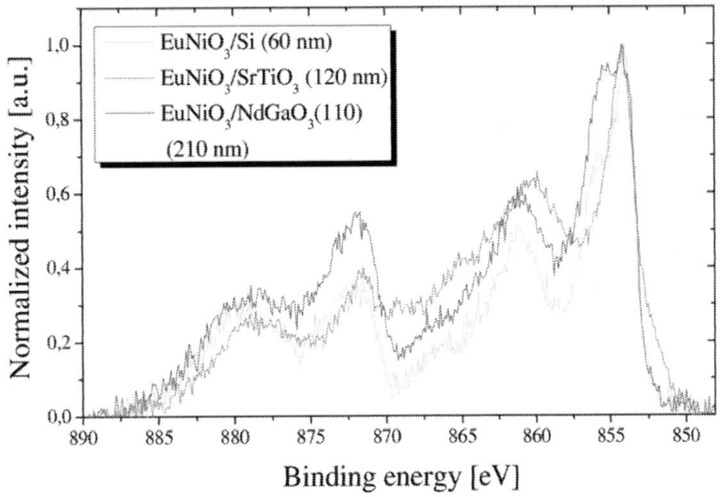

Figure 9. Ni2p spectrum in room temperature for thin films on various substrates.

The O1s spectra consist of at least two lines (figure 10). Mizokowa *et al.* [9] studied polycrystalline sample $PrNiO_3$ by XPS. They obtained a peak in O1s region at around 528.3eV. They found also a characteristic feature at around 531eV which was ascribed to the surface contamination. Our spectrum

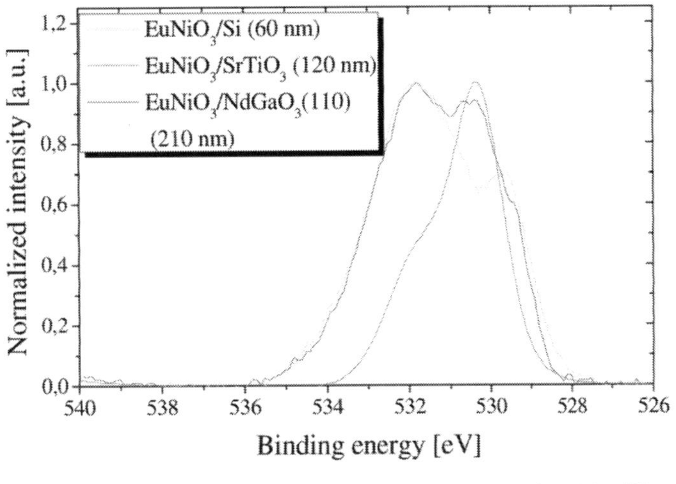

Figure 10. O1s spectrum in room temperature for thin films on various substrates.

shows also line at around 529-530eV. Such low energy line seems to be characteristic for nickelates with perovskite structure. Second line detected at 531eV (for $EuNiO_3$ on $SrTiO_3$ and $NdGaO_3$) or at 532.5eV (for $EuNiO_3$/Si) come from surface contamination. These conclusions were drawn on the basis of Galicka's thesis [8]. The highest level of contamination is present in $EuNiO_3$/Si and the lowest in $EuNiO_3/SrTiO_3$.

4. Conclusions

$EuNiO_3$ thin films were obtained by radio frequency sputtering. During the process a nickel deficiency was found and it was necessary to insert additional nickel pellets to the target so that the atomic ratio Eu/Ni would be close to 1. In order to obtained the right oxygen stoichiometry the films were annealed at approximately 800°C and oxygen pressure of around 170 bar.

After sputtering process and before oxygen annealing, the X-ray diffraction showed three important points. Films were still a mixture of Eu_2NiO_4 and NiO, they were poorly crystallized (for example for the film on $NdGaO_3$(110) with 210nm lines related to NiO are very large) and strongly oriented. It is evident since the intensity of the (200) and (101) peaks are strong in comparison with the XRD pattern of the target for example.

After annealing some films were still highly textured. A careful examination of the films after annealing shows the evolution of expected $EuNiO_3$ phase. Film on $SrTiO_3$(100) with 120nm thickness shows a line around 23° which is displaced towards higher angles after annealing. This is related to a decrease of the unit cell of perovskite subcell, from 99.06nm to 93.27nm. This line can be indexed as the (110) of the $EuNiO_3$ structure. We also observed that some of the films show unidentified lines probably due to reaction with substrate during annealing.

XPS measurements revealed that the films can be inhomogeneous in their bulk. Depth profiling of one of the samples confirmed that the film deposited on the Si consists not only of $EuNiO_3$ layer but also of a much thicker Eu_2NiO_4 layer present just above SiO_2. It indicates that annealing is a very surface sensitive process and does not allow to obtain desired composition in whole film.

Analysis of Ni2p photoemission line indicates that spin-orbit doublet of $Ni2p_{3/2}$ and $Ni_{1/2}$ consists of mixed-valence states of 2+ and 3+ valence. A careful examination of those spectra through phase transitions will be reported in future works.

References

[4] Alonso J A, Martínez-Lope M J and Rasines I 1995 *J. Solid State Chem.* **120** 174
[5] Laffez P, Zaghrioui M, Retoux R and Lacorre P 2000 *J. Magn. Magn. Mater.* **211** 111
[6] Barman S R, Chainani A and Sarma D D 1994 *Phys. Rev.* B **49** 8475
[7] Galicka K, Szade J, Ruello P, Laffez P and Ratuszna A 2009 *Appl. Surf. Sci.* **255** 4355
[8] Galicka K 2005 *Contribution to the study of the electronic states of rare earth oxide thin films by XPS and the Raman spectroscopy* Physics Co-tutelle Doctor's Thesis (Katowice: University of Silesia, Le Mans: University of Maine)
[9] Mizokawa T, Fujimori A, Arima T, Yokura Y, Mōri N and Akimitsu J 1995 *Phys. Rev.* B **52** 13865

Journal of Physics: Conference Series **289** (2011) 012015 doi:10.1088/1742-6596/289/1/012015

Synthesis and properties of doped $HgBa_2Ca_2Cu_3O_{8+\delta}$ superconductors

O Babych, Ya Boyko, I Gabriel, R Lutciv, M Matviyiv, B Sadovy, M Vasyuk

Electronics Department of Ivan Franko National University, Lviv, Ukraine

E-mail: orestbabych@gmail.com

Abstract. Synthesis of $HgBa_2Ca_2Cu_3O_{8+\delta}$ (Hg-1223) with higher chemical homogeneity and large reactivity of precursor powder $Ba_2Ca_2Cu_3O_{8+\delta}$ was carried out. Pb, Fe and Cd doped $HgBa_2Ca_2Cu_3O_{8+\delta}$ were prepared with using Hg-free precursor $Ba_2Ca_2Cu_3O_{8+\delta}$ obtained by the sol-gel method. After synthesis the samples were in the state overdoped with oxygen and reach the optimum state in flowing argon. The temperature dependences of the resistivity and magnetization measurements for different kinds of doped elements are presented.

1. Introduction

The Hg–based superconducting materials (HTSC) are compounds of the homologous series with the general composition $HgBa_2Ca_{n-1}Cu_nO_{2n+2+\delta}$, where n=1,2,3....Among them the highest superconducting transition temperature T_c = 133 K was observed for n=3 [1]. Unfortunately, there are still problems concerning the phase stability, especially the presence of CO_2 and humidity. Several reports show that the phase formation and superconducting properties of Hg-1223 are enhanced by means of cation substitutions. The critical current density and phase formation of Hg-1223 can be improved at doping by high-valence type Re, Pb or other elements [2-4].

The impurities, for example Re, Pb, are catalyzers for superconducting phase creating. This follows from comparision of the equilibrium mercury partial pressure temperature dependences for Hg-1223, Hg-1223 doping, and nonsuperconducting phase $HgCaO_2$, which are products of the synthesis reaction. It was shown, that the equilibrium mercury pressure for Hg,Re-1223, Hg,Pb-1223 near the HTSC synthesis temperature (860 C) is lower than that for $HgCaO_2$ what is not observed for the cleaned Hg-1223 [3].

On the other hand, creating the pinning centers with the destroying order parameter for critical current increase is necessary. Electric and magnetic states of defects play the main role in pinning vortex in superconducting materials. For example, the pinning enhancement in HTSC materials as a result of Fe ions influence on Y-, Bi- and Hg- HTSC has been shown [4,5].

At the same time, the solid state traditional synthesis process has several disadvantages such as high temperature and long time duration of synthesis. With the aim to activization of the solid state reaction the nanosized precursors were used. This leads to lowering the energy activation of Hg-HTSC phase formation.

2. Experimental

In this paper, by using the precursor method the synthesis of 15% Pb, 5% Fe and 5% Cd doped Hg-1223 was performed. The sol-gel method for preparing Pb,Fe and Cd doped precursors was used. This method has been also used for preparing nanosize powders of Y- 123 ceramic [6]. Fine homogeneous and reactive Hg-free precursor $Ba_2Ca_2Cu_3O_{8+\delta}$ samples were obtained by the sol-gel method using acetic acid as a complex-forming agent and the acetates of Ba, Ca, Cd, Cu, Pb and Fe were used.

The synthesis from the supersaturated solution of these acetates in water-acetone solution (1:1) with ammonium citrate gel was carried out. The nanogranules growth in the 250-350 C range was obtained as follows:

$$2BaO+2CaO+2{,}95CuO+0{,}05FeO+0{,}15PbO= Pb_{0,15}Ba_2Ca_2Cu_{2,95}Fe_{0,05}O_{7+\delta}\ .$$

This material was pressed and annealed at flowing oxygen at 900 C for 120 hours.

The obtained precursor was milled along with HgO and palletized at a pressure of 75 MPa. The pellets in sealed quartz tube were heated at 860 C for 30 h and furnace was cooled down with 10 C/h.

Samples of Pb, Fe and Cd doped $HgBa_2Ca_2Cu_3O_{8+\delta}$ superconductor were prepared by means of the following reactions (for example of Pb,Fe):

$$0{,}85HgO + Pb_{0,15}Ba_2Ca_2Cu_{2,95}Fe_{0,05}O_{7+\delta} = Hg_{0,85}Pb_{0,15}Ba_2Ca_2Cu_{2,95}Fe_{0,05}O_{8+\delta},$$
$$0{,}85HgO + Pb_{0,15}Ba_2Ca_2Cu_3O_{7+\delta} = Hg_{0,85}Pb_{0,15}Ba_2Ca_2Cu_3O_{8+\delta}.$$

Microstructure of the obtained samples was observed by scanning electron microscope. Electron probe microanalysis was also carried out for determining the composition of the superconducting phase and impurity phases in the samples.

The temperature dependences of the resistivity of Pb, Fe and Cd doped $HgBa_2Ca_2Cu_3O_{8+\delta}$ ceramics after synthesis and annealing at 350 C for 20-50 h in flowing oxygen and argon atmosphere were measured. The superconducting transition temperature T_c (on the basis of the derivative of the resistivity with temperature $d\rho/dT$) was presented. The magnetization versus temperature (4.2- 140 K) and magnetic field (up to 15 kOe) measurements for different kinds of the doped elements using a Quantum Design PPMS device were carried out.

3. Results and discussion

According to the microstructure and microprobe analyses fine homogeneous Hg-free precursor was obtained (for example of Fe doping, Fig.1a).

There are three main phases in the precursor material: the grey granules of $BaCu_{1-x}Fe_xO_2$, dark granules of $Ca_2Cu_{1-x}Fe_xO_3$ and white granules of $Ba_3Pb_2O_5$.

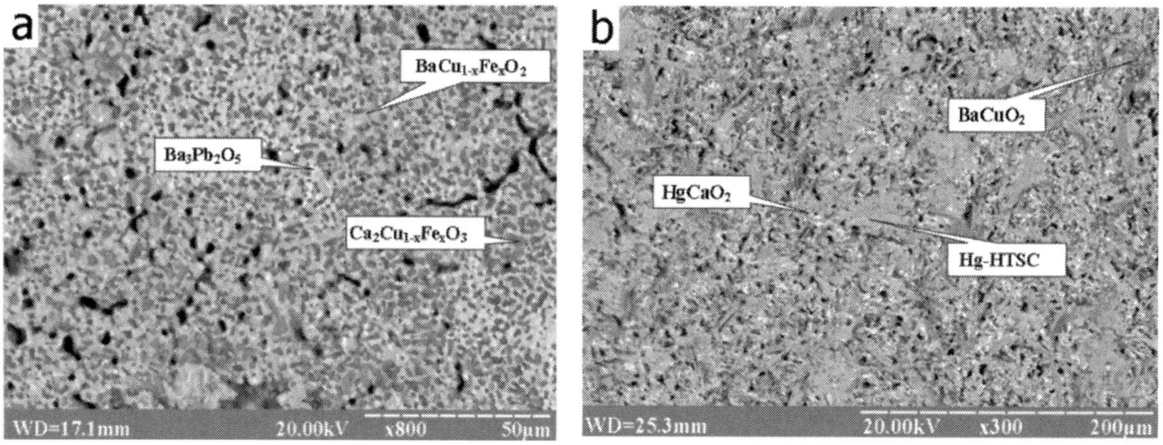

Figure 1 SEM image of $Pb_{0,85}Ba_2Ca_2Cu_{2,95}Fe_{0,05}O_{7+\delta}$ precursor-a, $Hg_{0,85}Pb_{0,15}Ba_2Ca_2Cu_{2,95}Fe_{0,05}O_{8+\delta}$ –b

Figure 1b shows the typical picture of doped Hg-1223 ceramics surface. The superconducting phase is shown in Fig.1 by the grey color. According to the microstructure analysis there are plate-like grains with a typical size of 10-20 μm. The Hg,Pb,Fe-1223 material obtained by the sol-gel method possessed high density and nonsuperconducting phases such as $HgCaO_2$ and $BaCuO_2$ were observed. The formation of $HgCaO_2$ by the interactions between HgO and oxides at 500C was achieved, however forming Hg-1223 phase was not observed. By the microprobe analyses it has been established that Fe is present in the superconducting phase near 5%. Probably, Fe ions replace Cu in the CuO_2 plane.

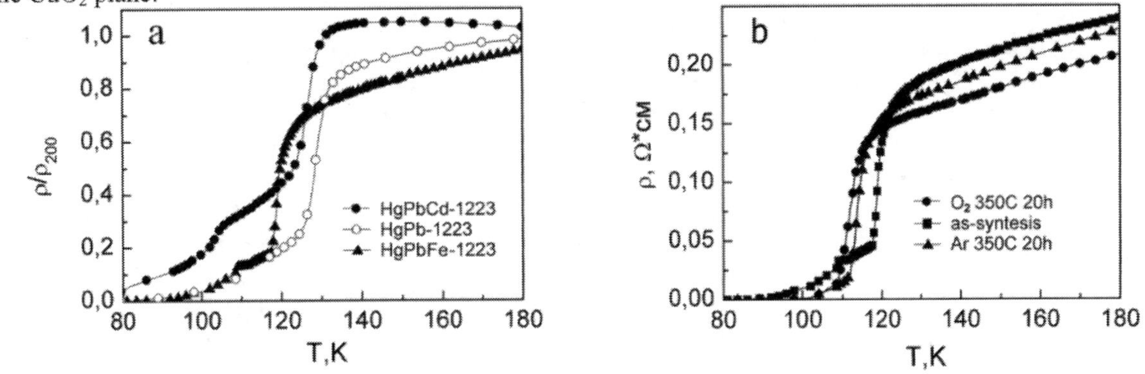

Figure 2 Temperature dependences of resistivity: ρ/ρ_{200} of Pb, Fe and Cd doped $HgBa_2Ca_2Cu_3O_{8+\delta}$ as-synthesized ceramics-a, and in flowing oxygen and argon for Fe, Pb doped $HgBa_2Ca_2Cu_3O_{8+\delta}$-b

The temperature dependences of the resistivity of Pb, Fe and Cd doped $HgBa_2Ca_2Cu_3O_{8+\delta}$ ceramics after synthesis and annealing in flowing oxygen and argon atmosphere for the example of Fe doping are shown in Fig.2a,b, respectively. There are two -step dependences on the superconducting transitions. The low temperature part of the transitions is very sensitive to magnetic field and depends on the intergranular media. The values of T_c (determined at $d\rho/dT$ maximum) for Pb doped, Hg,Fe,Pb-1223 and Hg,Cd,Pb-1223 samples were equal to 129 K, 118 K and 126 K, respectively. After synthesis the samples were in the oxygen overdoped state and reach the optimum state in flowing argon, similarly to the case of Fe doping. From Fig.2b we can see that T_c for the as-synthesized sample is equal to 118 K, while T_c decreases to 113 K at annealing in oxygen and grows up to 115 K in argon atmosphere. The presence of the superconductive state for Hg,Fe,Pb-1223 sample near 77K in contrary to Pb and Cd,Pb doping is connected with an increase of the transport critical current.

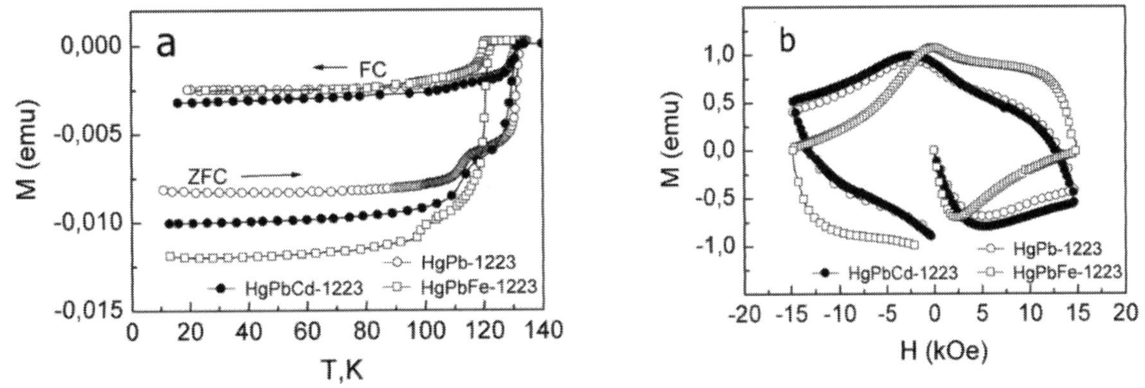

Figure 3 Magnetization of Pb, Fe and Cd doped $HgBa_2Ca_2Cu_3O_{8+\delta}$ ceramics: versus temperature – a, in magnetic field at 4,2 K – b

The results of DC zero field cooling (ZFC) and (FC) magnetization M measurement were shown in Fig.3a. There are two-step dependences on the superconducting transitions which correlate with the electric measurements. Magnetic hysteresis loops were measured at temperature 4,2 K in applied field up to 15 kOe for different kinds of the doped elements. The intragranular critical current densities were estimated using the Bean model [7]. These are J_c -6200 A/cm^2 for Hg,Pb-1223 and Hg,Cd,Pb-1223. The doping of Fe,Pb caused an increase of J_c to 6800 A/cm^2.

4. Conclusions

Samples of 15% Pb, 5% Fe and 5% Cd doped mercury superconductor $HgBa_2Ca_2Cu_3O_{8+\delta}$ were prepared employing highly homogeneous and reactive Hg-free precursor $Ba_2Ca_2Cu_3O_{8+\delta}$ obtained by the sol-gel method. According to microstructure and microprobe analyses the fine homogeneous Hg-free precursor was obtained. On the basis of the temperature dependences of the resistivity of doped $HgBa_2Ca_2Cu_3O_{8+\delta}$ ceramics (for the example of Fe doping) it was shown that after synthesis the samples were in oxygen overdoped state and reach the optimum oxygen content in flowing argon.

Measurements of the magnetization versus temperature (4.2- 140K) and magnetic field (to 15 kOe) for different kinds of doped elements were performed. The Fe doping of $HgBa_2Ca_2Cu_3O_{8+\delta}$ samples shows decreasing Tc however gives rise to pinning enhancement of the material and increasing the critical current.

References

[1] Schilling A, Cantoni M,.Guo J D and Ott H R 1993 Superconductivity above 130K in the Hg-Ba-Ca-Cu-O system. *Nature* **363** 56-58

[2] Passos C A C, Orlando M T D, Oliveira F D C,da Cruz P C M Passamai Jr J L ,Orlando C G P Eloi N A ,Correa H P S and Martinez L G 2002 Effect of oxygen content on the properties of the $Hg_{0.82}Re_{0.18}Ba_2Ca_2Cu_3O_{8+\delta}$ superconductors. Supercond. Sci. Technol. **15** 1177-1183

[3] Knizek K, Veverka M V,Pollert E,Sedmidubsky D and .Bryntse I 2002 Mercury pressure during synthesis of the $HgBa_2CaCu_2O_{6+\delta}$ superconductor. *Physica* C **371** N. 2 111–116

[4] Tian W, Shao H M, Zhu J S andWang Y N 2006 Depining lines of Fe - and Pb – doped High-T_c superconductors $HgBa_2Ca_2Cu_3O_{8+y}$ *Physica Status Solidi* (b) **196** 373-378

[5] Togulev P N, Bazarov V V, Khaibullin I B and Suleimanov N M 2002 Reinforcement of pinning by surface magnetic microparticles in high-Tc superconductors. *Low Temp. Phys.* **28** 250-253

[6] Levytska K E, Voitenko T A and Nedilko S A 2007 Superconductivity in $Y_xYBa_{2-x}Cu_4O_{8\pm\delta}$ system. *Physica* C **460-462** 722-723

[7] Bean C P 1962 Magnetization of hard superconductors. *Phys. Rev. Lett* **8** 250-253

Journal of Physics: Conference Series **289** (2011) 012016 doi:10.1088/1742-6596/289/1/012016

Study of optical and physicochemical properties of colloidal silver nanoparticles as an efficient substrate for SERS

M Cyrankiewicz and S Kruszewski

Medical Physics Division, Biophysics Department,
Collegium Medicum of Nicolaus Copernicus University,
ul. Jagiellońska 13, 85-067 Bydgoszcz, Poland

E-mail: micy@cm.umk.pl

Abstract. The unique optical and physicochemical properties of the noble metal colloidal nanoparticles enable their use in a wide range of applications, especially as a substrate in SERS and MEF study. The aim of this work is to characterize the conditions for the enhancement of Raman scattering by molecules adsorbed on silver surface. Silver sol is prepared by slightly modified Lee-Meisel's method and rhodamine 6G is used as a probe adsorbate. Pure colloidal silver suspension containing isolated nanoparticles exhibits relatively poor SERS efficiency. The extremely large electromagnetic field is induced in the junctions between two or more metallic nanocrystalites so some degree of their aggregation is necessary. The influence of potassium chloride and nitric acid as the aggregating agents is investigated here. The experiments show that both of them can promote the controlled aggregation process but chloride anions, unlike nitrate, much more effectively affect both electromagnetic and chemical mechanisms contributing to SERS. Due to the co-adsorption with rhodamine 6G they allow the dye molecules to directly interact with metallic surface. Moreover, the results clearly indicate that chloride in the presence of silver particles can induce the dimerization of the dye molecules.

1. Introduction

The noble metal nanocrystals due to their unique properties have recently attracted considerable interest of scientific world. They still remain the subject of fundamental research but also have begun to be widely employed for microanalytical applications in many fields, especially in biotechnological and biomedical research.

The great interest in colloidal silver and gold is related mainly to their ability to amplify the scattered light and fluorescence emission from adsorbed species. The first experimental and theoretical reports on surface-enhanced Raman scattering (SERS) [1][2] and metal-enhanced fluorescence (MEF) [3] appeared in the 70's of 20th century. Both phenomena manifest themselves in signal enhancement which, in the case of SERS, can reach at least six orders of magnitude [4] or even more on specific "hot spots" [5]. The earlier reports focused on SERS on silver electrodes with modified roughened surfaces [6] but since 1979 when enhancement of Raman signal on silver and gold sols was observed the first time colloidal metal particles have become the most commonly used substrates for SERS measurements (more recently also for MEF applications).

The generally accepted explanation of SERS is based on the combination of two mechanisms. The main contribution to the SERS originates from electromagnetic (EM) mechanism [7] and is related to the resonance of surface plasmons localized on the metallic nanostructures (LSPR). LSPR arises when the wavelength of light couples oscillations frequency of free conduction electrons of metal. Molecules adsorbed on the surface of silver or gold nanoparticles are affected by relatively enormous electric field. The EM enhancement is not chemically selective but intensity of Raman bands depends on the orientation of the scattering molecules and an enhancement of vibrational modes normal to the surface is most efficient [8]. The second contribution to the enhancement is a short-range resonant mechanism of chemical nature leading to the increase of molecular polarizability, mainly due to the charge-transfer (CT) and the adsorbate-metal complex formation. The prerequisite for the chemical enhancement is the adsorption of molecule directly on the metal, in so-called "active sites" [9]. The contribution of CT to the enhancement depends on the work function of metal and the energy of the HOMO or LUMO (highest occupied and lowest unoccupied molecular orbital) levels of the molecule (for transfer to or from the metal, respectively) [10]. The relative contribution of EM and CT effects can considerably differ for various compounds [11].

There are many procedures for obtaining silver sol exhibiting the desired plasmonic properties dependent on particle size, shape and surface charge. The method proposed by Lee-Meisel [12] was applied in this work. The further improvement of SERS activity of metallic colloid is achieved by its aggregation. The reason of such proceeding is the fact that the extremely large EM field is induced in the junctions between two or more metallic nanocrystals.

Rhodamine 6G (Rh6G) is used here as a probe adsorbate. Rh6G is a cationic xanthene dye that has a variety of applications, especially in biochemical and biomedical fields. Moreover, extremely SERS activity of Rh6G allowed the first single-molecule SERS observation in 1996 [5].

2. Experimental

2.1. Materials

Silver nitrate ($AgNO_3$), trisodium citrate ($Na_3C_6H_5O_7.2H_2O$), nitric acid (HNO_3) and potassium chloride (KCl) were obtained from POCH S.A (Poland). Rhodamine 6G (Rh6G) was purchased from Fluka (Switzerland). Mili-Q grade water (18.2 MΩ/cm) was used for the preparation of all solutions.

2.2. Silver colloids production and samples preparation

The nanoparticles used in experiments were prepared by chemical reduction of silver nitrate with trisodium citrate using a modified Lee and Meisel method [12]. Mechanism of reaction can be expressed as follows:

$$4Ag^+ + Na_3C_6H_5O_7 + 2H_2O \rightarrow 4Ag^0 + H_3C_6H_5O_7 + 3Na^+ + H^+ + O_2\uparrow$$

Silver colloid was obtained by addition of 2 ml of a 1% (w/v) trisodium citrate solution to 100 ml of a 1 mM $AgNO_3$ aqueous solution. First (0.2 ml) portion of citrate was added when $AgNO_3$ solution was heated to 60°C. Then the mixture was warmed up and kept at about 95°C. Next three portions (each of 0.6 ml) were added at an interval of 15 min (the first of them was added about 20 min after the addition of the initial small portion). Finally, the mixture was stirred for 30 min (still at 95°C). The resultant mixture was cooled down to room temperature and pure water was added to compensate the losses due to evaporation. pH of the mixture was about 6.5. The obtained suspension was diluted with water to the final concentration of Ag atoms of 0.167 mM in each measured sample.

KCl and HNO_3 were used as aggregating agents. All samples containing colloids with added appropriate amounts of these compounds were shaken on vortex stirrer for 30 sec after addition. Then Rh6G was added and mixing was continued for next 30 sec.

2.3. Instrumentation

Raman spectra were recorded on DK480 (Spectral Products Inc., USA), 0.5 m spectrometer equipped with ST6 CCD camera (Santa Barbara Instrument Group Inc., USA). Both devices were controlled by KestrelSpec software (Catalina Scientific Corp., USA). The spectra were excited by 632.8 nm line of

35 mW He-Ne laser (Coherent Inc., USA) at 30 sec exposure. The absorbance was measured with a two-beam V550 (Jasco Inc., USA) spectrophotometer. The particle sizing was performed by photon correlation spectroscopy. Fluctuations of the dynamically scattered light (DLS) intensity were analyzed on the system composed of a photon counting unit Photocor-PC1 (based on Hamamatsu R6358-10 photomultiplier) and 128 channels Photocor-FC correlator working in linear spacing mode (both devices with dedicated software were purchased from Photocor Instruments Inc., USA). The incident beam (632.8 nm light from a He Ne laser source) was polarized perpendicularly to the scattering plane. The scattered light was accumulated for 15 sec. Obtained autocorrelation functions were analyzed using DynaLS software (Alango Ltd., Israel). The steady-state emission fluorescence spectra of Rh6G diluted in silver colloidal suspension were recorded with a PTI (Photon Technology International, USA) spectrofluorometer working in typical "L-format" configuration. In order to improve the spectral purity of the light outgoing from excitation monochromator (of the wavelength 500 nm) the short pass 520 nm filter was mounted in excitation channel.

All measurements were performed in standard four-sided optical glass cell (10x10 mm inside width) in room temperature. They were started immediately after preparation of samples

3. Results and discussion

3.1. Characterization of silver sol.

Fig. 1. shows the optical absorption spectra of pure silver sol and Rh6G/silver sol system both without and with KCl addition. Silver nanoparticles dispersed in aqueous medium exhibit plasmon resonance at 453 nm. On the basis of Mie theory and optical data taken from CRC Handbook [13] authors have found that the dominant radius of nanoparticles is about 42 nm but the width of the resonance band points out to some diversification in size. Its worth to notice that both mean size of Lee-Meisel's colloid and its polydispersity strongly depends on the temperature at which suspension was kept during preparation [4].

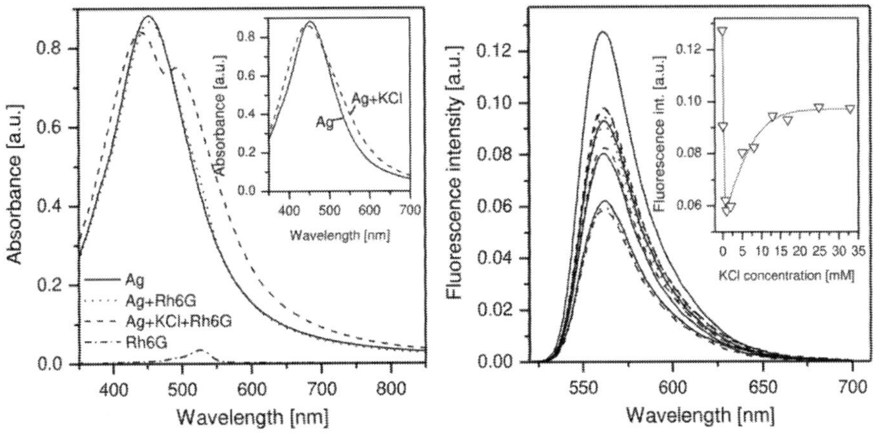

Figure 1. (Left panel) Aborbance spectra of the pure silver colloidal suspension and colloid with 1.7 mM KCl and 0.8 μM Rh6G added separately or together. (Right panel) Fluorescence spectra of 0.8 μM Rh6G on silver sol containing various amounts of KCl. The highest curve is related to the pure water solution of Rh6G.

According many authors particles obtained with the presented method are characterized by nearly spherical shape with smoothed multifaceted surface [14] whereas the others claim that up to 20% of them can be of different shape (mostly rods) [9]. However, the presence of elongated or non-ellipsoidal particles in the measured sample should manifest itself in the appearance of additional band in longer wavelengths [15] but in the presented case the absorbance beyond the plasmon resonance maximum smoothly and monotonically decays with the increase of wavelength.

As is shown in Fig. 1., after addition of the small amount of chloride ions the main absorption peak of silver sols is slightly suppressed, broadened and blue-shifted. This shift of the extinction maximum can probably be related to the changes in refractive index of medium surrounding the silver particles and formation Ag-Cl^- surface complexes.

3.2. The influence of small amount of chloride.

The used amount of Rh6G (the concentration of 0.8 μM) in the absence of chloride was insufficient to detect Raman signal from the dye molecules as well as it did not affect the resonance energy of surface plasmon on silver nanoparticles. The only change is the appearance of absorption peak of the dye at 527 nm. However, it should be mentioned here that Rh6G at higher concentration can alone induce the aggregation of colloid and considerable changes in its optical response. Author observed such effect for the Rh6G concentration of the order of 0.1 mM or higher.

The extinction spectrum changes significantly when Rh6G is added to the silver sol previously treated with chloride. After addition of the dye further decrease of main maximum is observed and new spectral band arises at about 500 nm, as one can see in Fig. 1.

It is remarkable that the presence of chloride in silver sol/Rh6G solution also results in the reduction of fluorescence emission from the dye molecules (see right panel in Fig. 1.). This effect is most evident for the concentration of KCl between 0.7 mM and 2 mM. Rh6G molecules due to their cationic properties easily adsorb on negatively charged metallic nanoparticles and tend to concentrate in their surroundings. Considering this fact the fluorescence quenching can be interpreted as a result of the resonant dye-metal energy transfer [16] but this kind of interaction should not have an effect on the absorption spectrum of the dye. The quenching of fluorescence emission can also be contributed by nonradiative deactivation of Rh6G dimmers (similarly as in reference [17]) due to dipole-dipole intermolecular interactions. The last mentioned relaxation process requires an appropriate distance and a specific mutual spatial orientation of the interacting molecules, i.e. a "sandwich" structure shown schematically in the Fig. 2. Theoretical calculations performed using a total energy minimization scheme [18] shows that the preferred orientations of adsorbed molecules with respect to the silver surface can promote the complexation of dye. One can suppose that the local increase of Rh6G concentration and partial ordering of molecules placed on or near the nanoparticles can result in the observed spectral effects.

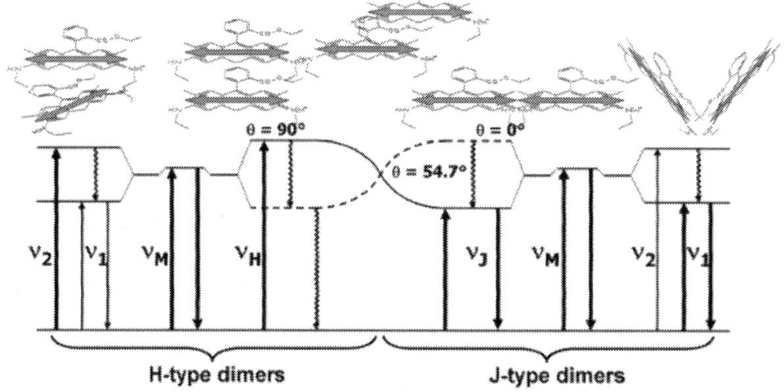

Figure 2. Simplified diagram for electronic transition in Rh6G dimers depending on the geometric arrangement of monomer units. Radiative transitions are represented by linear arrows, nonradiative - by waved ones. ν_M stands for a transition energy in isolated Rh6G molecules corresponding to about 527 nm. (diagram created on the basis of ref. [19])

According to the above diagram the appearance of new band (at about 500 nm) in the extinction spectra of colloidal silver/KCl/Rh6G system can be attributed to the formation of H-type dimers of Rh6G molecules.

3.3. Determination of size and aggregation extent of colloid.

The dynamic light scattering (DLS) technique allows to determine the particle size (more precisely: hydrodynamic radius) distribution of silver colloid and also provides information about the extent of its aggregation. The basic distribution obtained from a DLS measurement is an intensity - the other distributions can be (in the theory) generated from this using Mie theory. Nevertheless, the authors' experiences suggest that a posteriori modifications (reweighting) of regularized solution very often lead to the unrealistic results so only the primary distributions (i.e. size distributions by the scattering intensity obtained from DynaLS software) are presented here.

Despite the differences between absorbance spectra of pure silver sol and colloidal suspension with a small amount of Rh6G or KCl the particle sizing measurements gave almost identical results. It indicates that both Rh6G as well as KCl at low concentration does not cause distinct changes in the structure of colloid and a large majority of nanoparticles remain unaggregated.

3.4. Aggregation and SERS-activation of silver sol.

According to the fact that the extremely large EM field is induced in the junctions between silver nanocrystals some aggregation is desired in order to improve the SERS efficiency of silver sol. Initially the simple assemblies consisting of a few nanoparticles and then larger aggregates are formed with the increasing concentration of chloride ions. Due to the aggregation the decrease of extinction band related to the dipole plasmon resonance occurs (see Fig. 3.). Simultaneously, a new band arises beyond 600 nm and then increases and shifts towards longer wavelength. 1.7 mM concentration of chloride turned out to be insufficient to the effective SERS activation of silver sol and no Raman scattering of Rh6G was detected. As the content of chloride increased the SERS intensity of Rh6G on silver nanoparticles also rapidly increased and reached the maximum at 33.3 mM KCl concentration (see middle panel in Fig. 3.). The chloride at such high concentration strongly affected the stability of colloidal solution and after a few minutes a noticeable degradation of the sample, i.e. the gradual precipitation of colloid was observed. It has also resulted in considerable decrease of SERS intensity. From the other hand, the silver sol activated with smaller amount of chloride needs much more time to achieve the maximum of SERS efficiency (see [4]). In many cases this time are longer than the duration of measurements in the presented experiments.

As one can see in Fig. 3. the results of consecutive DLS measurements are correlated with the changes in absorption spectra. The increase of KCl concentration from 3.33 mM to 33.3 mM leads to the broadening of the particle size distribution function but the modal size varies relatively little. It suggests that most of the nanoparticles still remain unaggregated or form only small clusters consist of two or thee silver particles. One can notice that the coexistence of KCl and Rh6G in colloid results in a slightly greater extent of its aggregation than in the case of silver sol without Rh6G added.

Fig. 4. shows that also strong acids affect the Rh6G/silver sol system. The extinction and corresponding SERS spectra were studied at various concentrations of HNO_3 which was chosen to avoid the influence of chloride ions. The absorption spectra in Fig. 4. essentially differ from previously presented ones. The main difference is a smaller width of the blue band and the absence of secondary maximum in the red region. Instead, as pH decreases the absorption beyond 550 nm increases as the wide spectral wing monotonically decaying towards the longer wavelengths. For HNO_3 concentration changing stepwise from 0.83 mM to 5 mM the dipolar plasmon resonance band gets lower and shoulder peak attributed to the absorption of Rh6G (at 527 nm) becomes more distinct. It should be underlined that any spectral band indicating the complexation of dye molecules is observed. The sample containing 5 mM HNO_3 is very unstable and further addition of nitric acid leads to the fast (in just teens of seconds or even seconds) destruction of colloid what is indicated by darkening of the solution and sedimentation of large metallic aggregates.

It is noteworthy that not presented here spectra of KCl/silver sol system (without Rh6G) resemble spectra of silver colloid aggregated with nitric acid.

The results of DLS measurements of colloid aggregated by acidification with HNO_3 show that the range of pH values between 3 and 2 is critical for citrate reduced silver sol. This observation is consistent with [20]. As pH of the silver colloid containing 0.8 μM Rh6G decreases the average hydrodynamic radius of aggregates strongly increases. The right graph in Fig. 4. also shows that, unlike the case of the aggregation with chloride, the growth of aggregates leads to the fast decrease of the number of remaining isolated nanoparticles.

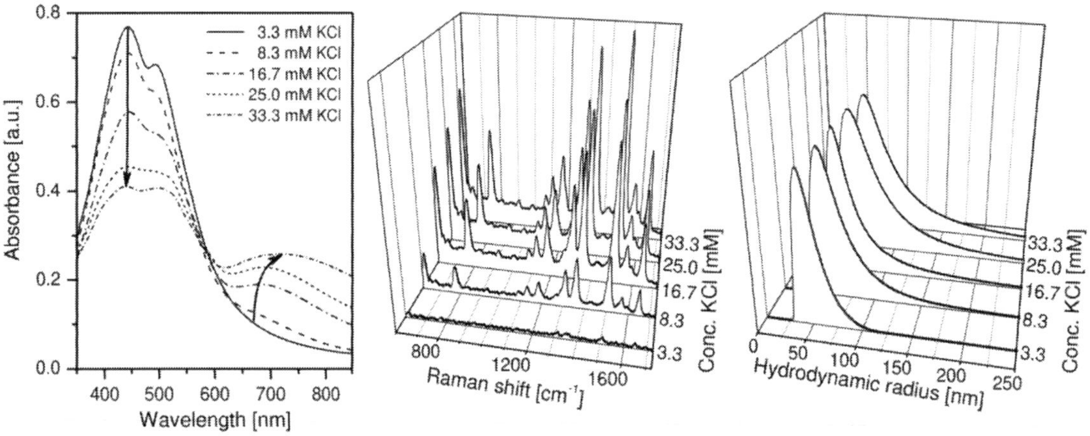

Figure 3. Extintion (left panel) and SERS (midle panel) spectra and the corresponding results of DLS measurements of 0.8 μM Rh6G on silver colloid with various amounts of KCl added to solution. The DLS intensity distribution function are normalized to equal maximum.

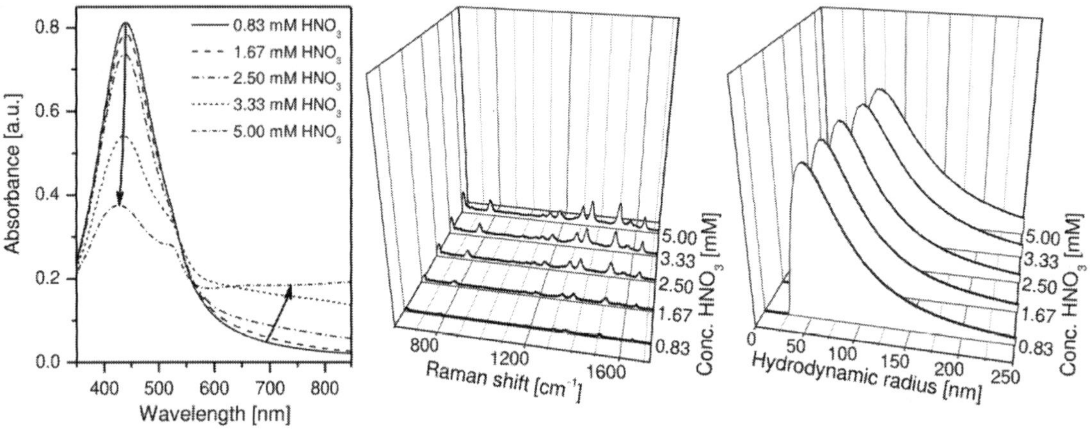

Figure 4. Extintion (left panel) and SERS (midle panel) spectra and the corresponding results of DLS measurements of 0.8 μM Rh6G on silver colloid after addition of various amounts of HNO_3. pH of the samples is equal to 2.92, 2.55, 2.26, 2.10, 2.05, respectively.

3.5. A discussion of the influence of chloride and acidity.

Chloride ions exhibit much larger complexation affinity to silver than nitrate and adsorb directly on the surface, replacing other anions. Detailed visual inspection of the SERS spectra of Rh6G on colloid aggregated with HNO_3 allows to perceive a weak hump at 1400 cm^{-1} (assignable to the carboxylate symmetric stretching vibrations in the citrate ion) which disappears in the presence of KCl (compare

SERS spectra shown in Fig. 5.). Also the changes in the intensity of fluorescence emission of Rh6G shown in Fig. 1. can suggest that KCl promotes chemisorption of Rh6G on silver only when the concentration of chloride ions does not exceed some critical value. The comparison of absorbance spectra of silver sol\Rh6G samples with the KCl concentration of 3.3 mM and 33.3 mM (see left and middle graph in Fig. 5., respectively) leads to the similar conclusion regarding the dimerization of dye molecules. The second derivative spectra clearly show that the band assignable to the H-dimers of Rh6G disappears at higher concentration of KCl.

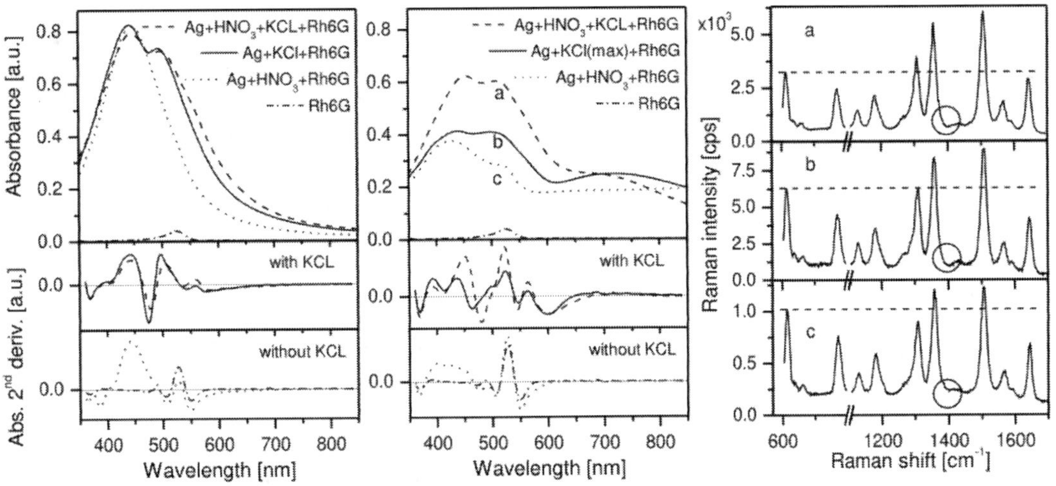

Figure 5. (Left panel) Extinction spectra of 0.8 μM Rh6G on silver colloid with 3.3 mM KCl and 0.83 mM HNO_3 added, both together or separately. Extintion (middle panel) and corresponding SERS (right panel) spectra of 0.8 μM Rh6G on silver sol with addition of 5 mM HNO_3 and 3.3 mM KCl (a), 33.3 mM KCl (b) and 5 mM HNO_3 (c).

Due to strong chemical interaction with silver, chloride ions can induce changes of surface morphology resulting in the creation "active sites" [9][21]. The co-adsorption of chloride with Rh6G allows the dye molecules to directly interact with silver surface and leads to the additional SERS enhancement. It is the main reason of the differences in the intensity between SERS spectra shown in Fig. 3. and Fig. 4. One can also suppose that the appearance of the second extinction band at the longer wavelengths is not only the manifestation of the aggregation of nanoparticles but also is an indirect evidence for the chemisorption of Rh6G and CT transition between dye molecules and silver surface [22]. From the other hand, most researchers agree that the aggregation of silver sol leads to the formation of fractal assemblies [21] and theoretical calculations predict that such pattern of aggregation is necessary for the occurrence of low-frequency plasmon modes [23].

Unlike nitric acid, chloride ions effectively affect both mechanisms contributing to SERS. Moreover, the comparison between the Raman spectrum marked as (a) in Fig. 5. and the spectra shown in Fig. 3. suggests that the influence of the small amount of chloride on SERS efficiency of silver sol increases when colloidal solution is previously acidified with HNO_3. The resultant effect of chloride measured as the ratio between intensity of the most pronounced Raman band at about 1505 cm^{-1} obtained for sample with KCl and without KCl addition decreases with the increasing of acidity. From the other hand, before the addition of chloride a role of chemical Rh6G-Ag interaction ("first-layer" effect) in the amplification of the SERS signal seems to be very limited probably due to the presence of the citrate coating on the surface. It is also worth to notice that the extinction spectrum of Rh6G on silver sol after 5 mM HNO_3 and subsequently 3.3 mM KCl addition, considerably differs from the ones obtained for the greatest amounts of both HNO_3 and KCl added separately. Despite the expected highest degree of aggregation and lowest stability, the mixture is more stable and exhibits high extinction coefficient below 600 nm and relatively low in infrared region. It is possible that in this case chloride ions partly reduce the influence of nitric acid.

All vibrational bands of molecules adsorbed on the silver sol are liable to the EM selection rules but some of them (stretching band at 614 and 775 cm^{-1} in the case of Rh6G) are additionally coupled to molecular and CT resonance. The excitation wavelength (632.8 nm) in presented experiments seems to bee too far away from resonance condition so the observed differences between SERS spectra shown in Fig. 5. (see right panel) should be rather attributed to the reorientation of Rh6G molecules with respect to surface of particles.

4. Conclusions

Results presented in this work show that the aggregation of silver nanoparticles plays the crucial role in SERS efficiency of colloid. The experiment shows that both potassium chloride and nitric acid can promote this process but among these two aggregating agents chloride exhibits better properties from point of view SERS of Rh6G on silver. Colloidal solution with the addition of small amount of chloride ions is characterized by greater stability in comparison to the silver particles aggregated with HNO_3. Moreover, unlike nitrate, chloride ions effectively affect not only electromagnetic but also chemical mechanism of SERS. It results in additional enhancement of Raman scattering. This work also indicates that chloride in the presence of silver particles can induce the dimerization of Rh6G molecules.

Rhodamine 6G as well as the other cationic compounds can easily adsorb on the negatively charged silver surface but in the case of neutral and anionic molecules further modification of the surface potential is needed. Our preliminary results show that it can be successfully achieved with the addition of appropriate amount of poly(L-lysine) but the further studies are needed.

Acknowledgments
The work was supported from Nicolaus Copernicus University Grant No. 36/2009.

References

[1] Fleishman M, Hendra P J and McQuillan A J 1974 *Chem. Phys. Lett.* **26** 163-166
[2] Janmaire D L and Van Duyne R P 1977 *J. Electroanal. Chem.* **84** 1-20
[3] Drexhage K H 1974 *Prog. Opt.* **12** 163-232
[4] Cyrankiewicz M, Wybranowski T and Kruszewski S 2007 *J. Phys.: Conf. Ser.* **79** 012013
[5] Kneipp K, Wang Y, Kneipp H, Perelman L T, Itzkan I, Dasari RR and Feld M S 1997 *Phys. Rev. Lett.* **78** 1667-1670
[6] Kruszewski S 1994 *Surf. Interface Anal.* **21** 830-838
[7] Moskovits M 2005 *J. Raman Spectrosc.* **36** 485-496
[8] Brolo A G and Irish D E 1996 *J. Electroanal. Chem.* **414** 183-196
[9] Hildebrandt P and Stockburger M 1984 *J. Phys. Chem.* **88** 5935-5944
[10] Haran G 2004 *Isr. J. Chem.* **44** 385-390
[11] Muniz-Miranda M, Cardini G and Schettino V 2004 *Theor. Chem. Acc.* **111** 264-269
[12] Lee P C and Meisel D 1982 *J. Phys. Chem.* **86** 3391-3395
[13] Weaver J H and Frederikse H P R 2004 Optical properties of selected elements *CRC Handbook of chemistry and physics, 84th Edition* ed D R Lide (Boca Raton: CRC Press) p 12-150
[14] Munro C H, Smith W E, Garner M, Clarkson J and White P C 1995 Langmuir **11** 3712-3720
[15] Noguez C 2007 *J. Phys. Chem. C* **111** 3806-3819
[16] Lakowicz J R, Shen Y, D'Auria S, Malicka J, Fang J, Gryczynski Z and Gryczynski I 2002 *Anal. Biochem.* **301** 261-277
[17] Chandrasekharan N, Hu J, Jones G, Kamat P V 2000 *J. Phys. Chem. B* **104** 11103-11109
[18] Noginov M A, Vondrova M, Williams S M, Bahoura M, Gavrilenko V I, Black S M, Drachev V P, Shalaev V M and Sykes A 2005 *J. Opt. A: Pure Appl. Opt.* **7** 219–229

[19] Martinez V M, Arbeloa F L, Prieto J B and Arbeloa I L 2005 *J. Phys. Chem. B* **109** 7443-7450
[20] Alvarez-Puebla, R A, Arceo E, Goulet P J G, Garrido J J and Aroca R F 2005 *J. Phys. Chem. B* **109** 3787-3792
[21] Otto A, Bruckbauer A, Chen Y X 2003 *J. Mol. Struct.* **661–662** 501–514
[22] Du Y and Fang Y 2004 *Spectrochim. Acta, Part A* **60** 535–539
[23] Karpov S V, Bas'ko A L, Popov A K and Slabko V V 2000 *Colloid J.* **62** 699-713

Journal of Physics: Conference Series **289** (2011) 012017 doi:10.1088/1742-6596/289/1/012017

Influence of SBN70 concentration in PVDF on dielectric and pyroelectric properties of nanocomposites

M Olszowy[1], E Nogas-Ćwikiel[2], K Ćwikiel[3]

[1]Institute of Physics, University of Zielona Góra, 4a Prof. Szafrana Str., 65-516 Zielona Góra, Poland
[2]Faculty of Computer Science and Materials Science, University of Silesia, 2 Śnieżna Str., 41-200 Sosnowiec, Poland
[3]Institute of Physics, University of Silesia, 4 Uniwersytecka Str., 40-007 Katowice, Poland

E –mail: M.Olszowy@if.uz.zgora.pl

Abstract. Strontium barium niobate $Sr_{0.7}Ba_{0.3}Nb_2O_6$ (SBN70) ceramic nanopowder was dispersed in a poly(vinylidene fluoride) (PVDF) matrix providing a composite with 0-3 connectivity. The SBN70-PVDF composites samples were obtained from ceramic and polymer powders by hot-pressing method. The SBN70 ceramic was prepared by a sol-gel method. The composite surface images were obtained by AFM tapping mode (NT-NDT Solver P47). The dielectric response of the composites was studied in the frequency range 100 Hz – 1 MHz and the temperature range 100 – 430 K. The dielectric properties of the composites display features originated from the PVDF polymer modified by those of SBN70 ceramics. The resulting pyroelectric currents were measured using a Keithley 6517 A electrometer and were used to calculate the pyroelectric coefficient *p*. The *p* of poled composites increases from ~24 $\mu C/m^2 \cdot K$ in pure PVDF to ~40 $\mu C/m^2 \cdot K$ in the composites of $\Phi = 0.2$ at room temperature.

1. Introduction

Ceramic-polymer composites show unique properties for a wide range of applications in the field of microelectronics, biomedical engineering and other modern technologies. These materials (i.e., polymer loaded with ferroelectric ceramic powder) have aroused increased interest due to their unique opportunities blending polymeric properties like mechanical flexibility, formability and low dielectric permittivity with high electro-active properties characteristic for ferroelectric ceramics. They were developed to be an alternative material to ferroelectric ceramics and to ferroelectric polymers. The ceramic-polymer composites are arranged in a specific configuration called connectivity. In composites with 0-3 connectivity a three-dimensionally connected polymer phase is loaded with isolated ceramic particles. The dielectric permittivity ε' and dielectric losses $tg\delta$ of composites are involved in "figures of merit" (FOM_I and FOM_{II}) of various physical quantities important for applications, therefore, the knowledge of dielectric response is essential for tuning their physical properties [1].

The ceramics type $Sr_xBa_{1-x}Nb_2O_6$ (abbreviated as SBN, where $0.25 \le x \le 0.75$) is a promising lead free ferroelectric material with tungsten bronze structure and it is a potential material for pyroelectric infrared detectors, piezoelectrics, electro-optics and photorefractive optics [2, 3]. It exhibits a ferroelectric phase with only one polar axis and a transition temperature depending on the Sr/Ba ratio. The SBN ceramics with $x = 0.7$ (SBN70), used to produce SBN70-PVDF composites, is characterized by the diffuse phase transition phenomena (i.e., they exhibit a broad Curie peak in the phase transition range) and is a well-known relaxor material [3, 4]. The SBN70 ceramics offer excellent piezo- and pyroelectric properties, high dielectric permittivity ($\varepsilon' \sim 3000$ at 1 kHz) [3] and low dielectric losses, albeit with low mechanical flexibility. Pure PVDF is a semicrystalline polymer, it has a monomer unit (CH_2CF_2) and at least four crystallite modifications [5]. The two most important ones are the nonpolar α phase and the polar β phase. PVDF film obtained from the melted material contains spherulites with nonpolar α phase. Mechanical stretching, annealing at high temperature or poled material in a high DC field is necessary to obtain the polar β phase. Such material has a dipole moment of 7.0 x 10^{-30} C·m perpendicular to the chain direction [6]. PVDF film (unpoled, radially oriented) offer low value of the dielectric permittivity ($\varepsilon' \sim$ 8-10) [1], high dielectric losses, but also a high flexibility and mechanical strength.

In this work, we present the dielectric response of the SBN70 ceramics obtained by a sol-gel method, pure PVDF radially oriented and PVDF loaded with nanosize particles of SBN70 ceramic powder with the volume fraction of the ceramics Φ = 0.1 and 0.2. The dielectric response was observed as a function of frequency and temperature. We investigated also the temperature dependence of pyroelectric coefficient with a selected DC electric field for composite samples. AFM tapping mode was used to perform SBN70-PVDF composite surface images.

2. Experimental

Dielectric and pyroelectric response was studied for SBN70-PVDF composite samples of 0-3 connectivity and volume fraction of ceramics Φ from 0.0 to 0.2. The composites were prepared from ceramic and polymer powders by hot-pressing (450 K, 3.2 MPa during 10 minutes) and cooled to the room temperature under pressure at a rate of 10 K/min. Powders of SBN70 used for fabrication of SBN70-PVDF composites were prepared by a sol-gel route using strontium acetate $(CH_3COO)_2Sr$, barium acetate $(CH_3COO)_2Ba$ and niobium ethoxide $(C_2H_5O)_5Nb$ as precursors. Powders obtained from dried gels were calcined at 870 K for 24 hours to burn off organics. Bulk ceramics was prepared from these powders by hot uniaxial pressing at 1470 K for 4 hours, under the pressure p = 5 MPa. The distribution of the ceramic grains on the surface of the PVDF matrix and the grain size were estimated by AFM tapping mode (NT-MDT Solver P47) in air at room temperature. The PVDF powder was delivered by Nitrogenous Concern (Tarnów, Poland). The samples had the form of discs with a diameter of 11 mm and a thickness of 50 – 100 μm. The samples were covered with gold electrodes on both surfaces.

Study of the dielectric properties of samples was carried out in the frequency range 100 Hz – 1 MHz using HP 4284A precision LCR meter. The measurements were performed on unpoled samples during heating from 100 to 430 K at a rate of 0.2 K/min. All the samples were aged for at least 72 h prior to measuring the dielectric and pyroelectric properties. The pyroelectric current response for each of the poled samples was measured by quasi-static method in the short-circuit regime. The samples were first cooled to the temperature of 100 K. Next they were heated to the temperature of 360 K and cooled to the temperature of 100 K in a DC electric field E_p = 5 MV/m. Heating and cooling was made at a rate of 0.2 K/min in the field applied. After poling, the electrodes were short-circuited for 0.5 h to eliminate surface charges at 100 K. Pyroelectric current was measured during heating of samples from 100 K to 360 K at a constant rate of β = 3 K/min using Keithley 6517 A electrometer coupled with a computer assisted cryogenic system. The pyroelectric coefficient is calculated from the formula $p = i_p/A\beta$, where i_p is the measured pyroelectric current and A is the area of the top electrode.

3. Results & discussion

SBN-PVDF composite surface image was obtained by AFM tapping mode. In the Figure 1 particles of ceramics are white and the polymer is dark. It is visible that value of ceramics grain size amounts < 200 nm. It is seen that the SBN particles are dispersed rather uniformly in the polymer matrix.

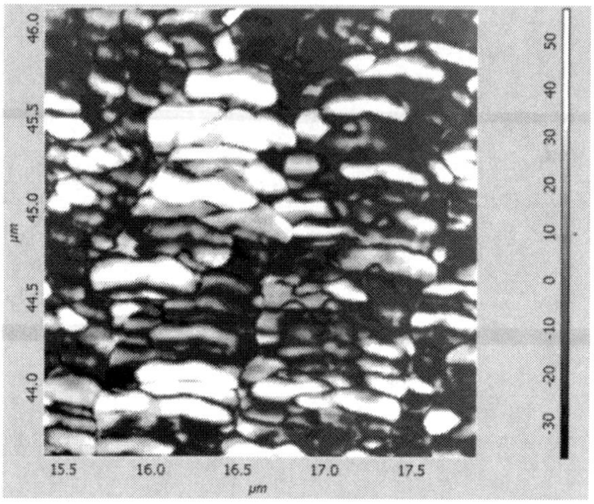

Figure 1. The micrograph of the surface of the SBN70-PVDF sample ($\Phi = 0.2$)

Dielectric permittivity ε' and dielectric losses tgδ were measured as a function of temperature and frequency. Figure 2 shows the ε' behaviour for the pure PVDF (a) and the bulk ceramic SBN70 (b). For PVDF three dielectric anomalies are observed. A dispersive ε' anomaly of the PVDF polymer (Figure 2a) in the temperature range 180 – 330 K is related to a freezing dipolar motion in the amorphous region. Dielectric behaviour in the range 330 – 380 K is ascribed to wide angle oscillations of dipoles attached to the chain, followed by their rotation with main chain co-operation appearing in the crystalline phase [1]. At ~420 K one observes a ε' anomaly characteristic of the ferroelectric-paraelectric phase transition. The diffused dielectric anomaly obtained for the SBN70 ceramic sample (Figure 2b) in 250 K – 350 K temperature range as well as the characteristic dispersion of the permittivity points to the relaxor properties of the ceramics used to produce SBN70-PVDF composites [4].

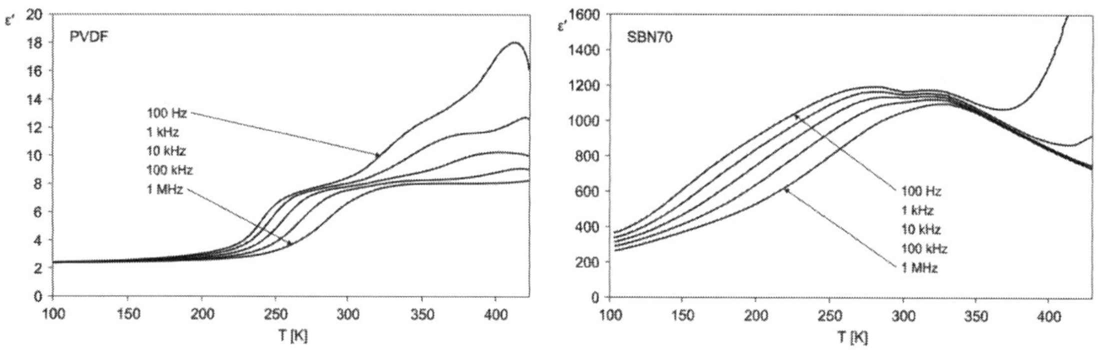

Figure 2. Temperature dependence of the dielectric permittivity ε' for the pure PVDF (a) and bulk ceramic SBN70 (b) measured at various frequencies

In Figure 3 the temperature dependencies of dielectric permittivity ε' and tgδ of SBN70-PVDF composites with volume fraction of ceramic Φ = 0.1 and 0.2 are presented. The values ε' of the composites lay between the values for the PVDF and values for the SBN70 as expected (see Figure 2). The addition of the SBN70 ceramics increases the value ε' of the composites in the whole temperature range, due to the high permittivity value of SBN70. The dielectric response of the composites is modified by the maxima originated from the PVDF polymer and the SBN70 ceramics.

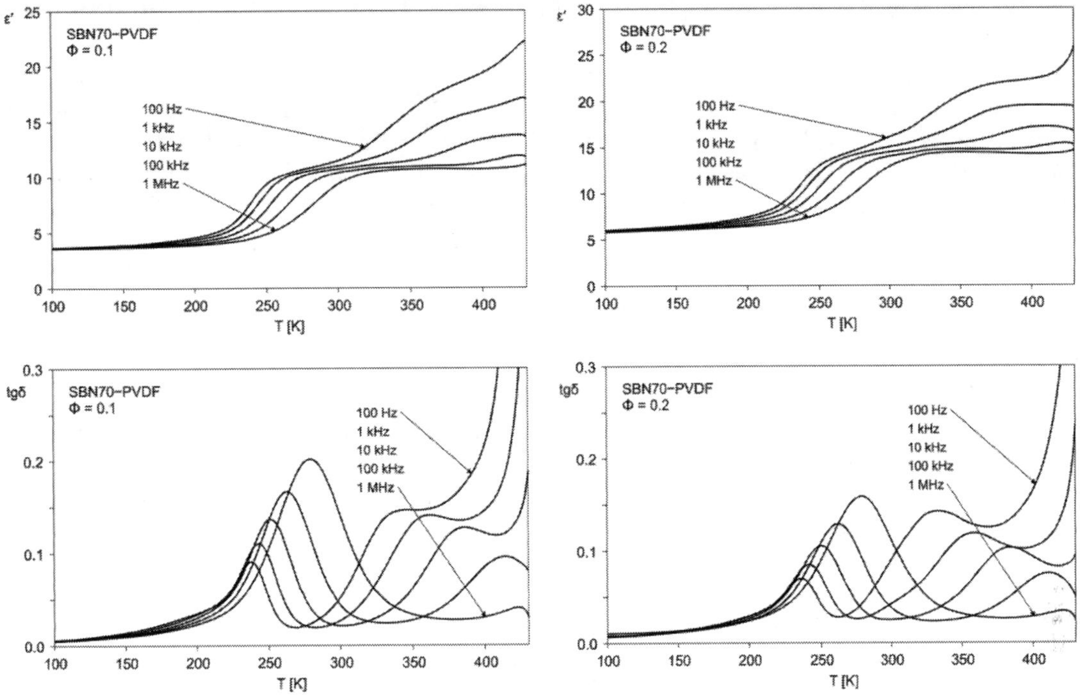

Figure 3. Temperature dependencies of the dielectric permittivity ε' and tgδ for SBN70-PVDF composites with two volume fraction of ceramic

The maxima of tgδ in the temperature range 230 – 280 K are attributed to the α-relaxation process, related to the glass transition in the polymer. In the temperature range 330 – 420 K the maxima of tgδ are attributed to dielectric losses in the crystalline phase of the polymer. In the low temperature range described maxima of tgδ increase and are shifting towards higher temperatures with increasing frequency, whereas in the high temperature the maxima decrease with increasing frequency. The addition of the ceramics to the polymer matrix decreases value of tgδ in the composite.

The pyroelectric coefficient p and figures of merit FOM_I and FOM_{II} of poled samples were studied. The pyroelectric current i_p were measured during heating samples and values of $FOM_I = p/\varepsilon'$ (significant for sensor with high impedance amplifiers) and $FOM_{II} = p/(\varepsilon' \cdot \mathrm{tg}\delta)^{1/2}$ (important in the case when the noises of the sensor are mainly due to the pyroelectric element) are calculated. Figure 4 shows the dependence of pyroelectric coefficient on temperature for PVDF and composites with two volume

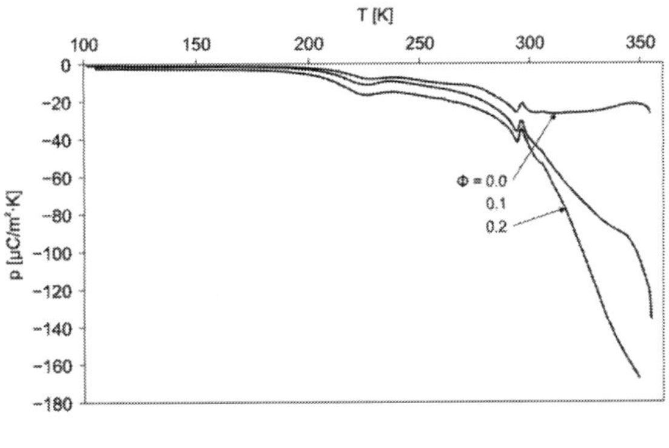

Figure 4. Dependence of pyroelectric coefficient on temperature for PVDF and SBN70-PVDF composites

fraction Φ of the ceramics: 0.1 and 0.2. The content of the SBN70 improves the pyroelectric properties of the investigated materials. The p of composites increases from ~24 $\mu C/m^2 \cdot K$ in pure polymer to ~40 $\mu C/m^2 \cdot K$ in the composites of $\Phi = 0.2$ at room temperature. The thermal variations of p (Figure 4) show anomalies corresponding to the dielectric behaviour mainly in the PVDF polymer. The calculated FOM_I and FOM_{II} of composites are higher in comparison with the PVDF polymer. The optimal of FOM_I and FOM_{II} equals to 3.5 $\mu C/m^2 \cdot K$ and 77.3 $\mu C/m^2 \cdot K$ for $\Phi = 0.1$, respectively. The increase of the Φ from 0.1 to 0.2 involved the reduction of the values of FOMs. We related this behaviour to the rapider increase of the value ε' than of the pyroelectric coefficient with the increase of the volume fraction Φ of investigated composites.

4. Conclusions

The SBN70-PVDF 0-3 composites samples with SBN70 volume fraction Φ up to 0.2 were fabricated by hot-pressing method and their dielectric and pyroelectric response was studied. The distribution of the ceramic grains on the surface of the polymer matrix and the grain size estimation were also studied. The SBN70 ceramics is characterized by the diffuse phase transition phenomena in 250 K – 350 K temperature range. The results of the ε' measurements for composites proved that the dielectric response of the ceramic-polymer composite is determined by the diffused dielectric anomaly coming from the ceramics and by relaxation processes in polymer. The addition of the ceramics to the polymer decreases dielectric losses in the SBN70-PVDF composite. Due to high permittivity value of the ceramics the ε' of the composite increases with the volume fraction of the ceramics. The pyroelectric coefficient of the SBN70-PVDF is higher than that of the pure polymer. The results of the pyroelectric and dielectric measurements indicated that the value of the $\Phi = 0.1$ is the optimum which provides the highest values of the FOMs. The SBN70-PVDF composites can be recommended for pyroelectric applications.

References

[1] Hilczer B, Kułek J, Markiewicz E, Kosec M and Malič B 2002 *J. of Non-Crystalline Solids* **305** 167
[2] Sakamoto W, Yogo T, Kikuta K and Hirano S, 1996 *J Amer. Ceram. Soc.* **79** 2283
[3] Kim M.-S, Lee J.-H, Kim J.-J, Lee H Y, Cho S.-H 2002 *J. of the Eur. Ceram. Soc.* **22** 2107
[4] Pawlaczyk C, Olszowy M, Markiewicz M, Nogas-Ćwikiel E, and Kułek J 2007 *Phase Transitions* **80** 177
[5] Rollik D, Bauer S, and Gerhard-Multhaupt R 1999 *J. of Appl. Phys.* **85** (6) 3282
[6] Das-Gupta D K 2000 Smart Ferroelectric Ceramic/Polymer Composite Sensor, in Polymer Sensors and Actuators, Osada Y, De Rossi D E, eds., (Springer) 109-146.

Influence of doping elements (Y, Fe, V) on magnetic properties of RM_2 (R= Gd, Er; M= Fe, Co, Ni) Laves phases and their hydrides

H Michor[1], B Kotur[2], O Myakush[2], G Hilscher[1]

[1]Institute of Solid State Physics, Vienna University of Technology, Wiedner Hauptstrasse 8-10, A-1040 Wien, Austria
[2]Department of Inorganic Chemistry, Ivan Franko National University of Lviv, Kyryla and Mefodia St. 6, UA-79005 Lviv, Ukraine

E-mail: kotur@franko.lviv.ua

Abstract. The ternary $R_{0.85}Y_{0.15}Ni_2$, $RNi_{1.85}Fe_{0.15}$ (R=Gd and Er), $ErM_{1.85}V_{0.15}$ (M=Fe, Co) and $Er(Fe,Co)_{1.85}V_{0.15}$ phases have been investigated for the influence of doping elements (Y, Fe, V) on magnetic properties of RM_2 (R=Gd, Er; M=Fe, Co, Ni) Laves phases and their hydrides. All investigated compounds can easily absorb hydrogen up to 3.3-3.9 at.H/f.u. at 0.1-0.12 MPa pressure without amorphization. The hydrides preserve the structure of the parent compounds. Rare earth substitution of Gd and Er by 5 at. % Y in RNi_2 based compounds reduces slightly both the ordering temperature and the magnetic moment due to the dilution effect of a magnetic rare earth element by a non-magnetic one. The corresponding 3*d* metal substitution of Ni by 5 at. % Fe increases T_C by a factor of 3-7 for Gd and Er, respectively. $Er(Fe,Co,V)_2$ compounds with a Fe content higher than 20 at. % exhibit ordering temperatures above room temperature. Hydrogen absorption reduces the saturation moment and suppresses long range magnetic order in majority of investigated alloys to the temperature below 2 K except for $ErFe_{1.85}V_{0.15}H_{3.9}$ and $ErFe_{0.925}Co_{0.925}V_{0.15}H_{3.6}$.

1. Introduction

Intermetallic compounds of rare earth elements with 3d-elements exhibit a large variety of interesting physical properties. Some of them also absorb large amounts of hydrogen which induces changes of their physical properties. Basic intermetallics and alloys and their hydrides attract interest as objects for fundamental investigations and for practical application in the production of magnetic materials, hydrogen accumulators, metal hydride electrodes, etc. [1-3].

During last years we studied a number of binary Laves phases RM_2 (R=rare earth; M=d-element) doped with a third component (R' or M') with respect to the influence of the doping element on the hydrogenation abilities of alloys and the crystal structure of their hydrides. These results are presented in detail in [4-11]. A literature and authors' review of influence of hydrogen absorption on the crystal structure of binary RM_2 and ternary $RM_{2-x}M'_x$, $R_{1-x}R'_xNi_2$ (R = rare earth; M = Mn, Fe, Co, Ni) Laves phase compounds is presented in [12]. Crystallographic parameters of the RM_2 (R=Gd, Er; M=Fe, Co, Ni), and related pseudo-binaries doped by a third component R or M, namely $R_{0.85}Y_{0.15}Ni_2$, $RNi_{1.85}Fe_{0.15}$, $ErM_{1.85}V_{0.15}$ (M = Fe, Co) and $Er(Fe,Co)_{1.85}V_{0.15}$ alloys and their hydrides are reviewed in Table 1. All investigated compounds can easily absorb a large amount of hydrogen up to 3.3-3.9 at.H/f.u. at 0.1-0.12 MPa pressure without amorphization. Hydrogen uptake leads to a lattice volume expansion for 14-26 %. The hydrides preserve the structure of the parent compounds.

As it is seen from the data presented in Table 1 substitution of Fe by V increases the hydrogen sorption capacity of alloys from 3.85 at.H/f.u. for $ErFe_2$ to 4.2 at.H/f.u. for $ErFe_{1.78}V_{0.22}$ whereas substitution of Co by V slightly influences the hydrogenation capacity of the samples. All V doped hydrides preserve the

Table 1. Crystallographic parameters of the RM_2, (R = Gd, Er; M = Fe, Co, Ni), $R_{0.85}Y_{0.15}Ni_2$, $RNi_{1.85}Fe_{0.15}$ (R=Gd and Er), $ErM_{2-x}V_x$ (M = Fe, Co) and $Er(Fe,Co)_{2-x}V_x$ alloys and their hydrides.

Alloys and their hydrides	Structure type	Lattice parameters			Ref
		a, nm	*V*, nm^3	Δ*V*/*V*, %	
$GdNi_2$	$TmNi_2$	1.44085(8)	2.991.2(3)	-	4
$Gd_{0.85}Y_{0.15}Ni_2$	$TmNi_2$	1.44017(7)	2.9870(2)	-	4
$Gd_{0.85}Y_{0.15}Ni_2H_{3.6}$	$TmNi_2$	1.5209(9)	3.5180(4)	17.8	4
$ErNi_2$	$TmNi_2$	1.42678(6)	2.9045(2)	-	5
$Er_{0.85}Y_{0.15}Ni_2$	$TmNi_2$	1.42920(6)	2.9193(2)	-	5, 6
$Er_{0.85}Y_{0.15}Ni_2H_{3.4}$	$TmNi_2$	1.5010(4)	3.3816(15)	15.8	5, 6
$GdNi_{1.85}Fe_{0.15}$	$MgCu_2$	0.72246(3)	0.37709(3)	-	4
$GdNi_{1.85}Fe_{0.15}H_{3.8}$	$MgCu_2$	0.7674(11)	0.452(2)	19.9	4
$ErNi_{1.85}Fe_{0.15}$	$MgCu_2$	0.71707(3)	3.6871(3)	-	5
$ErNi_{1.85}Fe_{0.15}H_{3.3}$	$MgCu_2$	0.7490(4)	0.4201(4)	13.9	5
$ErFe_2$	$MgCu_2$	0.7285(1)	0.3865(1)		7
$ErFe_{1.85}V_{0.15}$	$MgCu_2$	0.7298(1)	0.3887(1)		7
$ErFe_{1.85}V_{0.15}H_{3.9}$	$MgCu_2$	0.7877(1)	0.4887(1)	25.7	7
$ErFe_{1.78}V_{0.22}$	$MgCu_2$	0.73125(7)	0.3910(4)		8
$ErFe_{1.78}V_{0.22}H_{4.18}$	$MgCu_2$	0.7915(1)	0.4958(3)	26.8	8
$ErCo_2$	$MgCu_2$	0.71559(5)	0.36644(8)		8
$ErCo_{1.85}V_{0.15}$	$MgCu_2$	0.71773(5)	0.3697 (8)		8
$ErCo_{1.85}V_{0.15}H_{3.8}$	$MgCu_2$	0.7670(7)	0.4510 (1)	22,0	8
$ErCo_{1.67}V_{0.33}$	$MgCu_2$	0.71976(5)	0.37288(8)		8
$ErCo_{1.67}V_{0.33}H_{3.83}$	$MgCu_2$	0.7746(1)	0.4647(3)	24.6	8
$ErFe_{0.925}Co_{0.925}V_{0.15}$	$MgCu_2$	0.72560 (6)	0.3820 (9)		8
$ErFe_{0.925}Co_{0.925}V_{0.15}H_{3.6}$	$MgCu_2$	0.7774 (4)	0.4698 (7)	23,0	8

structure of the parent compounds, in contrast to the $ErFe_2H_{3.85}$ hydride, where hydrogenation is accompanied with the phase transition of the cubic structure ($MgCu_2$-type) into the rhombohedral one ($TbFe_2$-type) [6, 12].

The Fe-containing alloys $GdNi_{1.85}Fe_{0.15}$ and $ErNi_{1.85}Fe_{0.15}$ occur in the $MgCu_2$-type structure, however the $R_{0.85}Y_{0.15}Ni_2$ (R=Gd and Er) and RNi_2 (R=Y, Gd and Er) crystallize with the $TmNi_2$ cubic structure with doubled cell parameter *a* in comparison to the parent cubic $MgCu_2$-type structure. The above mentioned phases absorb 3.0-3.9 at.H/f.u. with the increasing cell volume up to ~26 %. Hydrides of these Laves phases maintain the crystal structure of the parent compounds (Table 1). Ternary substituted phases absorb more hydrogen in comparison with the binary compounds. The hydrogen storage capacity of the $GdNi_{1.85}Fe_{0.15}H_{3.8}$ hydride is the highest one among investigated solids and is equal to 1.27 H/M.

The magnetic properties of RMn_2 Laves phase compounds as well as their hydrides have been extensively studied in particular with respect to the interplay between the localized moments of the rare earth and the itinerant electron magnetism of the 3*d* sublattice leading to a wide variety of magnetic structures and associated magnetic properties [1-3, 13-15]. Also rare earth and 3d metal substitutions have systematically been investigated in several pseudobinary Laves phase systems and their associated hydrides where either the isoelectronic dilution effect of the localized 4f moment by "nonmagnetic" Y, La and Lu or the effect of electronic changes of the 3*d* metal subsystem was of primary interest. In the latter case mainly the magnetically ordered Mn, Fe, Co and Ni based Laves phase compounds have been studied by Fe/Co, Fe/Ni, Fe/Cu, Fe/Al and Co/Ni etc. substitutions. However substitutions of the above mentioned Laves phases with light 3*d* metals as Ti [16] or V are rather rare. As the hydrogen absorption capability can be enhanced in these systems by V substitution [12] without changing the crystal structure or that hydrogen induced amorphisation occurs at larger hydrogen content [17], it is the aim of this paper to present the relevant changes of the magnetic properties upon hydrogen absorption of the title compounds.

2. Experimental procedure

Alloys have been prepared by arc melting of the initial elements with a purity not less than 99.9 wt.% under argon. During arc melting the weight losses were less than 1 % of the total mass of the ingots. The alloys were annealed in quartz ampoules under vacuum at 600°C for 720 h.

All hydrides were synthesized at room temperature under 0.12 MPa hydrogen pressure in autoclave after preliminary activation of the samples in vacuum at 350-400°C. The amount of absorbed hydrogen was determined volumetrically.

The samples were examined by X-ray powder diffraction using diffractometers HZG-4a, (FeK_α-radiation) and DRON-3M (CuK_α-radiation). For crystal structure determination the diffraction data were collected using a θ-2θ scan mode with steps of 0.05° 2θ and exposition time 20 sec. at every point. All the crystal structure calculations were performed by the Rietveld method using the FullProf [18] and CSD [19] programs.

Magnetization measurements have been performed with a SQUID magnetometer in the temperature rangė 2-300 K in magnetic fields up to 6T.

3. Results and discussion

The available literature data and the present data on magnetic properties of the RM_2, $R_{0.85}Y_{0.15}Ni_2$, $RNi_{1.85}Fe_{0.15}$ (R = Gd and Er), $ErM_{2-x}V_x$ (M = Fe, Co) and $Er(Fe,Co)_{2-x}V_x$ alloys and their hydrides are collected in Table 2. All ternary compounds exhibit long range magnetic order, in particular compounds with more than 20 at. % Fe exhibit ordering temperatures above room temperature and are not accessible with the SQUID magnetometer used. As can be seen from Table 2 hydrogen absorption reduces the saturation moment and suppresses long range magnetic order in majority of the investigated alloys to below 2 K except for $ErFe_{1.85}V_{0.15}H_{3.9}$ and $ErFe_{0.925}Co_{0.925}V_{0.15}H_{3.6}$.

Table 2. Magnetic characteristics of the RM_2 (R=Gd, Er; M=Fe, Co, Ni), $R_{0.85}Y_{0.15}Ni_2$, $RNi_{1.85}Fe_{0.15}$ (R= Gd and Er), $ErM_{2-x}V_x$ (M = Fe, Co) and $Er(Fe,Co)_{2-x}V_x$ alloys and their hydrides.

Alloys and their hydrides	Structure type	T_C, K	μ_s/f.u., μ_B	μ_{eff}/R, μ_B	θ_p, K	Ref
$GdNi_2$	$TmNi_2$	75	6.9	7.92	73	20, *
$GdNi_2H_{3.9}$	$TmNi_2$	8	4.2	7.84	3	20, *
$ErNi_2$	$TmNi_2$	14-16	5.4	9.55	10	13,21
$ErNi_2H_x$	$TmNi_2$					21
$Gd_{0.85}Y_{0.15}Ni_2$	$TmNi_2$	60	5.89#	7.89	65	*
$Gd_{0.85}Y_{0.15}Ni_2H_{3.6}$	$TmNi_2$	<2	5.64#	7.89	9	*
$Er_{0.85}Y_{0.15}Ni_2$	$TmNi_2$	15	6.75#	10.1	15	*
$Er_{0.85}Y_{0.15}Ni_2H_{3.4}$	$TmNi_2$	<2	5.52#	10.35	4.5	*
$GdNi_{1.85}Fe_{0.15}$	$MgCu_2$	217	6.5#	8	224	*
$GdNi_{1.85}Fe_{0.15}H_{3.8}$	$MgCu_2$	<2	5.9#	7.3	25	*
$ErNi_{1.85}Fe_{0.15}$	$MgCu_2$	118	7.6#	9.88	113	*
$ErNi_{1.85}Fe_{0.15}H_{3.3}$	$MgCu_2$	<2	5.45#	4.0	- 5	*
$ErFe_2$	$MgCu_2$	590	4.85			22,25
$ErFe_2H_{1.2-2.5}$	$MgCu_2$	>300				23,29
$ErFe_2H_{3.5-3.9}$	$TbFe_2$	<300	~5.6			22-24
$ErFe_2H_{5.0}$	$ErFe_2H_{5.0}$		3.6			24
$ErFe_{1.85}V_{0.15}$	$MgCu_2$	>300	5.6#			*
$ErFe_{1.85}V_{0.15}H_{3.9}$	$MgCu_2$	>300	2.85#			*
$ErCo_2$	$MgCu_2$	32-33	7			25
$ErCo_2H_{3.5}$	$MgCu_2$	25	4.35			25
$ErCo_{1.4}Fe_{0.6}$	$MgCu_2$	510	4.85			25
$ErCo_{1.4}Fe_{0.6}H_{3.5}$	$MgCu_2$	>300	3.8			25
$ErCo_{0.66}Fe_{1.34}$	$MgCu_2$	670	4.65			25
$ErCo_{0.66}Fe_{1.34}H_{3.5}$	$MgCu_2$	>300	2.9			25
$ErCo_{1.85}V_{0.15}$	$MgCu_2$	33	6.8#	8.61	34	*
$ErCo_{1.85}V_{0.15}H_{3.80}$	$MgCu_2$	<2	4.9#	9.6	- 6	*
$ErFe_{0.925}Co_{0.925}V_{0.15}$	$MgCu_2$	>300	5.6#			*
$ErFe_{0.925}Co_{0.925}V_{0.15}H_{3.6}$	$MgCu_2$	25	5.05#			*

* - present paper; # - μ_s/f.u. taken at 3 K and 6 T

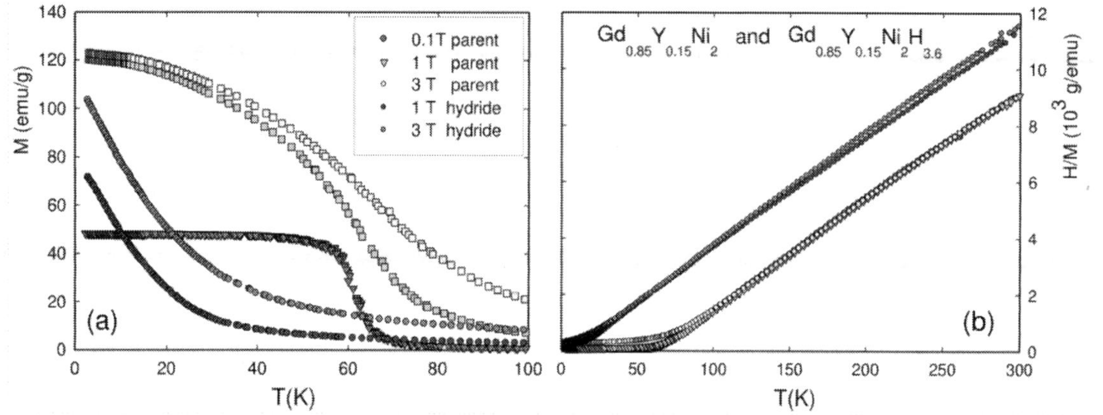

Figure 1. Magnetisation (a) and susceptibility (b) as a function of temperature of $Gd_{0.85}Y_{0.15}Ni_2$ and $Gd_{0.85}Y_{0.15}Ni_2H_{3.6}$ for various fields as labeled.

Shown in Fig. 1a is the temperature dependent magnetization of $Gd_{0.85}Y_{0.15}Ni_2$ and $Gd_{0.85}Y_{0.15}Ni_2H_{3.6}$. The dilution of the Gd and Er moment in RNi_2 by 5 at% Y substitution reduces T_C from 75 K for $GdNi_2$ to 60 K for $Gd_{0.85}Y_{0.15}Ni_2$ (Fig. 1a) while the analogous Er/Y substitution has only a minor influence upon T_C (Table 2). Previously it was suggested that Ni appears to have no or a rather small moment due to filled Ni 3*d* band [20] in $GdNi_2$, however, according to X-ray magnetic circular dichroism (XMCD) the Ni moment is 0.2-0.26 μ_B antiparallel to the Gd moment [26, 27]. The saturation moment for $Gd_{0.85}Y_{0.15}Ni_2$ at 3 K and 6 T is 5.89 μ_B/f.u. Taking account for the above mentioned Ni moment antiparallel to Gd yields a Gd moment of 7.4 μ_B being slightly larger than that expected for Gd^{+3} (7 μ_B), which can be attributed to the conduction electron polarisation. Hydrogen absorption reduces the saturation moment at 3 K and 6 T to 5.64 μ_B/f.u. for $Gd_{0.85}Y_{0.15}Ni_2H_{3.6}$ corresponding to a Gd moment of 6.63 μ_B. Long range order is suppressed to below 2 K in both hydrides $Gd_{0.85}Y_{0.15}Ni_2H_{3.6}$ and $Er_{0.85}Y_{0.15}Ni_2H_{3.4}$ according to an analysis of the M(H) curves of in terms of Arrott plots (M^2 versus M/H plots). The scaling behaviour of the magnetic isotherms as a function of the normalized external field with respect to the temperature (μ_0H/T) indicates that antiferromagnetic correlations are still preserved in the paramagnetic regime of $Gd_{0.85}Y_{0.15}Ni_2H_{3.6}$ which may be responsible for the reduced Gd moment mentioned above.

Figure 1b shows the temperature dependent magnetic susceptibility χ for $Gd_{0.85}Y_{0.15}Ni_2$ and the hydride plotted as $1/\chi$ versus T. The almost linear temperature dependence of $1/\chi$ above 20 K and 70 K for the hydride and parent compound, respectively, can well be described with a modified Curie-Weiss law $\chi = \chi_0 + C/(T - \Theta)$. $\chi_0 = 4\cdot10^{-8}$ m^3/kg represents a temperature independent Pauli susceptibility, C is the Curie constant yielding an effective moment 7.89 μ_B/Gd for both the parent compound and the hydride while the paramagnetic temperature Θ is reduced from 64 K for the parent compound to 9 K for the hydride. A similar behaviour is observed for $Er_{0.85}Y_{0.15}Ni_2$ and its hydride where the effective moment of 10.1 μ_B/Er is slightly raised to 10.35 μ_B/Er while the paramagnetic Curie temperature is reduced from 15 K to 4.5 K under hydrogen absorption. Both effective moments are close to those expected for Gd^{+3} (7.94 μ_B) and Er^{+3} (9.58 μ_B).

As shown in figure 2 substitution of 5 at. % Ni by Fe in RNi_2 (R = Gd, Er) increases T_C to 217 K and to 118 K for $GdNi_{1.85}Fe_{0.15}$ and $ErNi_{1.85}Fe_{0.15}$, respectively, while hydrogen absorption suppresses long range order to below 2 K. The saturation moment for $GdNi_{1.85}Fe_{0.15}$ (6.5 μ_B) indicates that the 3*d* moment is antiparallel to the Gd moment and can be estimated to be 0.25 μ_B/(Fe, Ni) if one assumes a Gd 4*f* moment of 7 μ_B. The irreversibility of the magnetization for zero field cooling and field cooling of the $ErNi_{1.85}Fe_{0.15}$ which is preserved in fields up to 1 T can be attributed to domain wall pinning. This is also seen in a metamagnetic like behaviour of the magnetic isotherms associated with a pronounced hysteresis at low temperatures. The effective moments of parent compounds are 8 μ_B/Gd and 9.88 μ_B/Er while those of the hydrides are 7.3 μ_B/Gd and 4 μ_B/Er.

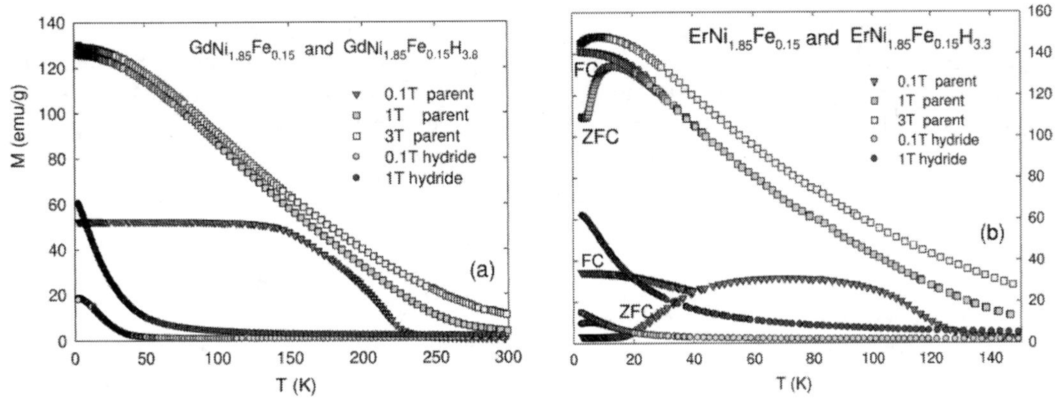

Figure 2. Magnetisation as a function of temperature of (a) $GdNi_{1.85}Fe_{0.15}$ and $GdNi_{1.85}Fe_{0.15}H_{3.8}$ for various fields and (b) $ErNi_{1.85}Fe_{0.15}$ and $ErNi_{1.85}Fe_{0.15}$ $H_{3.3}$ for various fields for zero field cooling (ZFC) and field cooling (FC).

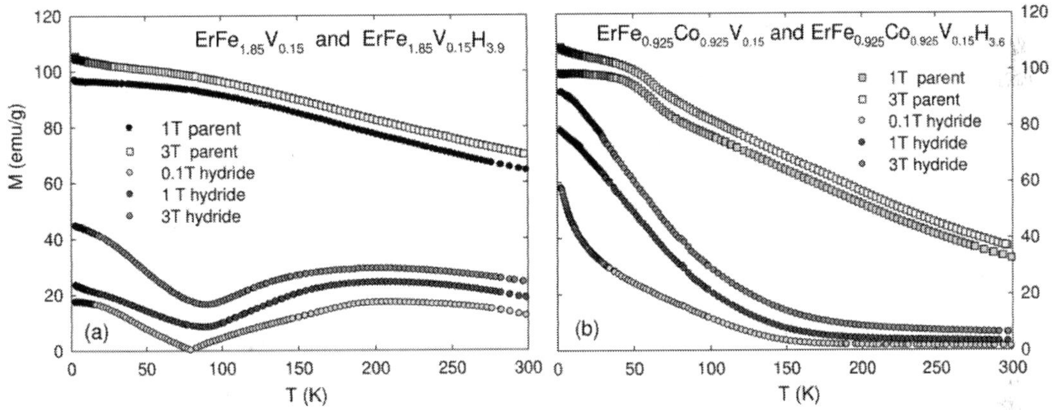

Figure 3. Magnetisation as a function of temperature of (a) $ErFe_{1.85}V_{0.15}$ and $ErFe_{1.85}V_{0.15}H_{3.9}$ and (b) $ErFe_{0.925}Co_{0.925}V_{0.15}$ and $ErFe_{0.925}Co_{0.925}V_{0.15}H_{3.6}$ for various fields.

Co/V substitution in $ErCo_2$ (T_C=32-33 K) has hardly an effect upon the first order transition T_C (33 K) for a V concentration of 5at% (Table 2) in a similar way as Co/Ti substitution [16]. Hydrogenation suppresses long range order to below 2 K in $ErCo_{1.85}V_{0.15}H_{3.8}$. The data of the effective moment and paramagnetic Curie temperature for the parent compound of the hydride are summarized in Table 2.

Both Fe-containing alloys $ErFe_{1.85}V_{0.15}$ and $ErFe_{0.925}Co_{0.925}V_{0.15}$ exhibit long range order above room temperature (Fig. 3). Hydrogenation in the former gives rise to a significant reduction of the bulk moment from 5.6 μ_B/f.u. to 2.83 μ_B/f.u. associated with a weakening of the Er-Fe exchange interaction. This has been studied extensively in $ErFe_2H_x$ [22-24, 28] where the compensation point of the parent compound at 410 K linearly decreases with the hydrogen content to 170 K for $ErFe_2H_{2.97}$ [29]. Also for $ErFe_{1.85}V_{0.15}H_{3.9}$ a compensation point occurs at about 80 K which is shifted to higher temperatures with rising fields but long range order remains above room temperature which is in line with the literature data mentioned above. Domain wall pinning in $ErFe_{0.925}Co_{0.925}V_{0.15}$ gives rise to a metamagnetic like behaviour of the magnetic isotherms at low temperatures associated with pronounced hysteresis. The additional feature of the magnetization at about 60 K in higher fields may be indicative for a spin-reorientation. According to Arrott plots the metamagnetic transition associated with long range order is reduced to T_C=25 K in the hydride $ErFe_{0.925}Co_{0.925}V_{0.15}H_{3.6}$.

4. Conclusions

$RNi_{1.85}Fe_{0.15}$ (R=Gd, Er), $ErM_{1.85}V_{0.15}$ (M=Fe, Co) occur in the cubic $MgCu_2$-type structure, however $R_{0.85}Y_{0.15}Ni_2$ (R=Gd, Er) crystallize with the $TmNi_2$-type cubic structure with doubled cell parameter *a*. The above mentioned phases absorb 3.3-3.9 at.H/f.u. with increasing the cell volume up to 26 %. The structure of the parent compounds and their hydrides remains unchanged.

Gd by Y substitution (5 at. %) in $GdNi_2$ reduces T_C from 75 K to 60 K. The analogous Er/Y substitution has only a minor influence upon T_C (T_C = 14-16 K for $ErNi_2$, T_C = 15 K for $Er_{0.85}Y_{0.15}Ni_2$). Hydrogen absorption suppresses long range order to below 2 K in both hydrides $Gd_{0.85}Y_{0.15}Ni_2H_{3.6}$ and

$Er_{0.85}Y_{0.15}Ni_2H_{3.4}$. Ni by Fe substitution (5 at.%) increases T_C to 217 K and to 118 K for $GdNi_{1.85}Fe_{0.15}$ and $ErNi_{1.85}Fe_{0.15}$, respectively, and hydrogenation suppresses long range order to below 2 K. Co by V substitution (5 at.%) in $ErCo_2$ (T_C=32-33 K) has no effect upon the T_C (33 K) for this V concentration. Hydrogenation reduces long range order to below 2 K in $ErCo_{1.85}V_{0.15}H_{3.8}$. Both Fe-based alloys $ErFe_{1.85}V_{0.15}$ and $ErFe_{0.975}Co_{0.975}V_{0.15}$ exhibit long range magnetic order up to above room temperature and T_C remains above room temperature for the hydride $ErFe_{1.85}V_{0.15}H_{3.94}$.

This work was supported by the Ukrainian-Austrian WTZ project No. UA 06/2009.

References

[1] Buschow K H J 1984 Hydrogen absorption in intermetallic compounds, in: Gschneidner Jr K A and Eyring L (Eds) *Handbook on the Physics and Chemistry of Rare Earths*, Vol. 10, (North–Holland Publishing Co.) 1–111
[2] Yvon K and Fischer P 1988 Crystal and Magnetic Structures of Ternary Metal Hydrides: A Comprehensive Review, in: *Topics in Applied Physics, Vol. 63. Hydrogen in Intermetallic Compounds I. Electronic, Thermodynamic and Crystallographic Properties, Preparation*, (Springer–Verlag, Berlin) 87–138
[3] Wiesinger G and Hilscher G 2008 Magnetism of Hydrides, in: K-H-J Buschow (Ed.) *Handbook on Magnetic Materials* Vol. 17 (Elsevier B.V.) Ch 5 293-456
[4] Kotur B, Myakush O and Zavaliy I 2007 *J Alloys Compounds* **442** 17
[5] Myakush O, Verbovytsky Yu, Kotur B, Kovalchuk I, Beresovetz V and Zavaliy I 2007 *J Phys.: Conf. Ser* **79** 012018
[6] Myakush O V, Verbovytskyi Yu V, Berezovets V V, Ershova O H, Dobrovolskyi V D and Kotur B Ya 2007 *Materials Science* **43** 682-688
[7] Myakush O, Martyniuk G, Verbovytsky Yu and Kotur B 2006 *Visnyk Lviv Univ Ser Chem* **47** 25
[8] Myakush O, Verbovytskyi Yu, Saldan I, Kovalchuck I, Zavalii I and Kotur B 2004 *Materials Science* **40** 781
[9] Verbovytsky Yu, Myakush O, Soldak A and Kotur B 2005 *Visnyk Lviv Univ Ser Chem* **46** 48
[10] Myakush O V, Denys R V, Kovalchuk I V, Verbovytskyi Yu V, Zavalii I Yu and Kotur B Ya 2003 *Materials Science* **39** 849
[11] Kovalchuk I Myakush O Denys R and Kotur B 2007 *Visnyk Lviv Univ Ser Chem* **48** 194
[12] Kotur B, Myakush O and Zavaliy I 2009 *Croat Chem Acta* **82** 469
[13] Burzo E, Chełkowski A and Kirchmayr H-R in: H-P-J Wijn (ed.),1990 *Landolt-Börstein, Numerical Data and Functional Relationships in Science and Technology, New Series, Magnetic Properties of Metals* Vol. 19-D2,(Springer, Berlin)
[14] Hilscher G 1982 *J Magn Magn Mat* **25** 229
[15] Hilscher G 1982 *J Magn Magn Mat* **27** 1
[16] Öner Y and Kamer O 2008 *J Alloys Compd* **464** 51
[17] Mushnikov N V, Gaviko V S and Goto T 2005 *J Alloys Compd* **398** 36
[18] Rodrigues-Carvajal J Program: FullProf.2k Version 2.20 (France: Lab Leon Brillouin CEA-CNRS)
[19] Akselrud L G, Zavalii P Yu, GrinYu N, Pecharsky V K, Baumgartner B and Wolfel E 1993 *Mater Sci Forum* **133-136** 335
[20] Malik S K and Wallace W E 1977 *Solid State Comm* **24** 283
[21] Wang D, Li Y, Long Y, Ye R, Chang Y and Wan F 2007 *J Magn Magn Mat* **311** 6971
[22] Fruchart D, Berthier Y, de Saxce T and Vulliet P 1987 *J Solid State Chem* **67** 197-209
[23] Paul–Boncour V and Perchéron–Guegan A 1999 *J Alloys Compounds* **293-295** 237
[24] Paul-Boncour V, Filipek S M, Perchéron–Guegan A, Marchuk I, Pielaszek J 2001 *J Alloys Compounds* **317-318** 83
[25] Pourarian F, Wallace W E and Malik S K 1982 *J Magn Magn Mat* **25** 299
[26] Mizumaki M, Yano K, Umehara I, Ishikawa F, Sato K, Koizumi A, Sakai N and Muro T 2003 *Phys. Rev. B* **67** 132404
[27] Yano K, Umehara I, Miyazawa T, Adachi Y and Sato K 2005 *Physica B* **367** 81
[28] Paul–Boncour V, Giorgetti C, Wiesinger G and Perchéron–Guegan A 2003 *J Alloys Compounds* **356-357** 195
[29] De Saxce T, Berthier Y and Fruchart D 1985 *J Less Common Met* **107** 35

Influence of Gd, Dy and Fe doping on electrochemical properties of $Al_{87}Y_5Ni_8$ amorphous metallic alloy

L Bednarska[1], J Kubisztal[2], A Budniok[2], M Kovbuz[1], O Hertsyk[1] T Mika[1] and B. Kotur[1]

[1]Department of Chemistry, Ivan Franko National University of Lviv, Kyryla i Mefodiya St. 6, 79005, Lviv, Ukraine, e-mail: lbednarska@yahoo.com
[2]Institute of Materials Science, Silesian University, Bankowa 12, 40-007 Katowice, Poland

Abstract: Corrosive properties of some Al-based amorphous metallic alloys (AMA) and the relationship between the AMA composition and nature of protective layers on the surface of alloys were studied. The complex of physico-chemical methods, namely the potentiometry, voltammetry, electrochemical impedance spectroscopy and electronic microscopy, has been used. It was shown that the influence of Dy doping on the same characteristics of alloys is opposite to that of Gd. It increases corrosion resistance of Dy-containing alloys in comparison to the reference $Al_{87}Y_5Ni_8$ alloy. Partial substitution of 4 at. % Ni by Fe causes further increase of corrosion resistance of $Al_{87}Gd_5Ni_4Fe_4$.

1. Introduction

Amorphous metallic alloys (AMA) of Al–Y–Ni-system are used in the precious electrotechnical devices due to their specific physico-chemical distinctions, which are caused by the inhomogeneity of chemical and phase composition, and also of possibility of interaction between components,. The AMAs of such type, doped by Y or the elements of lanthanum group, exhibits the magnetic and semiconductive properties. The doping elements, namely both rare-earth and transition metals, increase the mechanical stability of AMA, exactly their surface protection layers essentially. The mechanism of atomic ordering in the multicomponent AMA is connected with the change in its electronic structure [1, 2]. It is known that 4-*f* state is the most energetically favorable for the lanthanides regarding proximity of 4-*f* i 5-*d* states. Therefore in all cases, except for Gd, the 5-d^1 electron rises on the 4-*f* sublevel. Such electronic structure leads to their reactivity close to 3-d-metals. Lanthanides (especially Gd) are able to reacts with different metals, including Al, what leads for the formation of intermetallides under the certain conditions [3]. Rare-earth and transition metals doped AMA lightly oxidizes in the weak acid media with formation of protection oxides layers. Therefore the higher corrosion resistance of Al-based AMA with additives of rare-earth metals can be foreseen in advance, but high reactivity of Gd can be stipulated the partial dissolution of alloy [4]. Therefore the aim of this work is to study corrosive properties of some Al-based amorphous metallic alloys and a relationship between the AMA composition and nature of protective layers on the surface of alloys.

2. Experimental details and principles

The Al-Y-Ni amorphous ribbons of 30 μm thickness were produced by melt spinning with the following compositions: $Al_{87}Y_5Ni_8$, $Al_{87}Y_4Gd_1Ni_8$, $Al_{87}Y_4Dy_1Ni_8$, $Al_{87}Gd_5Ni_8$, $Al_{87}Dy_5Ni_8$, $Al_{87}Gd_5Ni_4Fe_4$ and $Al_{87}Y_4Gd_1Ni_4Fe_4$. The ribbons were produced in the Kurdyumov Institute of Metal Physics of the National Academy of Sciences of Ukraine (Kyiv).

The oxide films were formed on AMA surface during polarization of electrodes in 0.1 M NaCl aqueous solution. The process of the formation of passive layers was performed in Pyrex® (Radiometer No.1734) glass cell. The experiments were carried out in 0.1 M NaCl aqueous solution at room temperature with platinum as a counter electrode and SCE as a reference electrode. The exposed geometric surface area of working electrode was 1 cm^2. Steady-state polarization measurements as well as EIS experiments were conducted using Autolab®/PGSTAT20 with Frequency Response Analyser (FRA) and Differential Electrometer Amplifier (Eco Chemie B.V. the Netherlands), combined with one of the software packages and controlled by IBM compatible PC. Micrographs of the surface were obtained with scanning electron microscope REMMA-102 using various magnifications.

3. Results and discussion

The investigation of AMA corrosion properties was carried out by the voltammetry method in the potential interval ±50 mV (figure 1) from the stable value of open circuit potential (φ) (figure 2) without failure of the structure of Stern's double electric layer (figure 3).

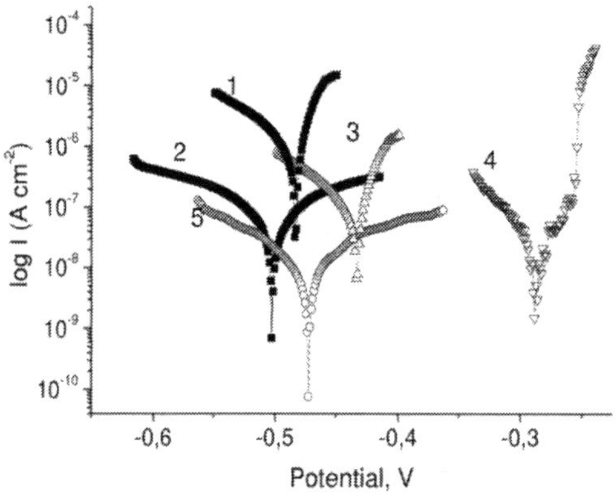

Figure 1. Stern voltammetric curves of the Al-based AMA in 0.1 M NaCl aqueous solution:1- $Al_{87}Y_5Ni_8$, 2- $Al_{87}Gd_5Ni_8$, 3- $Al_{87}Dy_5Ni_8$, 4 $Al_{87}Y_4Dy_1Ni_8$, 5- $Al_{87}Y_4Gd_1Ni_8$

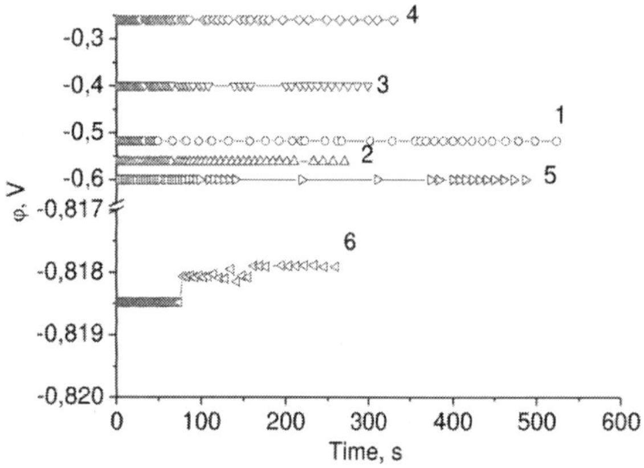

Figure 2. Potentiometric curves of the Al-based AMA in 0.1 M NaCl aqueous solution: 1- $Al_{87}Y_5Ni_8$ 2- $Al_{87}Gd_5Ni_8$, 3- $Al_{87}Dy_5Ni_8$, 4 -$Al_{87}Y_4Dy_1Ni_8$, 5- $Al_{87}Y_4Gd_1Ni_8$, 6- $Al_{87}Y_4Gd_1Ni_4Fe_4$

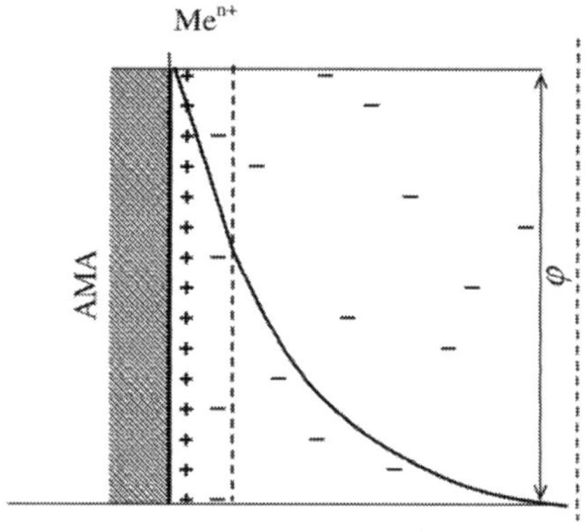

Figure 3. The decrease of electric potential in the DEL (accordingly to Stern)

Partial and total substitution of Y by Gd decreases corrosion potential ($E_{corr.}$) and increases of corrosion current ($j_{corr.}$), polarization resistance (R_p) (see table). The influence of Dy doping on the same characteristics of alloys is opposite to that of Gd. It increases corrosion resistance of Dy-containing alloys in comparison to the reference alloy $Al_{87}Y_5Ni_8$. Partial substitution of 4 at. % Ni by Fe causes further increase of corrosion resistance of $Al_{87}Gd_5Ni_4Fe_4$ (see table). This effect is suppressed by partial substitution of Gd by Y in $Al_{87}Y_4Gd_1Ni_4Fe_4$. The corrosion resistance depends on layer thickness and especially on its sealing, therefore good the quality control of both parameters has maximum significance [5].

Table. Corrosion characteristics of the Al-based AMA in 0.1 M NaCl aqueous solution

AMA	$E_{corr.}$, V	$j_{corr.}$, A·cm^{-2}	R_p, Ω
$Al_{87}Y_5Ni_8$	–0.502	$5.79\cdot10^{-7}$	$1.81\cdot10^{2}$
$Al_{87}Y_4Gd_1Ni_8$	–0.612	$1.34\cdot10^{-7}$	$2.10\cdot10^{2}$
$Al_{87}Y_4Dy_1Ni_8$	–0.259	$1.16\cdot10^{-9}$	$1.55\cdot10^{3}$
$Al_{87}Gd_5Ni_8$	–0.481	$3.40\cdot10^{-8}$	$6.47\cdot10^{3}$
$Al_{87}Dy_5Ni_8$	–0.433	$1.94\cdot10^{-9}$	$7.74\cdot10^{4}$
$Al_{87}Gd_5Ni_4Fe_4$	–0.472	$5.39\cdot10^{-10}$	$1.26\cdot10^{4}$
$Al_{87}Y_4Gd_1Ni_4Fe_4$	–0.643	$1.35\cdot10^{-5}$	$4.53\cdot10^{3}$

The processes of electrochemical behaviour and of AMA have been studied by the cyclic voltammetry method in a 0.1 M NaCl aqueous solution during electrode polarization (figure 4).

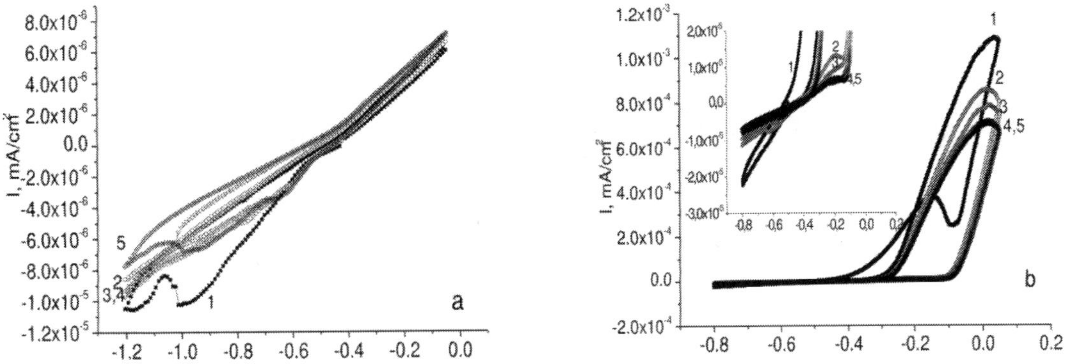

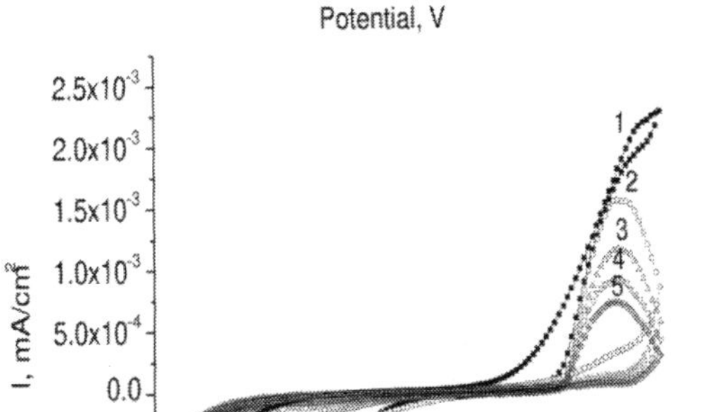

Figure 4. Cyclic voltammograms of the Al-based AMA in 0.1 M NaCl aqueous solution: a- $Al_{87}Y_5Ni_8$, b- $Al_{87}Y_4Gd_1Ni_8$, c- $Al_{87}Gd_5Ni_4Fe_4$

The results of electrochemical studies, namely the forms of voltammetric curves (figure 4 b), indicated that the most stable to the electrode polarization is the partial Gd–Dy-doped (to 1%) Al–Y–Ni AMA-systems. AMAs doped by 4% of iron were the most sensitive for the polarization influence. The values of the currents of oxidation and reduction of $Al_{87}Gd_5Ni_4Fe_4$ and $Al_{87}Y_4Gd_1Ni_4Fe_4$ AMA-electrodes approach 10^{-3} mA·cm^{-2} (figure 4 c). However, one can see from the voltammograms on the Fig. 4, the increase of the number of electrodes polarization cycles leads to the decrease of the values of oxidation current. The reason for such effect is the formation of stable oxides of metals, exactly aluminum (because its concentration on the surface is the largest) forms on the AMA surface (figure 5). The basic oxides of lanthanides Me_2O_3 are characterized by high enthalpy of formation (ΔG^o=~1600 kJ·mol^{-1}). Therefore the formation of passivation oxide layers and the rate of their formation can be explained by the ionization energy E_{ion} for $Me^o \rightarrow Me^+$ transition, which is equal 5.986 eV, 7.635, 6.217, 6.12 and 6.82 eV for $Al^o \rightarrow Al^+$, $Ni^o \rightarrow Ni^+$, $Y^o \rightarrow Y^+$, $Gd^o \rightarrow Gd^+$ and $Dy^o \rightarrow Dy^+$ transitions respectively. The decrease of ionic radius in the Y – Gd – Dy sequence leads to the compaction of amorphous surface layers (figure 5).

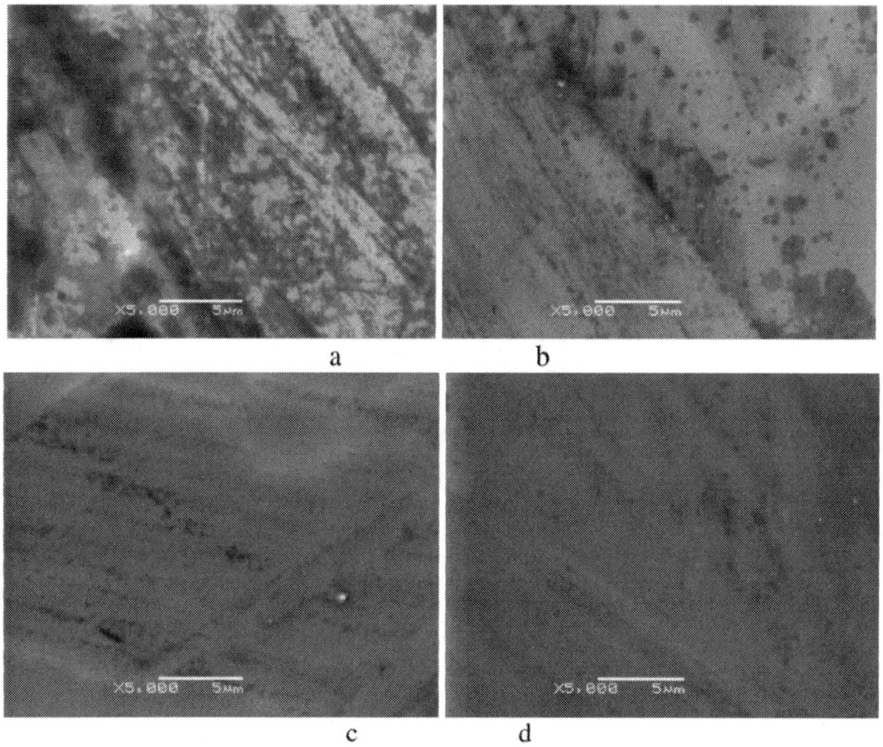

Figure 5. Micrographs of the AMA surface after polarization of samples in the 0.1 M NaCl aqueous solution: $Al_{87}Y_5Ni_8$ (a), $Al_{87}Y_4Gd_1Ni_8$ (b), $Al_{87}Y_4Gd_1Ni_4Fe_4$ (c) and $Al_{87}Gd_5Ni_4Fe_4$ (d)

Passivation ability of oxide layers on the AMA-electrodes was studied by electrochemical impedance spectroscopy method in a 0.1 M NaCl aqueous solution. The impedance model for the formation of oxide (AMA) | 0.1 M NaCl (aq.) and AMA | 0.1 M NaCl (aq.) phase boundary was chosen. The Bode curves (impedance and phase angle) (figure 6) indicate that the lowest resistance is typical for the $Al_{87}Y_5Ni_8$ AMA. This effect can be explained by the big radius of Y^{3+} ion, which is equal to 0.097 nm and is the largest in comparison to all the radiuses of both atoms and other ions in the AMA composition. The porous oxide films with thickness <100 nm, where the ionization of metals proceeds, are formed on the three-component alloys surface | 0.1 M NaCl aqueous solution boundary.

It was confirmed by the EIS method that ($Al_{87}Y_5Ni_8$) | 0.1 M NaCl (aq.) boundary can be described by the simplest electric scheme (figure 7 a). The charge transfer resistance increases due to the substitution of Y on Gd or Dy from 2.9 to 7.3 kOhm·cm^{-2}.

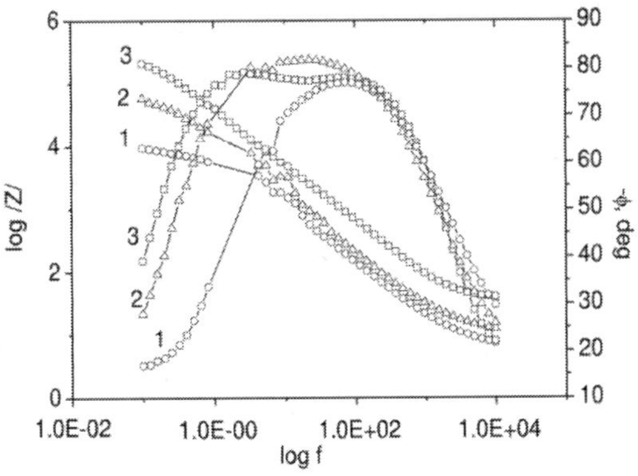

Figure 6. Bode plots of $Al_{87}Y_5Ni_8$ (1), $Al_{87}Dy_5Ni_8$ (2), $Al_{87}Y_4Dy_1Ni_8$ (3) amorphous alloy in the 0.1 M NaCl aqueous solution

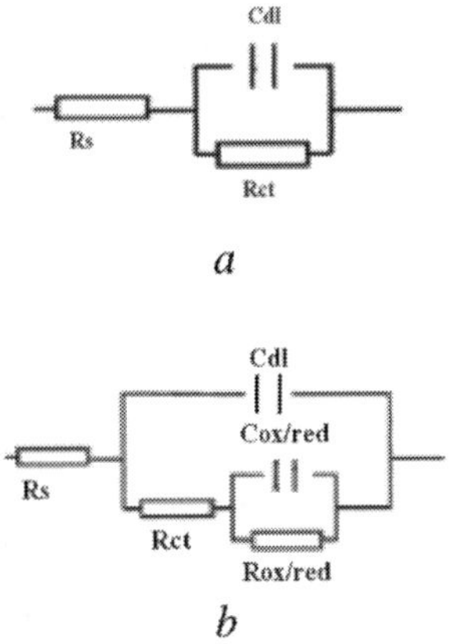

Figure 7. Equivalent circuit for plots in Figure 6: circuit a – curves 1, 2; circuit b – curves 3

The corrosion resistance properties of surface layers upgrade with the decrease of atomic radius of rare-earth metal. Moreover, the double-stage process of charge transfer on the oxide (AMA) | 0.1 M NaCl aqueous solution / boundary (figure 6, curve 3) indicates that the oxide layer, which forms in the presence of the ions of two different rare-earth metals, generates a diffusion barrier, that isolates the AMA surface. As a result, the electric scheme of this boundary can be described by Warburg element (figure 7 b).

Conclusion

The Dy-doped AMA has the highest corrosion resistance. Simultaneous doping by two metals, namely by the Y and Gd, Y and Dy, Ni and Fe leads to the improvement of the protective properties of AMA surface layers. The mixed oxides of Al and doping elements take part in the formation of the protective surface layers. It was confirmed by the EIS method that mixed oxides of Al, Ni, Dy or Gd forms the dense surface layers, where the reversible diffusion of the electrolyte ions takes place, and limits the contact of the aggressive ions of the electrolyte with metallic AMA surface. The studied AMAs were disposed in the following order according to the increase of their corrosion resistance: $Al_{87}Y_4Dy_1Ni_8$> $Al_{87}Dy_5Ni_8$> $Al_{87}Gd_5Ni_4Fe_4$> $Al_{87}Gd_5Ni_8$> $Al_{87}Y_4Gd_1Ni_8$>$Al_{87}Y_5Ni_8$>$Al_{87}Y_4Gd_1Ni_4Fe_4$

References

[1] Inoue A, Shen B and Takeuchi A 2006 *Mat. Sci. Eng.* **441** 18
[2] Brown G M, Shimizu K and Kobayashi K 1998 *Corr. Sci.* **40** 1579
[3] Xingwen Y, Chunan C and Zhiming Y 2003 *Corr. Sci.* **43** 1283
[4] Suay J J, Gimenez E, Rodriguez T, Habbib K and Saura J J 2003 *Corr. Sci.* **44** 611
[5] Bednarska L, Mika T, Kotur B, Kovbuz M and Mudry S 2008 *Chem. Met. Alloys* **1** 149

Relations between the cohesive energy, atomic volume, bulk modulus and sound velocity in metals

S Wacke[1], T Górecki[2], Cz Górecki[1], K Książek[2]

[1]Department of Physics, Opole University of Technology, 75 Ozimska str., 45-370 Opole, Poland
[2]Department of Physics, Opole University, 48 Oleska str., 45-052 Opole, Poland

e-mail: kasiak@uni.opole.pl

Abstract: By analysing the experimental data available in the literature, it has been found that the bulk modulus B of metals is proportional to the cohesive energy density E_c/V. For metals which start to melt having the close packed structure A1 or A3 the proportionality factor in the forementioned correlation is distinctly greater than that for metals melting from the A2 type structure. The existence of the correlation between the bulk modulus and the cohesive energy density leads to another, hitherto unrevealed correlation between the sound velocity, cohesive energy and the molar mass of metals: $u^2 \sim E_c/\mu$.

1. Introduction

The cohesive energy as well as the compressibility, or its reciprocal – bulk modulus, are important quantities determining the stability and other physical properties of condensed phases of substances. In the past several authors attempted to develope, both theoretically and semiempirically, formulas relating the cohesive energy and bulk modulus to various physical properties of solids, such as the melting temperature, atomic volume, lattice constants, vacancy formation energy, Debye temperature, valence of constituent atoms, etc. Attempts to find correlation between the cohesive energy and bulk modulus have been made, too [1-3]. All of the above mentioned relationships are, as a rule, restricted to some groups of substances of similar structure, bond type or other physicochemical property. A closer examination of these relationships suggests, that instead of searching for a direct relationships between the bulk modulus and the cohesive energy one should look for the relation between the bulk modulus and the volume density of the cohesive energy.

Cohesive energy density, defined as the ratio of the atomic energy of cohesion E_c, to the atomic volume V, may be expressed in units J/m^3, or, equivalently, in the pressure units – Pa. It has the meaning of some internal pressure preventing the atomization of the condensed phase.

The bulk modulus B (reciprocal of compressibility) of condensed phases defines their resistance against hydrostatic compression. At fixed temperature T, it is defined by the relation

$$B_T = -V_T \Big/ \left(\frac{dV}{dp}\right)_T$$

where V_T is the volume at temperature T. Its value may be expressed in units of pressure (N/m^2 = Pa) or in J/m^3. So, the bulk modulus defines the volumetric energy density connected with the compression work.

The available literature data, give, as a rule, the values of B determined at room temperature and under atmospheric pressure. In principle, when comparing the values of B *and* E_c for different substances, these values should be reduced to the temperature corresponding to the thermodynamically equivalent states, e.g. to the melting point, or to the absolute zero. Unfortunately, such reduction is often impossible because of the lack of appropriate data.

By changing the interionic or intermolecular distances, the compression work changes both the energy of their interaction (cohesion) as well as the volume of compressed material, consequently changing the cohesive energy density. So, intuitively, the bulk modulus should be related to the cohesive energy density. The purpose of the present study is to check this supposition.

2. Results and discussion

The available in the literature experimentally determined values of the bulk modulus B [4,5,10], cohesive energy (per atom) E_c [6] and of the atomic volume V [6] of 31 metals are collected in Table 1. On this basis the cohesive energy density E_c/V has been calculated and plotted against the values of bulk modulus of corresponding metals (Fig. 1).

At a first glance the distribution of experimental data points seems to confirm the prediction about the linear relationship between E_c/V and B – as a rule the metals with higher energy density are characterized by higher values of the bulk modulus. The scatter of the data points is, however, very large, and an attempt to fit them by a straight line gives a very unsatisfying result with a low value of the correlation factor $R \approx 0.85$.

A more detailed inspection of figure 1 revealed that there are two groups of data points scattered along two different straight lines. The first of them includes the first 15 metals (Cu – Os) listed in Table 1. The common feature of these metals is the close packed crystallographic structure (A1 or A3) in the whole temperature interval, ranging from absolute zero up to the melting point. The experimental data for these metals are well fitted with straight line described by relation

$$B = (4.068 \pm 0.193) E_c / V \tag{1}$$

The correlation factor for this empirical correlation is R = *0.947*, and the relative standard deviation amounts 4,7%.

The second group of metals (last 16 in Table 1, Ca – Yb) consists of elements showing the temperature – induced polymorphism. Under normal pressure all of them display a body centered cubic structure (A2) just below the melting point. For these metals the relation between the bulk modulus B and the cohesive energy density can be expressed by empirical relation

$$B = (2.010 \pm 0.116) E_c / V \tag{2}$$

with a correlation factor R = *0.914*, and the relative standard deviation amounts 5,8%.

The revealed relationships between the cohesive energy density and bulk modulus open new possibilities for searching for other correlations between various properties of condensed phases, often very important both for pure science and applications.

For example, as the sound velocity u is related to the bulk modulus B and mass density ρ by equation

$$u = f \sqrt{\frac{B}{\rho}} \tag{3}$$

where f is a constant proportionality factor then, from relations (1) or (2) it simply follows that

$$u^2 = c \frac{E_c}{\mu} \tag{4}$$

were μ is the molar mass, and *c* – proportionality factor, the same for a given group of metals.. The term E_c/μ on the right side of equation (4) has dimensions m^2/s^2 and has the meaning of the mass density of cohesive energy.

Table 1 Experimental data on the molar mass μ, atomic volume *V*, cohesive energy E_c and sound velocity *u* of metals

Metals	μ $[g/mole]$	$V \cdot 10^3$ $[nm^3]$	E_c $[eV]$	E_c/V [GPa]	*B* [GPa]	E_c/μ $10^{-6} \cdot [m^2/s^2]$	*u* $[m/s]$
Cu	63.55	11.81	3.49	47,28	142	5,30	4880
Ag	107.87	17.06	2.96	27,76	102	2,64	3830
Au	196.97	16.96	3.78	35,66	170.8	1,85	3361
Pb	207.2	30.33	2.04	10,76	40.4	0,95	2300
Ni	58.69	10.94	4.43	64,79	190	7,29	5894
Rh	102.91	13.77	5.75	66,81	287	5,38	6194
Pd	106.42	14.72	3.94	42,83	184.4	3,57	4630
Ir	192.22	14.14	6.93	78,42	370	3,47	5379
Pt	195.08	15.10	5.85	61,99	274.7	2,89	4075
Mg	24.31	23.23	1.53	10,54	34.5	6,06	5898
Zn	65.38	15.24	1.35	14,17	61.7	1,99	4170
Cd	112.41	21.58	1.16	8,60	47.7	0,99	2780
Co	58.93	11.13	4.39	63,11	167.1	7,18	5827
Ru	101.07	13.57	6.62	78.05	327.1	6,31	6534
Os	190.23	13.99	8.10	92,64	426	4,10	5478
Ca	40.08	43.48	1.83	6,73	18.26	4,40	4176
Sr	87.62	56.32	1.70	4,83	12.1	1,87	2782
Ce	140.12	34.37	4.77	22,21	19.8	3,28	2300
Ti	47.87	17.65	4.86	44,06	88.6	9,78	6263
Zr	91.224	23.27	6.32	43,46	75.6	6,67	4359
Hf	178.49	22.16	6.35	45,85	109	3,43	3671
Y	88.91	33.01	4.39	21,28	49.2	4,76	4217
La	138.91	37.12	4.49	19,35	26.8	3,11	2772
Pr	140.91	34.15	3.9	18,27	25.5	2,67	2660
Nd	144.24	34.18	3.35	15,68	29.2	2,24	2718
Sm	150.36	33.01	2.11	10,23	29.98	1,35	2702
Tb	158.93	31.14	4.1	21,07	40.7	2,48	2920
Dy	162.50	31.52	3.1	15,74	43.3	1,84	2958
Ho	164.93	31.12	3.0	15,42	40.8	1,75	3039
Er	167.26	30.64	3.3	17,23	47.4	1,90	3082
Yb	173.05	30.02	1.6	18,53	13.53	0,89	1820

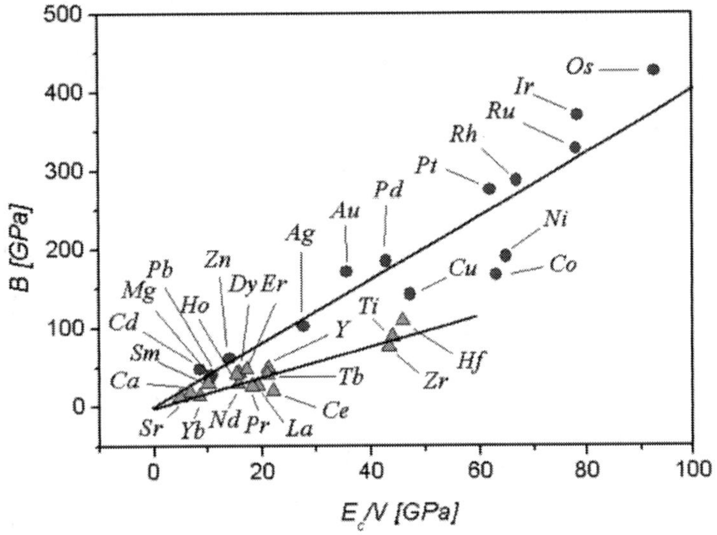

Figure 1. Linear correlation between the experimental data on the bulk modulus$B$ and cohesive energy density E_c/V. ● - structure A1 or A3, ▲ - structure A2, at the melting point.

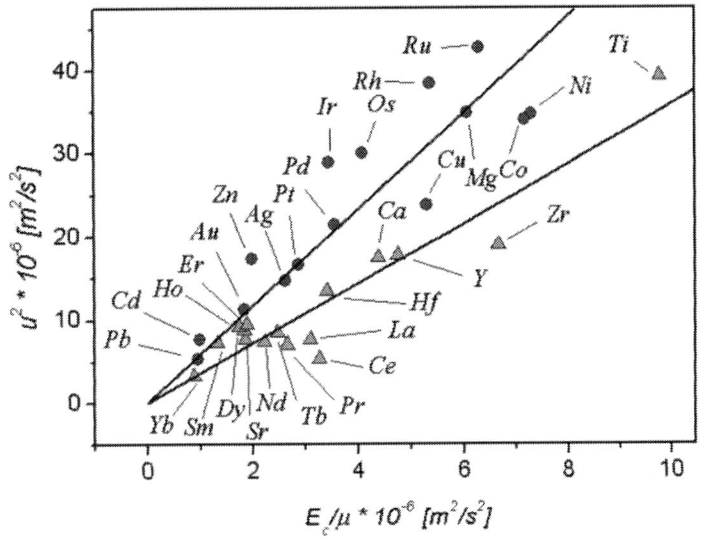

Figure 2. Linear correlation between the experimental data on the squared sound velocity u^2 of metals and their mass density of the cohesive energy, E_c/μ. ● - structure A1 or A3, ▲ - structure A2, at the melting point.

In order to check the validity of equation (4), in Table 1 the experimental data on the sound velocity u [7,8,9] are collected together with the values of their molar masses. On this basis the plot of the mass energy density E_c/μ versus the squared value u^2 of the sound velocity has been prepared (Fig. 2). Similarly as in Fig. 1, the experimental data points in Fig. 2 are scattered along two different straight lines. The data for metals melting from the close packed structure A1 or A3, are well fitted by a straight line described by relation

$$u^2 = (5.796 \pm 0.309) E_c / \mu \quad (5)$$

The correlation factor for this empirical correlation is R = *0.905*, and the relative standard deviation amounts 5,3%.

For metals showing the temperature – induced polymorphism and melting from the body centered cubic structure (A2), the relation between the squared sound velocity u^2 and the mass density of cohesive energy E_c/μ is well fitted by a straight line described by relation:

$$u^2 = (3.591 \pm 0.190) E_c / \mu \quad (6)$$

The correlation factor for this empirical correlation is R = *0.938*, and the relative standard deviation amounts 5,3%.

The existence of the relationships presented in the present communication suggests that similar relationships should be valid for other classes of substances, with other nature of the interionic or intermolecular bonds (e.g. condensed rare gases and covalently bonded solids).

As the values of the bulk modulus are related with other elastic moduli (shear modulus, Young modulus) [4], the values of the two later moduli should be also simply related to the volumetric energy density E_c/V, and the squared values of the longitudinal and transversal sound velocities in solids should correlate with the mass density of the cohesive energy, E_c/μ. The search for such correlations will be continued.

References

[1] Plendl J M Gielisse J M 1969 *Compressibility and polymorphism of solids* (Office of Aerospace Research, Bedford MA)

[2] Srivastava G P Weaire D 1987 The theory of the cohesive energies of solids *Advances in Physics* **36** 463-517

[3] Plendl J M Mitra S S Gielisse P J 1965 *Compressibility, cohesive energy, and hardness of non-metallic solids*

[4] Górecki T 1980 The relations between the shear modulus, the bulk modulus and Young's modulus for polycrystalline metallic elements *Materials Science and Engineering* **43** 225-230

[5] Bernsztejn M L Zajmowskij W A 1973 *Struktura i własności mechaniczne metali* (Wydawnictwo Naukowo-Techniczne,Warszawa)

[6] Górecki T 1979 Vacancies and generalized melting curve of metals High Temperatures-*High Pressures* **11** 683-692

[7] Shaikhiev G F Sulaimanova A S 1981 Rascet skorosti ultrazvuka v prostykh veshchestvakh po termiceskim konstantam *Zhurnal Fizicheskoi Khimii* **55** 3024-26

[8] Shaikhiev G F Sulaimanowa A S 1981 Temperaturnaia zavisimost skorosti ultrazvuka prostykh veshchestv v tverdom I zhidkom sostoianiyakh *Zhurnal Fizicheskoi Kimii* **55** 2584-88

[9] Shaikhiev G A Sulaimanova A S 1981 Vzaimosviaz skachkov entropii i skorosti ultrazvuka v tocke plavlenia *Zhurnal Fizicheskoi Khimii* **55** 2251-2256

[10] Kikoin I K 1976 *Tablicy Fizicheskikh Velichin. Spravocnik* (Atomizdat, Moscow)

Journal of Physics: Conference Series **289** (2011) 012021 doi:10.1088/1742-6596/289/1/012021

Kinetics of phase transitions in vitreous chalcogenide semiconductors $As_xSe_{100-x-y}Bi_y$ as studied by the differential thermal analysis and exoelectron emission methods

Cz Górecki[1] and T Górecki[2]

[1]Department of Physics, Opole University of Technology, 75 Ozimska str., 45-370 Opole, Poland
[2]Institute of Physics, Opole University, 48 Oleska str., 45-052 Opole, Poland

E-mail: c.gorecki@po.opole.pl

Abstract. Kinetics of glass transition (retrification) in chalcogenide semiconductors $As_xSe_{100-x-y}B_y$ (x = 20 or 30, and y = 0 and 1) has been investigated by parallel differential thermal analysis (DTA) and exoelectron emission (EEE) measurements. EEE is a surface effect accompanying the structural transformations in the surface layer, whereas the DTA technique gives information about the transformations occurring in the volume of the sample. Temperature dependencies of the DTA signal and of the EEE intensity have been determined and the values of the activation energy for both the volume and the surface retrification have been determined by the Ozawa method for each of the four investigated materials. It has been found that addition of Bi into the vitreous As_xSe_{100-x} glass changes distinctly the kinetics of both the surface and volume retrification. Addition of Bi causes a distinct decrease in the value of the activation energy for retrification process in both the surface layer and in the volume, i.e. reduces the thermal stability of investigated materials.

1. Introduction

Vitreous chalcogenide semiconductors remain to be a subject of intensive investigations because of unusual combination of their physical properties. The most stimulating factor for these studies steams from potential applications of these materials in numerous advanced technological applications (optoelectronics, microelectronics, etc.) [1]. These promising applications of chalcogenide glasses are, however, limited by the difficulties arising from their structural instability, resulting from the production process, in which the liquid phase is quenched to the ambient temperature [2]. The process of thermal ageing of the as-quenched material (e.g. long-lasting isothermal ageing at proper temperature) leads to the achievement of the metastable state characteristics of the overcooled liquid [1-5].

Amorphous materials tend to crystalize at a proper combination of temperature and time. The stability of the chalcogenide glasses may be characterized by the activation energies and temperatures of the thermally activated retrification and crystallization processes. There is an additional complicating factor caused by the fact, that the parameters determining the thermal stability of the volume and of the surface layer of amorphous materials may differ among themselves. About 20 years ago we elaborated a method [6] for determination of the thermal stability for both the surface and volume of amorphous materials by parallel measurements of the temperature dependencies of the DTA signal and the intensity of photostimulated EEE. This method has been already successfully applied in investigations of the kinetics of phase transformations in various chalcogenide glasses, including arsenic selenides [7-8] as well as in pure melt-quenched selenium [9].

Recently Golovchak and coworkers [10] reported on the effect of modification of the structure of arsenic selenide glasses by addition of bismuth. By analyzing the core-level XPS (X-ray photoelectron spectra) they have found that addition of Bi leads to formation of coordination defects and, with increasing concentration of Bi, to the formation of nanocrystallites in the vitreous matrix. The presence of such imperfections may lead to the lowering of potential barriers for structural transformations in investigated arsenic selenide glasses.

The purpose of the present study is to investigate the effect of Bi addition on the kinetics of the retrification process in the volume and in the surface layer of arsenic selenide glasses.

2. Experimental conditions

The investigated bulk chalcogenide glasses were prepared in Lviv Scientific Research Institute of Materials of SRC "Carat" by conventional melting of appropriate mixture of high purity (99,999%) precursors sealed in evacuated (~10^{-3} Pa) quartz ampoules, followed by air quenching to room temperature. The amorphness of the obtained materials was checked by X-ray diffraction and IR transmission techniques.

Measurements of the temperature dependencies of photostimulated exoelectron emission (EEE) intensity were carried out by means of the arrangement described in [12]. An open point counter with saturated ethanol quenching vapour was used for detecting the exoelectrons. The sample temperature, controlled using a Ni-CrNi thermocouple with an accuracy of about 5 K, was changed at four constant heating rates (2, 5, 10 and 20 K/min). The sample surface was irradiated during the measurements by unfiltered radiation from a quartz lamp with a Q-400 burner.

The calorimetric investigations of the volume retrification were performed at five heating rates (1, 2, 5, 10 and 20 K/min) using the NETZSCH DSC 404/3/F differential calorimeter with E-type thermocouple using an empty crucible made of high density Al_2O_3 as references.

All the EEE and DTA measurements were performed in air atmosphere under ambient pressure.

3. Results and discussion

Temperature dependencies of the intensity of photostimulated exoelectron emission and of the DTA signal were measured at different heating rates. The shape of the registered termograms and of the temperature dependencies of the EEE intensity for all the samples is very similar. All of them display single peaks corresponding to the retrification in the volume (DTA) or in the surface layer of investigated samples. The positions of these peaks, and their intensities for different chemical composition are different and depend on the heating rate.

As an example, on Fig. 1 the temperature dependencies of the EEE intensity and of the DTA signal for $As_{29}Se_{70}Bi_1$, measured at different heating rates, are presented. The temperature dependencies of the intensity of exoelectron emission presented in Fig. 1 display a maximum coinciding with the endothermal effect occurring on the DTA curves. The temperatures of these anomalies occurring on both the EEE and DTA curves systematically increase with increasing the heating rate. The process responsible for the effects displayed by the EEE (surface sensitive method) and DTA (volume sensitive method) is the retrification process occurring in the surface layer and in the volume of investigated sample, correspondingly. The temperature of the anomalies occurring on the DTA and EEE curves presented in Fig. 1 systematically shift toward higher values with increasing the heating rate. This confirms that the process responsible for the occurrence of these anomalies (retrification) is a thermally activated process.

Measurements of the temperature dependencies of EEE intensity at four different heating rates, and the DTA curves at five heating rates, enabled the determination of the activation energy for the retrification process on the surface and in the volume of the samples, responsible for the occurrence of the maxima on the EEE and DTA curves, correspondingly. Activation energy was determined using the well-known Ozawa relation [11]:

$$\ln V = A - E/kT$$

where V is the heating rate, A – constant, E – activation energy, k – Boltzmann's constant and T – transformation temperature.

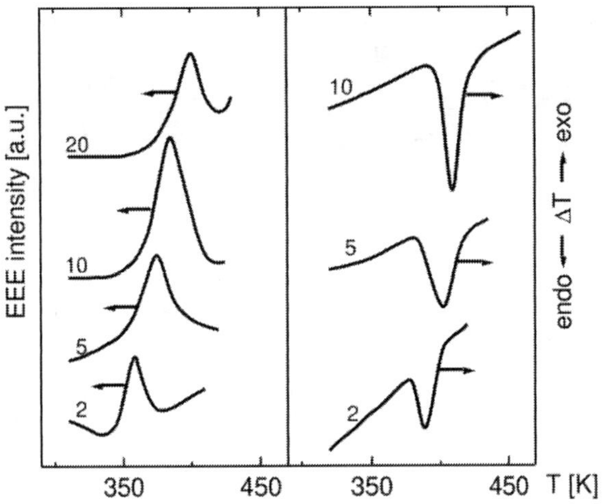

Figure 1. Temperature dependencies of the intensity of photostimulated exoelectron emission (EEE) and DTA signal for $As_{29}Se_{70}Bi_1$ glass. Parameter – heating rate in K/min.

The dependencies of the transformation (retrification) temperatures T, determined from the temperature dependencies of the EEE intensity and DTA signal, on the heating rate V, represented in the Ozawa coordinates ln V = f(1/T) - so called Ozawa plots for $As_{30}Se_{70}$ and $As_{29}Se_{70}Bi_1$ samples are shown on Fig. 2. The retrification temperature T has been determined from the position of the

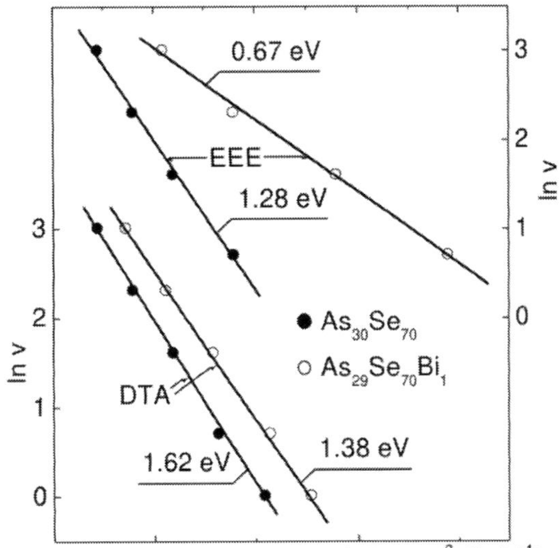

Figure 2. Ozawa's plots for the surface (EEE) and volume (DTA) retrification temperatures of $As_{30}Se_{70}$ and $As_{29}Se_{70}Bi_1$ glasses.

maximum on the EEE curves [12], and as the deflection point (from the base line) on the DTA curves [13]. The shapes of the Ozawa's plots for the second pair of investigated glasses ($As_{30}Se_{70}$ and $As_{29}Se_{70}Bi_1$) differs from those shown in Fig. 1 only by their slope.

Activation energy of the retrification process is a parameter characterizing well the thermal stability of amorphous materials. The values of the activation energy for the retrification process in the surface layer and in the volume of four investigated materials determined in present study on the basis of systematic EEE and DTA investigations by Ozawa method are collected in Table 1.

Table 1. Activation energies for the volume and surface retrification of investigated chalcogenide glasses as determined from the DTA and EEE measurements.

Composition	Activation energy in eV	
	Surface retrification	Volume retrification
$As_{20}Se_{80}$	1,10	1,61
$As_{19}Se_{80}Bi_1$	0,83	1,56
$As_{30}Se_{70}$	1,28	1,62
$As_{29}Se_{70}Bi_1$	0,67	1,38

The data collected in Table 1 permit to conclude that:

1. The activation energy for both the surface and the volume retrification of arsenic selenide glasses increases with increasing the arsenic content. This agrees well with the results of our previous studies [7,8].
2. Admixture of bismuth causes a distinct lowering of the activation energy for retrification of arsenic selenide glasses. This effect is most probably caused by formation of coordination defects and even mixed $As_2Se_3 - Bi_2Se_3$ nanocrystallites, observed in [10].
3. The combination of the EEE and DTA methods would be very useful and efficient in investigations of the crystallization kinetics of structural transformations and of the thermal stability of chalcogenide glasses.

Further studies are in progress.

4. Acknowledgments

The authors are very grateful to Prof. O. Shpotyuk and Dr R. Golovchak for kind donation of the samples of investigated materials.

References

[1] Saiter J M 2001 Physical ageing in chalcogenide glass *J. Optoelect. Adv. Mat.*, **3** 685-694

[2] Ohta T, Birukawa M, Yamada N and Hirao K 2002 Optical phase change and magneto-optical recording *J. Magn. and Magn. Mater.* **242-245** 108-115

[3] Golovchak R, Górecki Cz, Kozdraś A and Shpotyuk O 2006 Physical ageing effects in vitreous arsenic selenides *Sol. State Commun.* **137** 67-69

[4] Golovchak R, Shpotyuk O, Shpotyuk M, Górecki Cz and Kozdraś A 2005 Ageing effects in $As_{10}Se_{90}$ chalcogenide glasses induced by γ-irradiation *Ukr. J. Phys.* **50** .690-693

[5] Wagner T, Kasap S O and Petkow K 1997 Temperature-modulated differential scanning calorimetry studies of the structure of bulk and thin film $Ge_xAs_yS_{60}$ chalcogenide glasses *J. Mat. Sci.* **32** 5889-5893

[6] Górecki Cz, Górecki T and Michno Z 1987 Thermal stability of the $Fe_{40-x}Ni_{40}Cr_xB_{20}$ metallic glasses as studied by the EEE and DTA methods *Acta Phys. Polon.* **A72** 157-160 Thermal stability of the $Ni_{78}Si_xB_{22-x}$ metallic glasses as studied by the EEE and DTA methods *Acta Phys. Polon.* **A72** 161-163

[7] Górecki Cz, Górecki T, Golovchak R Żurawska A and Wacke S 2005 Kinetics of phase transitions in semiconducting amorphous chalco,enides As_xSe_{100-x} as studied by differential thermal analysis and exoelectron emission techniques *Visnyk Lviv Univ.* **38** 399-403

[8] Górecki Cz, Górecki T, Kozdraś A and Golovchak R. 2007 Kinetics of phase transitions in vitreous chalcogenide semiconductors As_xSe_{100-x} as studied by differential thermal analysis (DTA) and exoelectron emission (EEE) methods *Visnyk Lviv Univ.* **40** 254-260

[9] Górecki Cz and Górecki T 2007 Structural transformations in amorphous selenium as studied by the differential thermal analysis and exoelectron emission technique *J. of Phys.* **79** 1-5

[10] Golovchak R, Shpotyuk O, Kovalskiy A, Miller A C, Cech J and Jain H 2008 Coordination

defects in bismuth-modified arsenic selenide glasses; High-resolution x-ray photoelectron spectroscopy measurements *Phys. Rev.* **B 77** 172201 1-4
[11] Ozawa T 1970 Kinetics of non-isothermal crystallization of amorphous materials *J. Thermal Anal.* **2** 301-312
[12] Sujak B and Górecki T 1973 Egzoemisja elektronów podczas przemian fazowych i reakcji w ciałach stałych *Wiadomości Chemiczne* **37** 361-384
13] Schultze D 1974 Termiczna Analiza Różnicowa (PWN Warszawa)

Journal of Physics: Conference Series **289** (2011) 012022 doi:10.1088/1742-6596/289/1/012022

Effect of electrolytical hydrogenation on the thermal stability and crystallization kinetics of metallic glass $Fe_{79}Si_9B_{12}$

Cz Górecki[1] and T Górecki[2]

[1]Department of Physics, Opole University of Technology, 75 Ozimska str., 45-370 Opole, Poland
[2]Institute of Physics, Opole University, 48 Oleska str., 45-052 Opole, Poland

E-mail: c.gorecki@po.opole.pl

Abstract. The effect of electrolytical hydrogenation on both the surface and volume crystallization kinetics and thermal stability of amorphous alloy $Fe_{79}Si_9B_{12}$ has been investigated. The parameters of the surface and volume crystallization (temperature, activation energy) have been determined applying the exoelectron emission (EEE) and differential thermal analysis (DTA) methods, respectively.

It has been found that the surface crystallization of investigated material occurs at temperature much lower and with activation energy smaller than the volume crystallization. The determination of the activation energies for the volume and surface crystallization by the combination of DTA and EEE techniques enables the determination of activation energies for both the nucleation and growth of the crystalline phase in metallic glasses and other amorphous materials. Hydrogenation of the investigated metallic glass reduces its thermal stability, what is manifested by an decrease in the activation energies for both the surface and volume crystallization.

1. Introduction

For many years amorphous metallic alloys (metallic glasses) remain to be subject of intensive research activity [1-3], spurred by both the science and technology. The main reason for this permanent interest is the broad variety of potential applications of metallic glasses. Among others, the application for hydrogen storage are very intensively investigated [1,2].

The applications of metallic glasses are, however, limited by the difficulties arising from their thermodynamical instability. Amorphous alloys tend to crystallize at a proper combination of temperature and time. After crystallization they change drastically their properties and as a rule become useless [1-3].

The crystallization is a thermally activated process and in order to know the thermal stability of metallic glasses one has to determine the activation energy for this process and the crystallization temperature determined at well defined heating rates. There is an additional complication caused by the fact that the parameters describing the thermal stability (activation energies, crystallization temperature) of the volume and that of the surface layer of amorphous materials are, as a rule, different.

In the mid-eighties of XX century we elaborated a method for determination of the thermal stability for both the surface and the volume crystallization of amorphous materials by parallel differential thermal analysis (DTA) or differential scanning calorimetry (DSC) and measurements of the temperature dependencies of the intensity of photostimulated exoelectron emission (EEE). The results of such investigations for a variety of amorphous materials have been reported in a series of papers [4-9].

Considering the potential application of metallic glasses as the hydrogen storage materials, one has to take into account the possible effect of hydrogen charging on their thermal stability. The aim of the present communication is to report the results of investigation of the effect of electrolytic hydrogenation on the thermal stability and surface and volume crystallization kinetics of metallic glass $Fe_{79}Si_9B_{12}$, as determined by the parallel DTA and EEE measurements.

2. Experimental conditions

The investigated amorphous ribbon $Fe_{79}Si_9B_{12}$, prepared from high purity metals by liquid quenching on a rotating wheel, was produced in the Institute of Materials Science of the Warsaw Technical University.

The samples were by 2 and 5 hours cathodically charged with hydrogen in a 1 N aqueous solution of H_2SO_4 with 2 g of thiourea per litre. The current density was 0,5 A/dm^2.

Measurements of the temperature dependencies of photostimulated EEE intensity were carried out by means of the arrangement described in [10]. An open air point counter with saturated ethanol quenching vapour was used for detecting the exoelectrons. The sample temperature, controlled using an Ni-CrNi thermocouple with an accuracy of about 5 K, was changed at five constant heating rates (5, 10, 20, 40 and 70 K/min). The sample surface was irradiated during the measurements by unfiltered radiation from a quartz lamp with a Q-400 burner.

DTA measurements of investigated samples were performed at seven (0.5, 1, 2, 5, 10, 20 and 50 K/min) heating rates using the NETZSCH DSC 404/3F microcalorimeter, with Pt-PtRh measuring head and high-density Al_2O_3 sample pans. An empty high density Al_2O_3 crucible was used as the reference. All the EEE and DTA measurements were performed in air under normal pressure.

3. Results and discussion

Three sets of samples of metallic glass were investigated:: 0h - as delivered (uncharged), 2h – samples electrolytically charged with hydrogen by 2 hours and 5h – samples electrolytically charged with hydrogen by 5 hours.

The experimentally determined temperature dependencies of the EEE intensity and DTA curves for 0h and 2h type samples, all measured in the first heating run at the same heating rate of 10 K/min, are shown on Fig. 1. As it is seen, the DTA curves display two exothermal peaks at about 740 K and 790 K. This means that the volume crystallization of investigated material occurs in two stages.

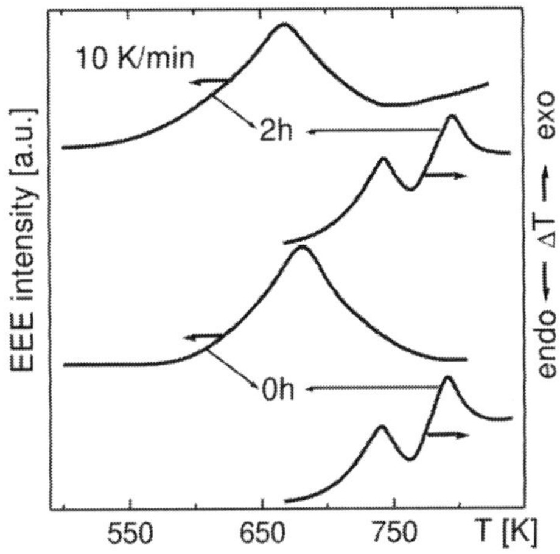

Figure 1. Comparison of the temperature dependencies of the EEE intensity and DTA curves for samples uncharged (0h) and charged with hydrogen by 2 hours (2h). Heating rate – 10 K/min.

The temperature dependencies of the intensity of photostimulated exoelectron emission display a maximum located at a temperature of about 670 K. There are at least two reasons for supposing that the peak of EEE intensity is caused by irreversible surface crystallization. First, the positions of EEE maximum depend on the heating rate (Fig. 3) in a manner similar to that of the DTA peaks connected with the volume crystallization (Fig. 2). This permits to explain the occurrence of the EEE maximum as the effect connected with some thermally activated process. Secondly, similarly as in the case of DTA, there are no peaks on the EEE curves registered in the second and further heating runs.

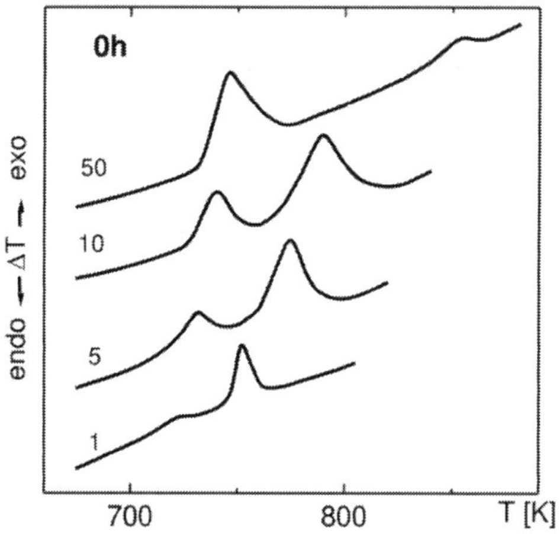

Figure 2. DTA traces for the 0h-type samples registered at different heating rates. Parameter – heating rate in K/min.

The measurements of the temperature dependencies of the DTA curves performed at seven heating rates ranging from 0.5 to 50 K/min together with the registration of the temperature dependencies of EEE intensity performed at five different heating rates ranging from 5 to 70 K/min, enabled the determination of the activation energies for both the volume and the surface crystallization processes governing the appearance of the maxima on the temperature dependencies of the DTA curves and of EEE intensity. Exemplary DTA traces for the 0h-type samples, registered at four different heating rates, are presented in Fig. 2. As it follows from the results shown in Figs. 3 and 4 the

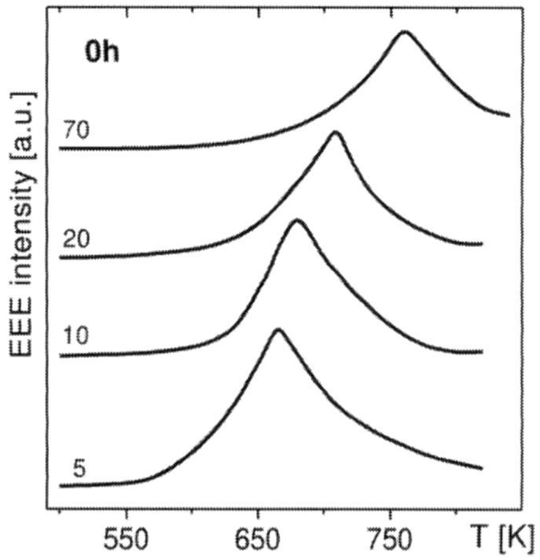

Figure 3. Temperature dependencies of the intensity of EEE from 0h-type samples.. Parameter – heating rate in K/min.

volume crystallization occurs in two stages (two peaks on the DTA curves).

The experimentally determined temperature dependencies of the EEE intensity for 0h-type samples, registered at four heating rates, are presented in Fig. 3. As it is seen on Figs. 2 and 3, with increasing the heating rate the peaks on the DTA and EEE curves systematically shift towards higher temperatures. This behaviour is typical of thermally activated phenomena.

From the shifts of the DTA and EEE peaks the activation energies for the volume and surface crystallization of investigated samples were determined using the Ozawa relation [11]:

$$\ln V = A - E/kT$$

where V is the heating rate, A – constant, E – activation energy, k – Boltzman's constant and T – crystallization temperature. The crystallization temperature T has been determined from the position on the maximum on the EEE curves [10] and as the deflection point (from the base line) on the DTA curves [11].

The dependencies of the crystallization temperatures T, determined from both temperature dependencies of the DTA signal and of the EEE intensity, on the heating rate V, represented in the Ozawa coordinates $\ln V = f(1/T)$ - so called Ozawa plots for 0h-type samples are shown on Fig. 4. The accuracy of the determination of the activation energies was better than 0,2 eV. The same procedure was applied for determining the activation energies of all the three types of investigated samples.

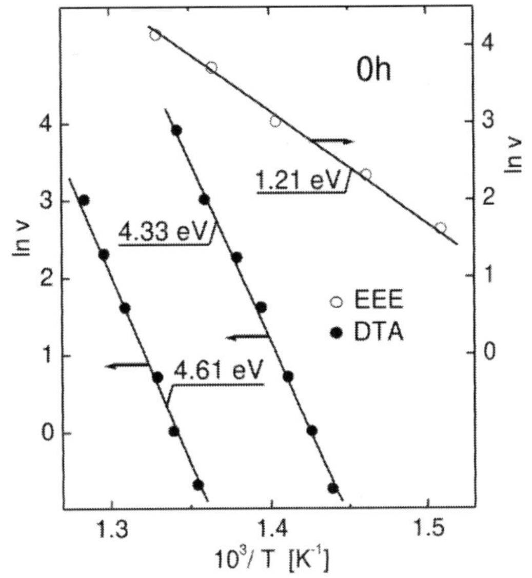

Figure 4. Ozawa's plots for the surface (EEE) and the volume (DTA) crystallization processes in 0h-type samples.

The value of the activation energy is an important parameter characterizing adequately the thermal stability of amorphous materials. The values of the activation energies for the surface and the volume crystallization of all types of investigated samples in the present work, determined by the Ozawa method on the basis of DTA and EEE measurements, are collected in Table 1.

The data presented in Table 1 indicate that hydrogenation causes an increase in the activation energies for both stages of the volume as well as of the surface crystallization. So, the hydrogenation not only does not deteriorate the thermal / temporal stability of investigated amorphous alloy $Fe_{79}Si_9B_{12}$, [13] but slightly enhances it. It seems very interesting to check if this statement holds also for other metallic glasses.

Parallel applications of both the EEE and DTA methods enable to follow the structural transformations in both the surface and in the volume of metallic glass. The surface crystallization of investigated metallic glass occurs at temperature much lower and with activation energy smaller than the volume crystallization. The difference between the activation energies for the volume and surface crystallization may result from the existence of crystal nuclei [12] formed in the surface layer already during the fabrication process. The bulk of the investigated material crystallizes by usual nucleation and growth mechanism, thus the difference between the activation energy for the first stage of the volume crystallization and the activation energy for the surface crystallization could be treated as a measure for the nucleation energy. The lowering of the thermal stability of the surface layer, with respect to that for the volume of material, can seriously restrict the potential applications of metallic glasses.

Table 1. Effect of the hydrogenation time on the activation energies for volume (DTA) and surface (EEE) crystallization of $Fe_{79}Si_9B_{12}$ metallic glass.

Hydrogenation time [h]	Activation energy in eV	
	Volume crystallization - DTA	Surface crystallization - EEE
0	4.33 4.61	1.21
2	3.74 4.37	0.80
5	3.85 4.03	0.84

References

[1] Luborsky F E 1983 *Amorphous Metallic Alloys* (Butherworths London)
[2] Adler D, Schwartz B B and Steele M C 1985 *Physical Properties of Amorphous Materials* (Plenum Press New York)
[3] Duhaj P, Mrafko P and Svec P 1997 Rapidly *Quenched and Metastable Materials* (Elsevier Amsterdam)
[4] Górecki Cz, Górecki T and Michno Z 1989 Crystallization of amorphous metals studied by means of the EEE and DTA techniques *Radiat. Protect. Dosim.* **4** 213.-215
[5] Górecki Cz, Górecki T and Michno Z 1987 Thermal stability of the $Fe_{40-x}Ni_{40}Cr_xB_{20}$ metallic glasses as studied by the EEE and DTA methods *Acta Phys. Polon.* **A72** 157-159, Thermal stability of the $Ni_{78}Si_xB_{22-x}$ metallic glasses as studied by the EEE and DTA methods *Acta Phys. Polon.* **A72** 161-163
[6] Górecki T and Górecki Cz 1989 Effect of electrolytical hydrogenation on the crystallization kinetics and exoelectron emission from metallic glass $Ni_{78}V_5Si_{10}B_{12}$ *Z. Phys. Chem.* **163** 361-365
[7] Górecki Cz 1981 Structural transformations in amorphous solids detected by the exoelectron emission technique *J. Non-Cryst. Solids* **32** 63.-67
[8] Górecki Cz and Górecki T 1993 Thermal stability and crystallization kinetics of metallic glasses $Fe_{80}T_xB_{20}$ as studied by EEE and DTA techniques *Surface Physics Chemistry Mechanics* (RAS) **7** 63-72
[9] Górecki Cz, Górecki T and Szymura S 2002 Surface and volume crystallization of metallic glass $(Ni_{50}Zr_{50})_{99,9}P_{0,1}$ as investigated by exoelectron emission (EEE) and differential thermal analysis (DTA) *Mod. Phys. Lett.* **16** 87-92
[10] Górecki T and Sujak B 1972 Photostimulated exoelectron emission during phase transitions of binary and ternary nickel alloys *Acta Phys. Polon.* **A42** 487-501
[11] Ozawa T 1970 Kinetic analysis of derivative curves in thermal analysis *J. Thermal Anal.* **2** 301-324
[12] Watanabe Y Kubozoe T Nakamura Y 1993 Crystallization – temperature dependence of exoelectron emission from metallic glasses *Surface Physics Chemistry Mechanics* (RAS Moscow) **8** 25-29
[13] Górecki Cz Górecki T 2007 Effect of electrolytical hydrogenation on the thermal stability and crystallization kinetics of METGLASS MBF-50 *J. of Physics Conf. Series* **79** 1-5

Investigation of the effect of intensive milling in a planetary ball mill on the thermal decomposition of basic nickel carbonate

K Książek[1], **S Wacke** [2], **T Górecki** [1], **Cz Górecki** [2]

[1]Institute of Physics, Opole University, 48 Oleska str., 45-052 Opole, Poland,
[2]Department of Physics, Opole University of Technology, 75 Ozimska str., 45-370 Opole,Poland,

E-mail: kasiak@uni.opole.pl; s.wacke@po.opole.pl

Abstract: The kinetics of thermal decomposition of basic nickel carbonate $NiCO_3 \cdot Ni(OH)_2 \cdot nH_2O$ and the effect of intensive milling in a planetary ball mill on its parameters, have been investigated. The values of the reaction heat and of the activation energy of thermal decomposition have been determined. Investigations of the thermal decomposition of the products of ball milling of investigated compound revealed a distinct effect of milling on the reaction temperature and heat consumed during the thermal decomposition of investigated compound.

1. Introduction

Studies of the effect of mechanical deformation on the chemical reactivity of solids, called mechanochemistry, are nowadays an important research area in the field of physical chemistry and chemistry of solids [1,2]. Mechanochemistry offers the possibility to conduct various reactions, including the synthesis of substances in the nanocrystalline and amorphous states, at low temperatures without dissolution or fusion of the reactants.

Mechanochemical treatment of substrates in producing new thermodynamically stable and metastable materials, often non achievable by traditional methods, is well established and received industrial applications. While the method is progressing at the industrial level, knowledge about the physical mechanisms operating during the mechanochemical treatment (eg. ball milling) is still very limited due to serious difficulties in quantifying this complex process [3].

A survey of the recent literature indicated that nowadays the research effort in mechanochemistry is mainly devoted to preparation of new metastable and stable metallic materials. In the field of nonmetallic inorganic compounds the mechanically induced polymorphic phase transition in many oxides were reported. Successful syntheses of many compounds, eg. copper and silver chalcogenides, have been described, too [3-6].

In our previous studies [7,8] a distinct effect of ball milling on the thermal decomposition of lead and cadmium carbonates, as well as of the basic zinc carbonate, has been observed. The purpose of the present study was to observe the eventual effect of ball milling on the thermolysis of basic nickel carbonate $NiCO_3 \cdot Ni(OH)_2 \cdot nH_2O$. The thermal decomposition of basic nickel carbonates is of particular interest, since pure nickel and several nickel catalysts are prepared by decomposition of these carbonates and reduction of the decomposition product. The activity and service life time are dependent on the prehistory of parent materials, calcinations conditions, doping with foreign oxides, etc. Despite of intensive research effort (see e.g. [9-12]), including even the action of ionizing radiation, there is still not quantitatively substantiated model of thermal decomposition of basic nickel carbonate. Till now the effect of ball milling on the thermal decomposition on the kinetics of thermal decomposition has not been investigated.

2. Experimental conditions

Chemically pure basic nickel carbonate $NiCO_3{\cdot}Ni(OH)_2{\cdot}nH_2O$, supplied by POCh Gliwice, was used in the experiments. Thermal decomposition of starting material, as well as of products obtained by intensive ball milling, was investigated by differential thermal analysis, performed using the NETZSCH DSC 404/3F differential scanning calorimeter with the S-type Pt-PtRh measuring head and standard platinum sample pans. An empty sample pan was used as the reference. The masses of the samples in each DSC experiment series were nearly the same ~19 mg.

The investigated compound was milled in a planetary ball mill Pulverisette 7 produced by FRITSCH. Milling was carried out in the two grinding vials of 25 ccm volume containing 5 balls with diameter of 12 mm. Both the vials and balls were made of stainless CrNi steel. Each of the two milling vials containing 5 balls were charged with 2 g of the investigated basic nickel carbonate. The rotation speed of the plate supporting the milling device was 370 turns per minute. The milling conditions were changed only by changing the milling time.

All the milling procedures and DTA experiments were performed in atmospheric air under normal pressure.

3. Results and discussion

The DSC traces for unmilled $NiCO_3{\cdot}Ni(OH)_2{\cdot}nH_2O$ as well as of the product of ball milling lasting 2 hours, all registered at the same heating/cooling rate of 10 K/min, are shown on Fig.1. In the first heating run distinct endotherms, revealing the two main stages of thermal decomposition of investigated compound, were observed. On the DSC traces registered in the cooling run as well as during the second heating run (with the same sample) no thermal effects were observed. This means that the investigated carbonate samples have been completely decomposed in the first heating run.

As it seen on Fig.1, the thermal decomposition of investigated carbonate occurs in two stages. The low temperature stage consists of evolution of crystalline and hydroxilic water. The solid product of

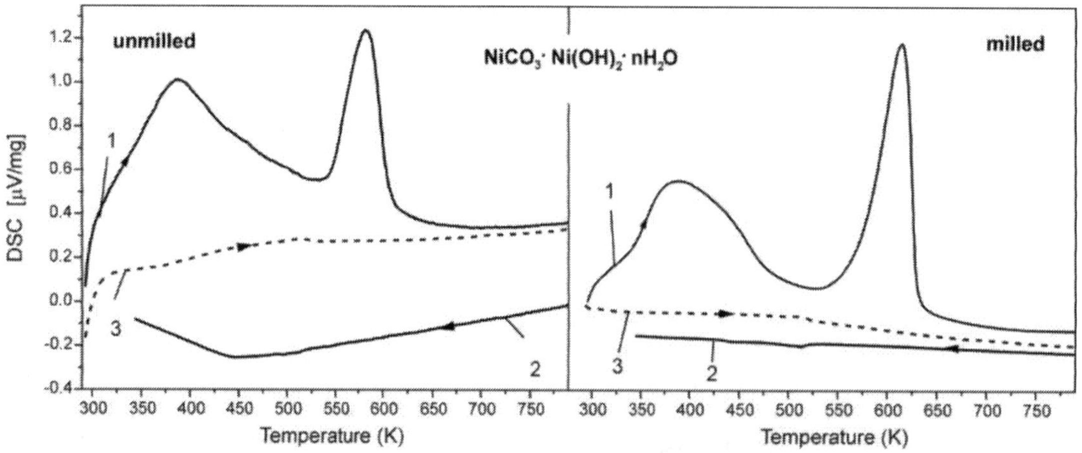

Figure 1. DSC traces for $NiCO_3{\cdot}Ni(OH)_2{\cdot}nH_2O$ registered at the heating/cooling rate of 10 K/min. 1 – first heating run, 2 - cooling run, 3 – second heating run.

this process is NiO. The DSC endotherm corresponding to this stage is anomalously brood. Most probably, the loss of crystalline water occurs at temperature slightly differing from that for the decomposition of $Ni(OH)_2$. These two processes produce two overlapping DSC peaks, and the low temperature endotherm in the registered DSC traces is a result of their superposition.

The high temperature endotherm peaking at about 590 K corresponds to the decomposition of $NiCO_3$ into NiO and gaseous CO_2. So, the only solid product of heating of investigated basic nickel carbonate up to ~ 770 K is NiO.

On the DSC traces registered in the first cooling run as well as in subsequent heating and cooling runs a small discontinuity (jump) is observed at temperature of about 510 K . This effect can be ascribed to the antifferromagnet-paramagnet phase transition (Neel point) of NiO, being the final solid product of thermolysis of the investigated carbonate. The Neel temperature for pure stoichiometric NiO is 520 K, and this second order phase transition is accompanied by a well know λ - anomaly in the temperature dependence of the heat capacity.

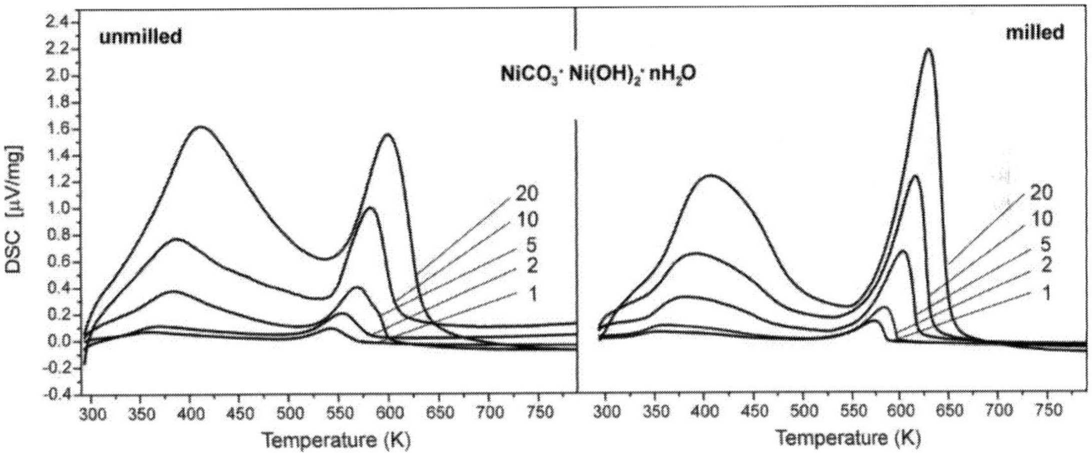

Figure 2. Exemplary DSC traces for unmilled and milled by 3 hours $NiCO_3 \cdot Ni(OH)_2 \cdot$samples registered at different heating rates. Parameter – heating rate in K/min.

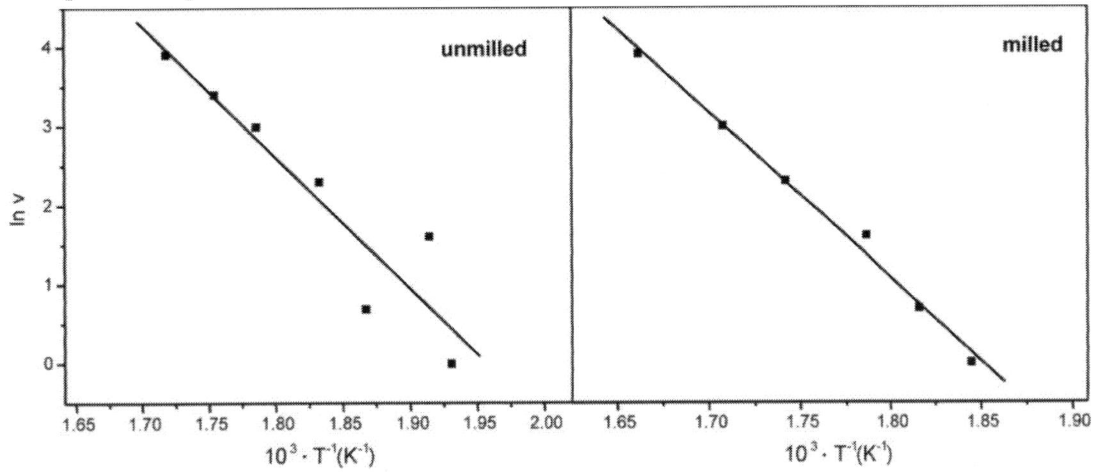

Figure 3. Ozawa's plots for reaction temperatures of unmilled and milled basic nickel carbonate. Milling time – 2 hours.

With increasing the heating rate in the DSC experiments, the positions of endotherms corresponding to the two stages of thermal decomposition of investigated compound systematically shift towards higher temperatures (Fig. 2). This effect is typical of thermally activated processes, and allows for determination of the value of activation energy.

The activation energies for thermal decomposition of $NiCO_3$ (high temperature stage) has been determined by the method proposed by Ozawa [13]. For this purpose the so called "Ozawa plots" $\ln v = f(1/T)$ were constructed (Fig. 3), where T denotes the reaction temperature determined at the heating rate v. Reaction temperatures T were determined from the intersection of tangent to the low-temperature (left) side of endotherm with the base line of recorded DSC trace. As it is seen on Fig. 3, the data points for reaction temperatures registered at different heating rates are located along straight lines. The slopes of these lines multiplied by the universal gas constant R give the value of the activation energies of investigated processes.

The obtained this way value of the activation energy for the high temperature stage of thermal decomposition of unmilled basic nickel carbonate (thermolysis of $NiCO_3$) is 142.46 kJ/mole. Activation energy for the same stage of thermolysis of compound submitted to 2 hour ball milling is distinctly higher and amounts 167.29 kJ/mole. This leads to the conclusion that the milling procedure increases the thermal stability of investigated compound.

Determination of the activation energy for the low temperature stage of thermolysis of investigated basic nickel carbonate was impossible, because of the complexity of shape of the low temperature endotherm on the registered DSC traces.

The conclusion about enhancement of the thermal stability of basic nickel carbonate by milling has been confirmed by investigations of the effect of the milling time on the reaction temperatures and on the heat consumed during the two stages of he thermal decomposition of investigated material.

The area under the peaks on the DSC traces is proportional to the heat of observed chemical reaction or structural transformation. Making use of the calibration experiments in which the endotherms corresponding to the melting of high purity tin and zinc were registered, the heat consumed during the thermal decomposition of investigated nickel carbonate has been determined.

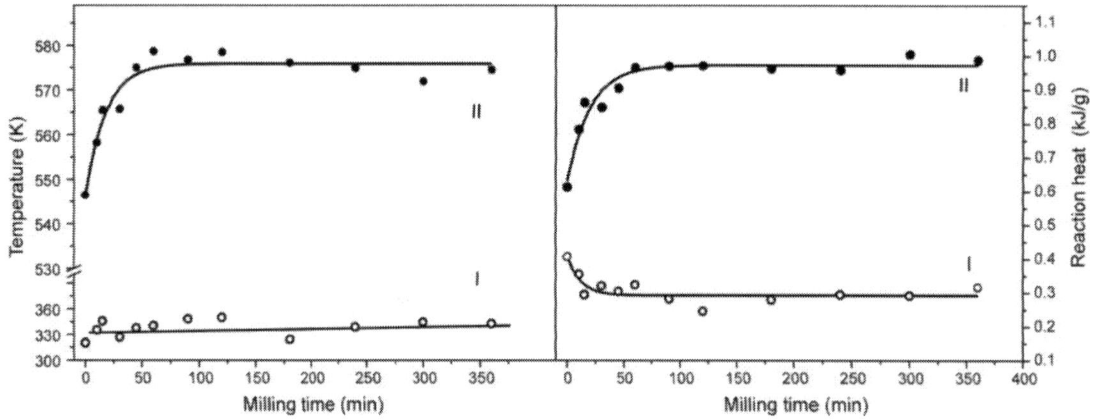

Figure 4. Effect of the milling time on the decomposition temperature (left) and the reaction heat (right) for the two stages of thermolysis of basic nickel carbonate.

Figure 4 represents the experimentally determined dependencies of the temperatures (left) and heats (right) consumed during the two stages of thermolysis of basic nickel carbonate on the milling time. As it is seen, milling causes a distinct increase of temperatures and heat of both the stages of thermolysis. The enhancing effect of milling on the thermal stability of investigated basic nickel carbonate, documented by the results of the present study, is rather surprising – in our previous studies [7,8] on the lead carbonate and basic zinc carbonate an opposite effect has been observed and no substantial effect of milling on the thermal decomposition of cadmium carbonate has been found. The elucidation of this discrepancy requires further experimental and theoretical studies.

4. References

[1] Butyagin P Yu 1994 Problems in mechanochemistry and prospect for its development *Russ. Chem. Revs* **53** 965-976

[2] Schulz R 2001 Electrochemical and Chemical Reactivity of Amorphous and Nanocrystalline Materials (*Trans Tech Publications*, Aedermannsdorf – Zürich)

[3] Shinghu P H (ed.) 1992 Mechanical alloying (*Trans Tech Publications*, Aedermannsdorf – Zürich)

[4] Ohtani T Maruyama K Ohshima K 1997 Synthesis of copper, silver and samarium chalcogenides by mechanical alloying *Material Research Bulletin* **32** 343 – 350

[5] Ohtani T Nonaka T Araaki M 1988 Sonochemical synthesis of coper and silver chlcogenides *J. Solid State Chemistry* **138** 131-134

[6] Kozdraś A Górecki Cz Żurawska A Górecki T 2000 Mechanochemical synthesis of silver telluride as controlled by exoelectron emission, photoacoustics and differential scanning calorimetry Latvian *Journal of Physics Supplement* **6** 160-163

[7] Książek K Wacke S Górecki T Górecki Cz 2007 Investigation of the effect of intensive ball milling in a planetary ball on the thermal decomposition of cadmium carbonate and basic zinc carbonate *Journal of Physics: Conf. Series* **79** 012022

[8] Książek K Górecki T Żaba M Górecki Cz Wacke S 2005 Effect of ball milling on the kinetics of thermal decomposition of lead carbonate as observed by the differential scanning calorimetry *Chemia i Ochrona Środowiska* **10** 167-170

[9] Lvov BV 2002 Mechnism and kinetics of thermal decomposition of carbonates *Thermochimica Acta* **386** 1-16

[10] Rhamdani MA Jak E Hayes PC 2008 Basic nickel carbonate:Part II. Microstructure evolution during industrial nickel production from basic nickel carbonate *Metallurgical Materials Transactions* **39** B 234-245

[11] Fried D Dollimore D 1994 Isothermal and rising temperature kinetic studies of doped nonstoichiometric basic nickel carbonate *Journal of Thermal Analysis* 41 323-336

[12] Buzdov E A 1966 Sintez i termiceskoe razlozhenie karbonatov nikela i kobalta *Zhurnal Neorganicheskoi Khimii*

[13] Ozawa T 1970 Kinetic analysis of the derivative curves in thermal analysis *J. Thermal Anal.* **2** 301-324

Journal of Physics: Conference Series **289** (2011) 012024 doi:10.1088/1742-6596/289/1/012024

Synthesis and optical behaviour of mesoporous silica functionalized by organometallic molecules

L Laskowski[1,2], **A Kassiba**[1] , **M Makowska-Janusik**[2], **N Errien**[1], **A Mehdi**[3] and **J Swiatek**[2]

Laboratoire de Physique de l'Etat Condensé LPEC, UMR CNRS n° 6087, Université du Maine Avenue Olivier Messiaen 72085 - Le Mans CEDEX 9 France.
Institute of Physics, Jan dlugosz University, Al. Armii Krajowej 13/15, 42-200 Czestochowa, Poland
Institut Charles Gerhardt, UMR 5253 Chimie Moléculaire et Organisation du Solide, CC 1701 Université Montpellier II Place E. Bataillon, F-34095 Montpellier Cedex 5, France.

E-mail:m.makowska@ajd.czest.pl

Abstract. Mesoporous silica SBA-15 functionalized by (1,4,8,11-tetraazacyclotetradecane) cyclam groups chelating nickel ions (Ni-cyclam) were synthesized by two different approaches. Characterizations by transmission electron microscopy TEM and UV-VIS absorption spectroscopy were performed to monitor the structure and optical properties of the material with regard to the used synthesis methods. The assignment of the experimental UV-VIS absorption spectra is carried out by using the support of suitable numerical simulations based on quantum chemistry DFT codes developed on the modelled (Ni-cyclam) system as free molecule and also constrained in the pores of mesoporous silica matrices.

1.Introduction

Mesoporous silica functionalized by suitable active vectors are attractive architectures for fundamental understanding of physical phenomena at the interfaces or for potential magneto-optical, controlled drug delivery, bio-sensing, catalytic applications or other molecular engineering [1-7]. A wide synthesis activity has contributed to control the synthesis routines which are now well improved to obtain mesoporous silica with variable pore sizes (2- 20 nm). Additionally, architectures based on several functional groups located either in the channels or in the silica network, were successfully obtained [7]. The incorporation of active groups inside the pores can be performed either by grafting the active molecules to the silica channels walls using organosilica chains or by direct synthesis methods. In this case, the guest groups are expected to be more regularly distributed in the channels leading to a homogeneous media. As matter of fact, the physical properties such as electronic, magnetic or optical, of mesoporous silica based composites can be tuned following the nature of the active groups.

The present work is devoted to cyclam moieties (1,4,8,11-tetraazacyclotetradecane) chelating nickel ions and used as functional groups in the mesoporous silica (SBA-15). Two main synthesis methods were developed based on sol-gel procedure. The performed physical investigations are devoted to shed light on the key features of the different classes of materials with the main criterion of homogeneous materials based on good dispersion of active groups in the host matrices. Experimental investigations of structure and optical properties were carried out by using TEM and UV-VIS absorption. Numerical simulations methods were introduced for the assignment of the absorption bands. The spatial organisation of Ni-cyclam groups inside the mesoporous silica matrices including agglomeration, dispersion or Ni-cyclam pairing, have also a great incidence on the probed properties. This fact justifies our approach based on comparative investigations of two mesoporous silica families.

2.Material Chemistry

Two mesoporous silica functionalized by Ni-Cyclam group were synthesized and referred below as SA and SB series. Both classes of samples had originally the same structure, but the procedure of preparing pre-functionalized silica-chlorine material (Cl-SBA, where SBA is the mesoporous silica) is different. For SA-series samples, chloropropyl groups were attached to mesoporous silica matrix by grafting in already prepared backbones whereas for SB material class, the pre-functionalized silica Cl-SBA samples, were obtained by a direct synthesis.

2.1 Procedure A : pre-functionlized [Cl-SBA$_9$] samples (SA)

The triblock copolymer ($EO_{20}PO_{70}EO_{20}$) Pluronic P123 (5.0 g) was dissolved in 150.0 mL of comparatively high acidic aqueous solution of HCl (2M). After about 1h stirring the clear solution was obtained. Amount of 134 g from the resulting solution was added to a tetraethylorthosilicate (7.50 g or 36.0 mmol). The mixture was vigorously stirred for 45 minutes at temperature of 40°C in hot oil bath in order to start polycondensation of silane groups. After this time a white suspension appeared indicating that a mesoporous material was formed. The second step deals with the grafting of the active groups through the 3-chloropropyltrimethoxysilane (0.79 g, 4.0 mmol) added to a suspension which contains hexagonal arranged mesoporous silica. In order to initialize grafting process, tightly closed flask with liquid was put into oven (90°C) for one night. The resulting white powder was filtered off and the surfactant was selectively removed by soxhlet extraction over ethanol for 24 hours. After drying at 120°C overnight under vacuum , **Cl-SBA$_9$** were obtained as white powder according the scheme depicted in figure 1.

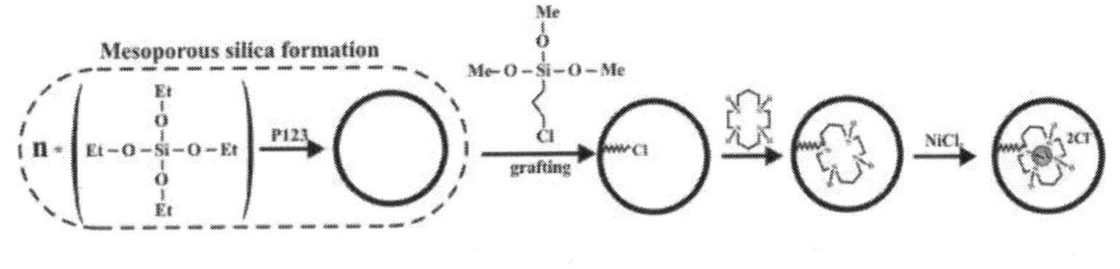

Figure 1: Steps of the mesoporous silica synthesis by procedure A

2.2 Procedure B: Preparation of Cl-SBA$_9$ (SB)

According the the well defined protocol [7], the triblock copolymer ($EO_{20}PO_{70}EO_{20}$) Pluronic P123 (4.0 g) was dissolved in 160 mL of an aqueous solution of HCl (pH=1.5). The resulting clear solution was then added to a mixture of 3-chloropropyltrimethoxysilane **1** (0.84 g, 4.49 mmol) and tetraethylorthosilicate (8.41 g, 40.41 mmol). The mixture was vigorously stirred for 3 hours at room temperature until a transparent solution appeared. The solution was transferred in a hot oil bath at 60°C and NaF (76.0 mg, 1.80 mmol) and then immediately added to induce the polycondensation. A white precipitate appeared within a few minutes and the resulting suspension was further stirred for 2 days at 60°C. The resulting white powder was filtered off and the surfactant was selectively removed by soxhlet extraction over ethanol for 24 hours. After drying at 120°C overnight under vacuum , 2.89 g (94 %) of **Cl-SBA$_9$** were obtained as white powder. The main steps of the procedure were depicted on figure 2.

2.3 Reactivity of chloropropyl groups toward cyclam.

After the above steps, the phase of functionalization aims to locate cyclam groups grafted in the mesoporous silica channels. The substitution reactions within the different solids, obtained by procedures A and B, have been carried out and the preparation of cycl-SBA$_9$ is given below as an example.

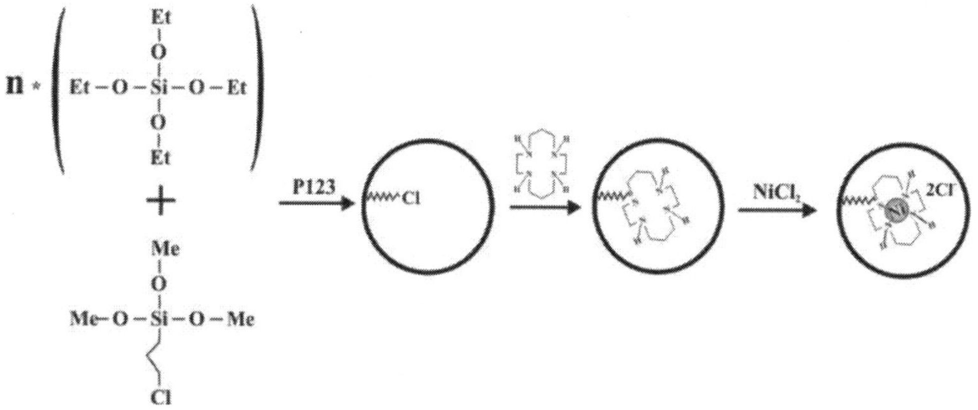

Figure 2. Synthesis of the functionalized mesoporous silica by using the procedure B

2.4 Preparation of cycl-SBA

A mixture of cyclam (0.42 g, 2.09 mmol), Cl-SBA$_9$ (1.00 g, 1.49 mmol), 90 mL of acetonitrile and triethylamine 1.1 g, (10.9 mmol) was heated under reflux and stirring for 2 days. The white solid was quantitatively recovered by filtration and washed 5 times with hot chloroform and 3 times with hot ethanol to remove the excess of cyclam. After drying at 120°C overnight under vacuum, cycl-SBA$_9$ were obtained as a white powder.

2.5 Complexation of cycl-SBA$_9$ by nickel salts

1.00g (1.49 mmol) of cycl-SBA$_9$ was added to a solution of $NiCl_2$ (0.017 M, 123 mL) dissolved in ethanol. The resulting suspension was heated overnight under reflux with stirring. The solid was quantitatively recovered by filtration and washed by dry ethanol several times to remove the excess of nickel salt. After drying at 120°C overnight under vacuum, functionalised mesoporous silica as cycl-SBA$_9^{Ni}$ were obtained as a green powder.

3.Physical characterizations

3.1 Transmission electron microscopy (TEM)

TEM observations were carried out in order to check the homogeneity and morphology of the mesoporous silica. JEOL 2010 equipment with accelerating voltage of 200 kV was used as well as liquid nitrogen anticontamination trap devoted to improve the vacuum around the sample. All TEM images were taken by using a contrast diaphragm. The sample preparation consists in small amount of powder, grinded in pure ethanol. One droplet of the suspension was then deposited on copper grid covered by amorphous carbon membrane. The sample resists quite well to the electron beam and the shape of the mesoporus silica can be resolved in very precise way. As shown in figure 3, elongated grains are clearly seen and testify the achievement of the synthesis of the appropriate mesoporous silica structures.

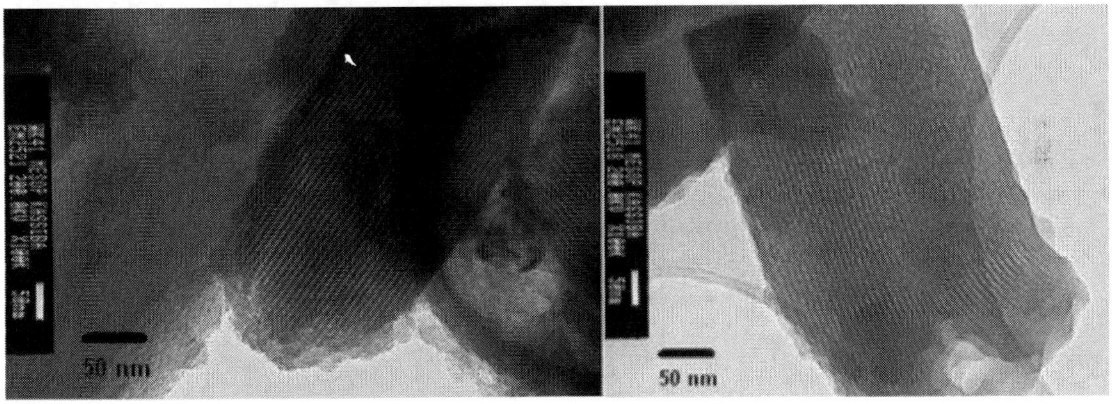

Figure 3. TEM images of mesoporus silica

The morphology of the mesoporous grains is marked by alternating clear and dark fringes with a periodicity about 10 nm. Such contrast reveals the porous structure of the sample with parallel and well aligned pores. However, with regard to the functional (molecular) group dimensions (Ni-cyclam), it wasn't easy to evaluate their presence in the host silica.

3.2 UV-VIS reflectance measurements and band assignment

Room temperature UV–visible diffuse reflectivity spectra were collected on a finely grinded sample with a Cary 5G spectrometer (Varian). This instrument was equipped with a Cary Praying Mantis Diffuse Reflectance accessory and computer control using the "Scan" software. The absorption in diffuse reflectance versus wavelength measurements were made in the 250–800 nm range with a 2 nm step using a mirror as reference (100% reflectance).

Fig. 4. shows the absorption reflectance measurement recorded for the SA and SB samples, pure mesoporous silica as well as bare Ni-Cyclam powders.

Journal of Physics: Conference Series **289** (2011) 012024 doi:10.1088/1742-6596/289/1/012024

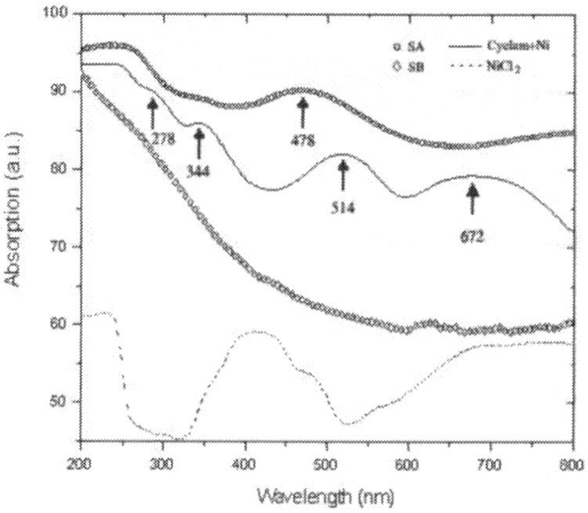

Figure 4. UV-Vis spectra recorded for the mesoporous samples SA and SB, pure mesoporous silica and bare Ni-Cyclam powders.

The bare mesoporous silica did not show any absorption band in the 250 nm – 800 nm region. SB sample exhibits more or less resolved bands located at 250 nm, 377 nm and 463 nm whereas the SA sample is characterized by a large absorption background without clearly defined peaks. For the bare Ni-cyclam powders, the absorption bands are relatively narrow and well resolved at wavelength positions 278 nm, 344 nm, 514 nm and 672 nm. The assignment of the experimental absorption requires an efficient tool as it is briefly outlined below.

3.3 Assignment of the UV-VIS absorption bands

The experimental UV-VIS absorption curves are analyzed by using the support of of numerical simulations. The calculations of the excitation energies and vibrational frequencies were obtained using the RESPONSE module [8] of the Amsterdam Density Functional program (ADF) [9]. Three different molecules were taken into consideration, namely pure cyclam ring, Ni-cyclam with the nickel ion bounded to one chlorine (pyramidal-like configuration) and to two chlorine ions (bi-pyramidal-like). In the first stage of computation, the geometry optimization of the structures was performed assuming C_1 symmetry of all molecules. At the first stage, the Generalized Gradient Approximation (GGA) was employed applying the Becke-Perdew (BP) exchange-correlation (XC) functional [10,11] using a Slater-type orbitals triple-□ (TZ) basis set without the frozen core approximation.

In a second stage, the molecules were rotated to align their ground-state dipole moment along the Z-axis. The UV-vis absorption spectra calculations were carried out by means of the potential model based on statistical averaging of orbital potentials (SAOP) [12]. The SAOP potential was recently used in order to yield improved estimates for the second and third order optical susceptibilities □ and □, as well as excitation energies for prototype molecules such as N_2, C_2H_4, and CO in comparison with the LDA and LB94 potentials. For all considered molecules, the excitation parameters of UV–VIS spectra were calculated by using the time dependent (TD-DFT) method [13-15] implemented in the ADF-RESPONSE code and limiting all the carried out simulations to the singlet–singlet state excitations.

According to this numerical support illustrated in figure 5, it is shown that the pure cyclam molecules without nickel atom exhibit an absorption band near 250 nm while Ni-cyclam bonded to one and two chlorine absorbs respectively at 320 nm and 450 nm. From these results, it is worth noticing that, as the nickel ion is chelated by cyclam and bonded to chlorine ions, the main absorption band is shifted to high wavelengths and exhibits a net broadening.

These facts underline the possibility to finely tune the UV-VIS absorption in the Ni-cyclam based material by a stabilization of relevant configurations defined by the number of bonded chlorine ions. As illustrated in fig. 5, a combination of the three computed spectra obtained on pure cyclam, and Ni-cyclam bonded to chlorine ions give rise to adjustable shape of the absorption spectra. The obtained curve consists in a superposition of contributions about 89.29 % from cyclam-$NiCl_2$, 1.78 % of cyclam+NiCl and 8.93 % pure cyclam without complexation by nickel ions. On the other hand, though the calculated wavelength positions are shifted with respect to the measured ones, the numerical procedure seems sound and relevant. Indeed, the calculation which consider free and isolated molecules in vacuum contrast with a real powder where molecular interactions are involved.

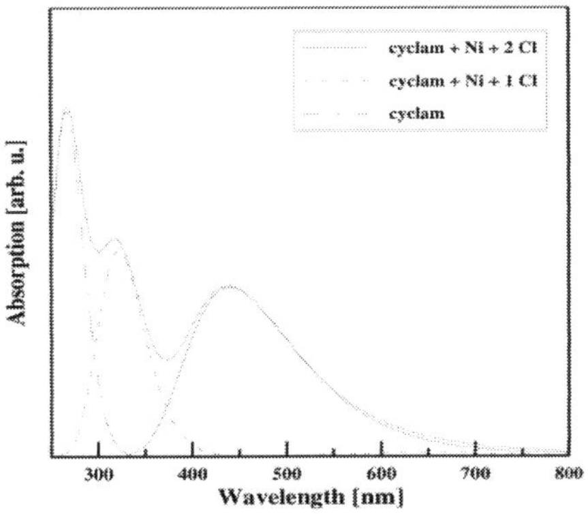

Figure.5. Calculated spectra for bare cyclam – nickel powder.

Furthermore, when the active molecules are grafted inside the mesoporous silica, constraints on the molecular configurations are induced as well as a clustering which occurs in the case of inhomogeneous media. For the SB samples it seems that three configurations of the active groups (pure cyclam, Ni-cyclam bonded to one and two chlorines) are involved. As clearly seen in the SB sample, configurations such as Ni-cyclam bonded to two chlorines dominate the features of the absorption spectrum. Finally, the comparative investigations of the two sample series (SA, SB) and the carried out experiments point out a more homogeneous distribution of active groups in SB sample series compared to SA samples.

4.Conclusions

Two families of mesoporous silica containing chloropropyl groups were prepared by two main approaches in to realize a homogeneous distribution of the functional groups inside the pores. Beyond checking the realization of mesoporous structures, the UV-VIS investigations give rise to characteristic bands which can be used to monitor the achievement of the synthesis process and the effective grafting of cyclam moieties within the channel pores. Computational work was performed and serves as a support for the assignment of the UV-VIS absorption bands. Tests of magneto-optical functionalities constitute challenging applications for all these synthesis work. This task is under development.

Acknowledgment. We are grateful to Dr.S.Kodjikian for TEM images performed on mesoporous silica matrices.

References

[1] Beck J S, Vartuli J C, Roth W J, Leonowiez M E, Kresge C T, Schimitt K D, Chu C T –W, Olson D H, Sheppard E W, McCullen S B,Higgins J B and Schlenker J L 1992 *J. Am. Chem. Soc.* **114** 10834.
[2] Inagaki S, Fukushima Y and Kuroda K 1993 *J. Chem. Soc. Chem. Commun.* 680.
[3] Moller K and Bein T 1998 *Chem. Mater.* **10** 2950.
[4] Yang C M,,Sheu H S and Chao K J 2001 *Adv. Funct. Mater.* **12** 143.
[5] Corriu R 2003 *J.Organometal.Chem.*,**686**, Issues 1-2 32
[6] Corriu R, Mehdi A and Reyé, C 2004 *J.Organometal.Chem*,**689**, 4437
[7] Qunli Tang' Yao Xu, Dong Wu and Yuhan Sun, 2006 *Chem.Lett.* **35,** 474
[8] X.Ji, Hu Q, Hampsey J E, QQiu X, Gao L, He J and Lu Y, 2006 *Chem.Mat.***18**(9) 2265
[9] Slowing I, Trewyn B G, Giri S and Lin V S –Y 2007 *Adv. Fun. Mat* **17,** 1225
[10] Consul J M D, Peralta C A, Benvenutti E V, Ruiz J A C, Pastore H O and Baibich I M, 2006 *J.Mol.Cat.*A **246** 33
[11] Jal P K, Patel S and Mishra B K 2004 *Talanta* **62** 1005
[12] Van Gisbergen, S. J. A.; Snijders, J. G.; Baerends, E. 1999 J. *Com. Phys. Com.* **118,** 119
[13] Baerends E J 1998 *Theor. Chem. Acc.* **99,** 391
[14] Becke A D 1988 *Phys. Rev.* A **38**, 3098;
[15] Perdew J P 1986 *Phys. Rev.* B (1986) **33**, 8822.]
[16] Schipper P R T ,Gritsenko O V, van Gisbergen S J A and Baerends E J 2000 *Chem. Phys.* **112**, 1344
[17] Runie E and Gros E K U 1984 *Phys. Rev. Lett.* **52**, 997.
[18] Gross E K U and Kohn W 1990 *Adv. Quantum Chem.* **21**, 255.
[19] Van Leeuwen R 2001 *Int. J. Mod. Phys.* ***B*** **15**, 1969.

EPR in functional structures based on doped (nano, meso)-porous silica and titanium dioxide

A Kassiba[1], M Makowska-Janusik[2], A Mehdi[3]

[1] Laboratoire de Physique de l'Etat Condensé LPEC, UMRCNRS 6087
Université du Maine, Avenue Olivier Messiaen 72085 Le Mans Cedex 9 France
[2] Institute of Physics, Al Armii Krajowej 13/15, Akademina Im Jana Dlugosz w Czestochowa, 42200, Poland
[3] Institut Charles Gerhardt, UMR 5253 Chimie Moléculaire et Organisation du Solide, Université Montpellier II Place E Bataillon, 34095 Montpellier Cedex 5, France
E-mail: kassiba@univ-lemans.fr

Abstract. EPR investigations are performed on mesoporous silica (SBA15) functionalized by Nickel-cyclam complexes (1,4,8,11-tetraazacyclotetradecane groups chelating nickel ions) and on mesoporous titanium dioxide with nitrogen doping. For functionalized silica, the magnetic behaviour of organometallic groups, their mutual interactions and dispersion in the host matrices are compared with respect to the doping rates and the synthesis procedures. The relaxation processes were analyzed from the thermal evolution of the paramagnetic spin susceptibilities and EPR line-widths. Particularly, some samples show the formation of clusters where phonon assisted one dimensional (1D) ferromagnetic ordering occurs below 45 K. For the mesoporous TiO_2, systematic EPR investigations were performed on two main classes of materials with regard to the porosity degrees. The EPR experiments point out the efficiency of EPR method to probe the degree of functionalization of mesoporous silica or the nitrogen doping achievement in TiO_2, and in general to give a valuable feedback to improve the synthesis routes of smart materials.

1. Introduction

Nanostructures based on functional mesoporous silica or titanium dioxide are attractive materials for versatile applications in catalysis [1,2], bio-sensing [3], drug delivery [4, 5], photovoltaic conversion, CO_2 sequestration or degradation of organic pollutants. Several strategies for the synthesis routes were developed to obtain the well known SBA-15 mesoporous silica with variable pore size ranging from 2 nm – 50 nm. Also, nanoparticles or active molecules with original magnetic, optic or charge transfer properties, can be grafted in the channels or even incorporated in the host matrix backbone [6]. The high specific surfaces ensure a grafting of a large amount of active groups required for enhanced physical responses leading to effective applications. However, the main problem encountered in the functionalization process lies in the formation of aggregates or clusters of the active groups. In this context, by using paramagnetic organometallic molecules as functional vectors in mesoporous silica, the EPR technique was successfully applied to probe these active groups including their local environment, mutual interactions, dispersion in the host matrices and the correct achievement of the material processing. Analyses of the relaxation mechanisms and thermal behavior of the spectral EPR parameters were performed to give information on the intrinsic electronic and magnetic of organometallic groups as well as the degree of functionalization of the silica matrices.

The second class of investigated mesporous materials is based on TiO2 with or without doping agents. By using $TiOCl_2$ as precursor of mesoporous matrix, thiourea for nitrogen doping and P123 as the structuring agent, TiO_2 powders with different degree of ordering and porosity were synthesized [7]. The efficiency of doping and the determination of the electronic features consecutive to the presence of doping agents were investigated by EPR technique. Different paramagnetic species mainly due to oxygen radicals were revealed with the nature and stability dependant on the calcinations temperature up to 500°C. The nitrogen based radicals were identified and their molecular structure and stability are discussed with regard to the sample features from the synthesis and treatment conditions.

2. Experimental

2.1. EPR spectroscopy

EMX Bruker spectrometer, equipped by X-band, was used with variable temperature equipment using an Oxford Instrument cryostat which allows a temperature stabilisation within ± 0.02K from room temperature down to 4 K. Due to the high concentration of paramagnetic species, low micro-wave power (< 2 mW) and low modulation field (< 0.5 Gs) were typically used to record the EPR spectra far from saturation conditions.

2.2. Mesoporous materials

2.2.1. Functionalized silica matrices

Two families of mesoporous silica functionalized by Nickel-cyclam molecules were synthesized according to the process described in references [6]. The sample features is depicted briefly in figure 1 with the main steps such as the realization of mesoporous silica backbones and the grafting of the Nickel-cyclam molecules inside the pores.

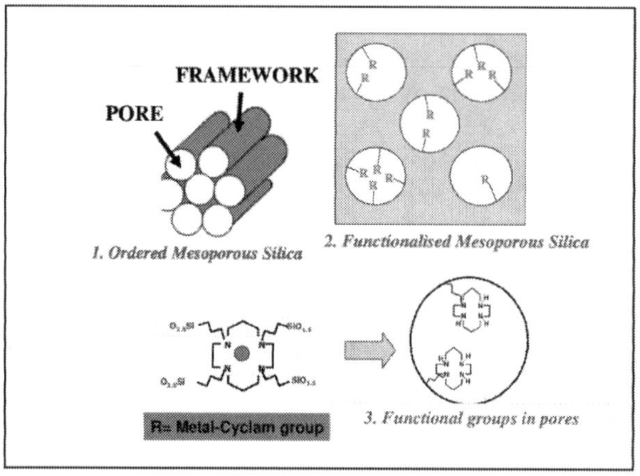

Figure 1. Mesporous silica with the functionalization of the pores by metal-cyclam groups.

Two synthesis procedures, namely A and B, were developed to obtain the functional materials. The main differences lie in the dispersion of active groups (Ni-cyclam) in the host matrices and then different magnetic behaviors are expected for the nickel based molecules. Particularly, the procedure (A) give rise to non- homogeneous grafting while procedure (B) allows obtaining homogeneous distribution of active groups inside the pores. The investigated samples are hereafter referenced as (SA5) from procedure (A) and (SB2, SB5, SB11) from the procedure B. The doping rates by Ni-cyclam are (2%) in SB2 sample, 5% in SA5 and SB5 samples, and 11% in SB11.

2.2.2. TiO_2 based materials

The investigations were carried out on two families of titanium dioxide materials synthesized by sol-gel method. One class is dedicated to mesoporous network (or structured) by using P123 as surfactant and the second one without any porosity is referenced below as non-structured materials. For all samples, the nitrogen incorporation in the TiO2 network was from thiourea. For the main relevant details on the sample, we may below summarize briefly the material features.

Structured and nitrogen doped mesoporous TiO_2. The synthesis of structured samples is based on mixing in ethanol solution, the precursor of TiO2 such as titanium oxychloride, P123 as surfactant and thiourea for nitrogen doping. The formation of gel followed by a calcination at several temperature and time durations lead to powders samples with crystalline phases of TiO2 mainly such as anatase and rutile with the nomenclature SN-TiO_2-400-0.5, SN-TiO_2-400-4.5 and SN-TiO_2-500-12. The first letter "S" denotes that we deals with structured TiO_2 , the second letter "N" is related to Nitrogen doping and the rest standing for the treatment temperature and time (h).

Non-structured and doped mesoporous TiO_2. Similar synthesis procedures were performed to obtain TiO2 without any porosity. In this aim, no P123 surfactant was required and titanium tetraisopropoxide was used as the titanium dioxide precursor. All reactants were dissolved in ethanol which is removed under reduced pressure lading then to the formation of a gel. Calcination of the products was performed at 400°C and 500°C to obtain the samples in the same conditions as for the structured ones. These non-structured samples will be designated hereafter by the following nomenclature N-TiO_2-400-0.5, N-TiO_2-400-4.5 and N-TiO_2-500-12.

Comparative X-ray diffraction investigations were performed on same representative samples. Structured doped or undoped samples exhibit different crystalline orders even under the same calcination conditions [7]. Thus, undoped samples show a coexistence of crystalline polytypes for a short time duration (0.5 h) of calcination at 400°C. At the opposite, nitrogen doped samples under the same treatment conditions reveal only one polytype (anatase) and the superposition of polytypes occurs only for longer time of calcinations (4.5 h). The behavior of the non-structured TiO_2 is also quite different and even for long time of calcinations (4.5 h), only rutile-like order is realized in the sample with only anecdotic anatase content. Also, what can be addressed from such analysis are the different crystalline orders involved in the samples after similar treatments.

Representative samples of the investigated series were annealed at 500°C and their structural features were compared. Thus, the structured or non-structured samples which were treated at 500°C overnight are pure anatase despite doping or structuring. Thus, calcination at 500°C during 12 h was adequate to produce pure anatase out of the various preparations. It should be noted here that during this treatment, the mesostructure is completely destroyed [7].

3. EPR investigations : experiments and discussion

3.1. Mesoporous silica functionalized by Nickel-cyclam

3.1.1. Representative Sample from synthesis procedure A

CW-EPR spectroscopy is used to characterize the effective achievement of the sample synthesis and its correct functionalization by the active groups. Figure 2 reports the thermal evolution of EPR spectra recorded on the sample (SA5) in the temperature range [200K, 5K]. A single EPR line is clearly shown below 200 K and its intensity increases when the temperature is varied from 200 K to 60 K. additionally, a net narrowing of the line width is observed down to 60 K. At lower temperatures, the intensity loss and the development of broad features are indicative of magnetic ordering which will be discussed below.

The EPR signal of nickel ions in the SA5 sample, consists in a single and symmetric line located at the effective g-factor = 2.20(3) at 100 K. This value correspond to nickel ions in the valence state Ni^{3+}, with an effective spin S = 1/2 and the electronic configuration ^2E_g as a ground state. The absence of anisotropy, hyperfine or superhyperfine couplings with nearest ligands such as chlorine or nitrogen ions, suggests the formation of Ni-cyclam clusters in the SA5 sample.

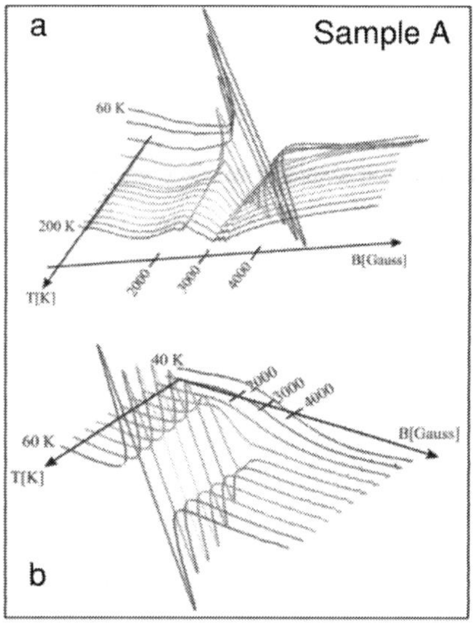

Figure 2. Thermal behavior of EPR spectra [8] in the sample (SA5) obtained by the procedure (A) (see the text) and containing a rate of doping about 5% by Nickel-cyclam groups.

The thermal evolution of the EPR spectra of SA5 sample brings the mark of the relaxation phenomena and the interaction between paramagnetic species. The results concerning the line intensity and line width are summarized in figure 3. The thermal evolution of the EPR line intensity was fitted according to the Curie-Weiss law $\chi(T)=\frac{C}{T-\theta}$. The parameter $(\theta=38K)$ indicates that ferromagnetic exchange interactions hold between spins which can be also represented by a magnetic exchange integral (J) through the relation $\frac{J}{K_B}=-\frac{3.\theta}{2.z.S(S+1)}\approx -38K$. S = 1/2 refers to Ni^{3+} ion and z represents the number of interacting ions which was fixed at z = 2 with regard to the expected molecular arrangements.

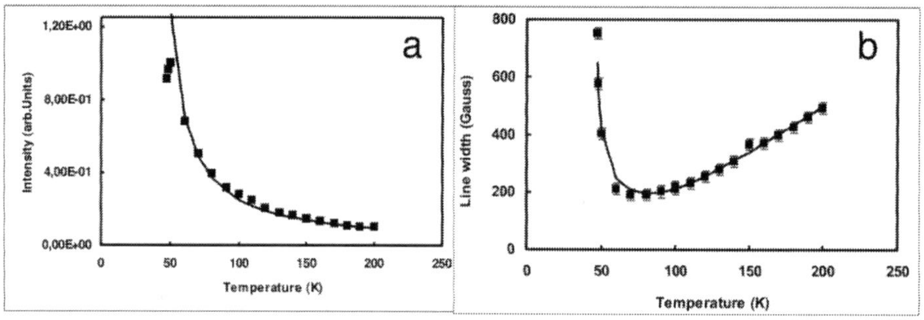

Figure 3. Thermal evolution of the EPR spectrum line intensity (a) and line width (b).

In the temperature range [200 K, 60 K], a monotonous decrease of the line width (ΔH) with the temperature is observed. Below a temperature around 45 K, a divergence of the line width is supported by the occurrence of magnetic transition. Indeed, a magnetic ordering, ferromagnetic or anti-ferromagnetic seems involved in the Ni-Cyclam based clusters and induces a divergence of the correlation length of spin-spin interactions at the critical temperature ($T_c \approx 45K$). The fitting curve of the EPR line width requires two main terms. One is related to a critical contribution behind the line width divergence due to a magnetic ordering and the other from dynamic Jahn-Teller effect mediated by the spin-spin exchange interaction according to the following expression [8, 9]:

$$\Delta H(T) = A(T - T_c)^{-\nu} + B\exp(-\tilde{\Delta}/k_B T)$$

The obtained fitting parameters are:

$$A(Gauss) = 920, \quad T_c = 45\,K, \quad \nu = 0.5, \quad B\,(Gauss) = 2040 \;\; and \;\; \hat{\Delta} = 315\,K$$

The energy barrier $\tilde{\Delta}$, in the order of 220 cm^{-1}, makes us confident in the correct assignment of the relaxation process as monitored by Jahn-Teller effect on the Ni^{3+} ions. However, the amplitude B ~10^{10} s^{-1} cannot discriminate clearly between the phonon-controlled tunnelling or the Orbach-Aminov relaxation phenomena [10]. The relaxation mechanisms are monitored by the dynamic Jahn-Teller effect at high temperature mediated by the exchange interactions between Ni^{3+} ions. Cooperative phenomena seems involved and lead to a magnetic transition below 45K with the features of ferromagnetic-like order.

3.1.2. EPR in representative samples from synthesis procedure B
The EPR spectra recorded at 100K are summarized in figure 4 for SB 2,5,11 samples. At a first glance, the doping ratio is directly traduced on the central and well resolved EPR signal around g = 2.18. The existence of a superhyperfine structure on the EPR spectra indicates the dispersion of Nickel-cyclam groups inside the silica pore with the molecule conformation depicted in figure 5d. The broad resonance absorption background observed in the low magnetic field range is a consequence of the existence of agglomerated groups. From the thermal evolution of the EPR spectra, no critical behaviour was observed on the line intensity, line width or line positions. The isolated Ni-cyclam groups involved in the samples from the procedure (B) didn't show any critical behaviour or magnetic interactions.

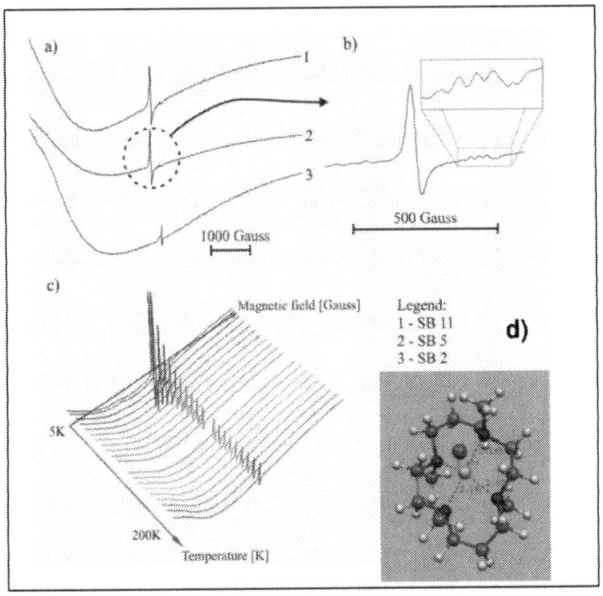

Figure 4. EPR spectra [8] from the samples synthesized by the procedure (B). (a) comparative spectra with different doping rates by nickel-cyclam, (b) magnification of the central part of spectra to resolve a superhyperfine structure, (c) thermal evolution of central part of the spectra and d) the nickel-cyclam molecule.

3.2. Mesoporous titanium dioxide

Journal of Physics: Conference Series **289** (2011) 012025 doi:10.1088/1742-6596/289/1/012025

The performed EPR investigations compare the paramagnetic centres (PC) signatures in the structured and non-structured N-TiO_2 powders. The expected effect is related to changes on the electronic and optical properties of the host matrix. In the synthesized samples, examples of the EPR spectra are shown in figures 5 for the structured sample (figure 5a) and the non-structured ones (figure 5b). As a main PC in the structured samples, O_3^- radicals or alternatively F centres give rise to the characteristic axially symmetric g-tensor whose components are ($g_\perp$ = 2.004, $g_{//}$ = 2.007). The discrimination between these centres is not so easy with regard to previous reports in TiO_2. However, the evolution of the EPR signal with the calcination of the sample at high temperature suggests the relevance of O_3^- radicals. On other hand, no EPR signal was identified as resulting from nitrogen based radicals. This possibility is realized only when the samples are annealed at 500°C during 12 h. In the structured sample, the development of an important EPR signal characteristic of NO radicals occurs in detriment of PC such as O_3^- species. The tendency of nitrogen to be associated to Oxygen and the disappearance of O_3^- suggest the occurrence of NO radicals on the surfaces of voids or pores in the TiO_2 host matrix. The nature of the stabilized NO centres indicate that oxygen molecules can be evacuated from the mesoporous structure after calcination at high temperature (500°C).

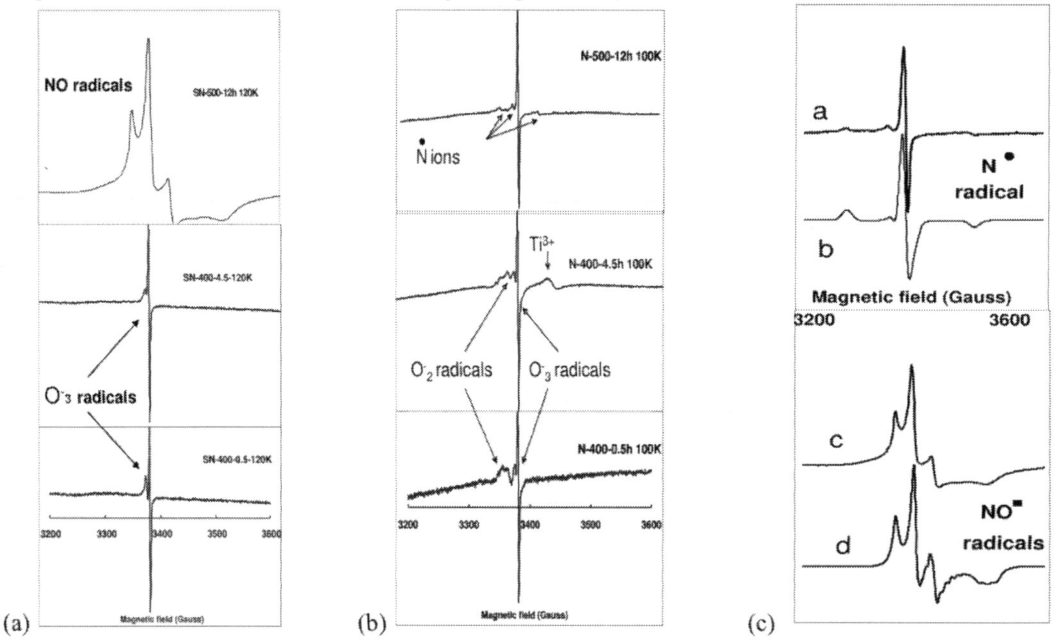

Figure 5. EPR spectra recorded on the structured samples (a), non-structured samples (b) at different calcinations temperatures and time durations. In figure (c) the labels a,d designate the simulated spectra for samples calcinated at 500°C by using spectral parameters related to nitrogen $N^{\bullet}$ and NO^- radicals as explained in the text.

Table 1. Spectral EPR parameters in the nitrogen doped structured and non-structured samples submitted to calcinations at 500°C.

	g_1	g_2	g_3	A_1 ($10^{-4}cm^{-1}$)	A_2 ($10^{-4}cm^{-1}$)	A_3 ($10^{-4}cm^{-1}$)	$\Delta H_1 = \Delta H_2$ (Gs)	ΔH_3 (Gs)
Non-structured	2.003	2.003	2.001	2.3	4.2	30.0	2	3
Structured	2.003	2.000	1.935	~0.0	30.0	10.0	6	20

In the non-structured samples (figure 5b), the EPR investigations show quite different PC with respect to the structured samples. Thus, O_3^- radicals, peroxide radicals O_2^- and Ti^{3+} paramagnetic ions coexist in the non-structured samples annealed at 400°C whatever the time duration (0.5 h, 4.5 h). Nitrogen doping effects is shown in the non-structured samples only when the calcination is performed at 500°C during 12 h [11]. The shape of the EPR spectra, marked by quite narrow lines; points out the good crystalline quality around the nitrogen paramagnetic centres involved in the bulk TiO_2. The simulated spectra (figure 5c) was achieved by using $S = 1/2$ as electronic spin coupled to nuclear spin $I = 1$ as expected for paramagnetic centres involved in the vicinity of nitrogen ions. The following parameters deduced from adjustment of the spectra are summarized in the table 1.

4. Conclusion

The performed EPR experiments give valuable information on the organisation of active groups in mesoporous silica or the doping efficiency in mesoporous TiO_2. For silica based samples, the magnetic behaviour of nickel-cyclam molecules depends on the synthesis procedure. For the procedure named (A), agglomeration of active groups lead to one dimensional magnetic ordering of nickel-cyclam molecules. In the opposite, isolated nickel-cyclam molecules were clearly obtained by the procedure B. For TiO_2 samples, the carried out experiments show the general tendency of the nitrogen doping effect to be efficient only when the calcination is performed at temperature as high as 500°C. The nonstructured samples show the relevance of nitrogen centres being incorporated in the bulk TiO_2 while only NO radicals are involved in the structured samples. These results point out the role of the mesostructure (structured network) to stabilize the effective doping agents in the bulk structure.

References

[1] Cónsul J M D, Peralta C A, Benvenutti E V, Ruiz J A C, Pastore H O and Baibich I M 2006 *J. Mol. Cat.* A **246** 33
[2] Taguchi A and Schüth F 2005 *Microporous Mesoporous Mater.* **77** 1
[3] Slowing I, Trewyn B G, Giri S and Lin V S Y 2007 *Adv. Funct. Mater.* **17** 1225
[4] Tang Q, Xu Y, Wu D and Sun Y 2006 *Chem. Lett.* **35** 474
[5] Ji X, Hu Q, Hampsey J E, Qiu X Q, Gao L, He J and Lu Y 2006 *Chem. Mater.* **18** 2265
[6] Corriu R 2003 *J. Organomet. Chem.* **686** 32
[7] Mei P, Henderson M, Kassiba A and Gibaud A 2009 (to be published)
[8] Laskowski L, Kassiba A, Makowska-Janusik M, Mehdi A, Gibaud A, Errien N and Swiatek J 2009 *J. Phys.: Condens. Matter* **21** 076004
[9] Kassiba A, Hrabanski R, Bonhomme D and Hader A 1995 *J. Phys.: Condens. Matter* **7** 3339
[10] Hoffmann S K, Hilczer W, Gosiar J, Massa M M and Calvo R 2001 *J. Magn. Reson.* **153** 92.

Journal of Physics: Conference Series **289** (2011) 012026 doi:10.1088/1742-6596/289/1/012026

Influence of the weak magnetic field on nanostructure and physical properties of the single crystal Si(111)

L Steblenko[1], O Koplak[1], S Mudry[2]

[1]Department of Physics, Taras Shevchenko National University of Kyiev, Volodimyrska Str.64, MSP 01601 Kyiv, Ukraine
[1] Physics of Metals Department, Ivan Franko National University of Lviv, Kyrylo and Mephodii Str. 8, 79005 Lviv, Ukraine

e-mail: matvieva@yandex.ru

Abstract. The possibility to modify nanostructure of silicon by applying of the weak (B-0.2T) constant magnetic field is proposed. It has been shown that structure and structure sensitive properties of Si undergo essential changes due to the action of a magnetic field. It has been found that relaxation of investigated experimental characteristics after magnetic treatment correlates with a relaxation of metastable states of nanoclusters, which commonly consist in structural defects and restoration of initial structure. Observed changes of physical characteristics are supposed to be connected with occurrence of structure relaxation processes.

1. Introduction

Silicon, as a basic material for microelectronics, is well studied, but taking into account new demands of more high technologies, caused by various reasons and first of all due to the transition of planar production of higher level, its further investigation, especially in single crystal state is needed. On that reason an actual problem is a study of the structure and physical properties of Si/SiO_2 system in order to control their properties by various external influences, in particularly, by weak magnetic field . Unfortunately there are available only few works [1-4] devoted to studies of a weak magnetic field influence on various physical properties of the diamagnetic crystals, in spite of suggestion that magnetic fields of B~ 1T can have effects just by an order of magnitude $(\mu_B/kT)\sim10^{-3}$. Especially, it should be noted that a little of publications consist results on changes of the structurally-dependent properties in semiconductors effected by external magnetic field [5]. On that reason the studies of Si structure and structure-dependent properties changed by an applied magnetic field are important both from fundamental and practical use viewpoints.

2. Exprimental

The n-Si(111) samples under investigation have been doped by P in order to reach the specific resistance ρ=4.5Ω·cm. The surface of sample was covered with SiO_2 oxide layer, of the tick ~2nm. Magnetic treatment (MT) was performed in the weak magnetic field (*B*~0.2T) at room temperature during 7-10 days.

For diagnostics of the crystal structural changes the several methods were used: the X-ray diffraction (Cu*K*α - radiation) measuring the lattice parameter of near surface region: the physical-chemical state of the Si surface was investigated by laser ellipsometry and X-ray photoelectron spectroscopy (XPS) method. Measuring of the energy and intensity of photoelectrons was carried out with using of EC-2402 electron spectrometer, equipped with energy analyzer PHOIBOS-100SPECS (Germany). For excitation of photoelectrons the Mg*K*α X-ray radiation was used (*E*=1252.6 eV).

3. Results and discussion

The magnetic treatment of n-Si(111) causes the non-monotone changes of the inter layers distances *d*, and significant changes of the diffraction peak intensities (Fig.1).

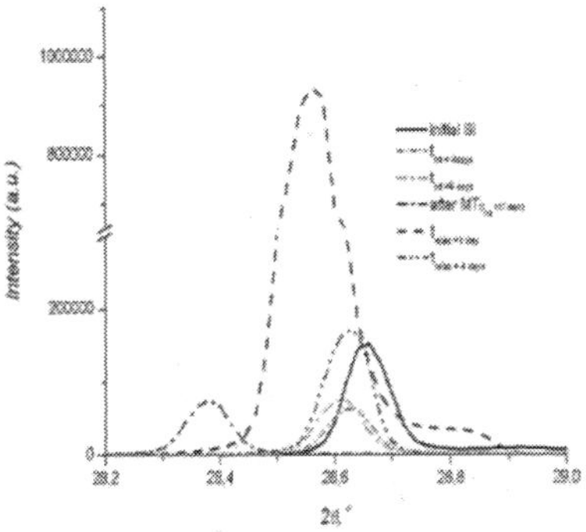

Figure 1. Diffraction peak profile in n-Si(111)

As it can be seen from this figure, the diffraction peaks shift to the less diffraction angel values with increasing magnetic treatment duration. While the magnetic treatment is completed, the considerable increasing of the diffraction peak intensities and corresponding shift of their angular positions to large angle values, i.e. returning to their initial positions. The experimental result is the evidence of the change of resiliently strained state of the near surface layers in Si single crystal. On Fig.2 the dependence of Si relative deformation change as function of magnetic field treatment duration is shown.

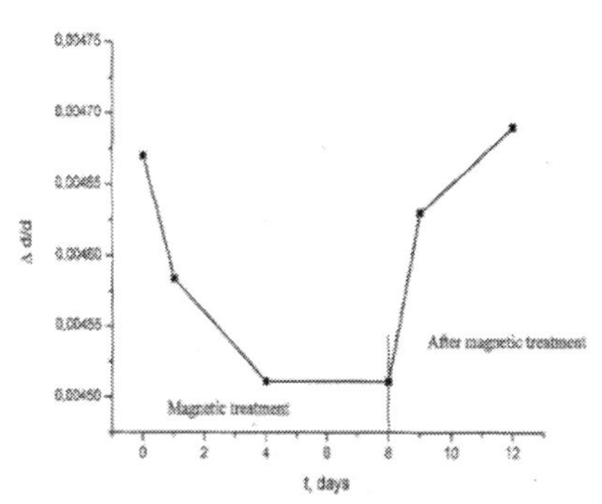

Figure 2. Change of relative deformation Δd/d as function of magnetic treatment duration

Magnetic treatment results in the decreasing of the relative deformation value (3%). After finishing of the magnetic field influencing the relaxation of interplanar distance Δd/d was denoted. This fact indicates that structural changes causing the changing of the relative deformations, and as a result the internal tensions, are reversible. In this work a changing of the lattice parameters when magnetic field treatment duration varies, was also observed. It is important to note, that after finishing of the magnetic field influence, the lattice parameter shows the non-monotone relaxation returning nearly to the initial value. We suppose that changing of the resiliently – strained state in the near surface layers of Si due to magnetic treatment is directly related to the modification of the imperfectly-structural subsystem in the SiO_2 oxides as well as in Si/SiO_2 interface.

The analysis of the experimental data, obtained by means of X-ray diffraction method, reveals the structural changes of the Si crystal in near surface region under act of the weak magnetic field, and on that reason it is of interest to study in more details the physical-chemical state of the Si surface modified by MT.

Using the XPS method the change of native oxide layer thickness was observed. These results allowed us to establish the following experimental facts (Table 1): the oxide layer thickness of the samples which were not under MT (reference samples) had the size of 0.89 nm. After MT during 10 days the SiO_2 layer thickness grew and attained 3.48 nm. Similar results were obtained in [6], where it was shown by means of laser ellipsometry, that action of the impulsive magnetic field results the increasing of the oxide layer thickness.

Table 1. Results of XPS investigation for Si/SiO_2 system

Si crystal	(Si) E_B/I	(SiO_2) E_B/I	Thickness of oxide, d (nm)
n-Si(111) initial	99.8 eV/86.8%	103.5 eV/13.2%	0.89
n-Si(111) t_{mt}=10 days	99.5 eV/50.8%	103.2 ev/49.2%	3.48
n-Si(111) t_{relax}=12 days	99.6 ev/82.1%	103.4 eV/17.9%	1.11

Twelve days later, since the magnetic field treatment was finished, there was decrease of the oxidized layer thickness to the size 1.17 nm. These experimental results show that magnetic treatment activates the Si-surface, increasing the surface chemical activity.

It is known that main factor, responsible for the increasing of magnetic chemical activity is the forming of more defect Si surface. Influence of the weak magnetic field on the real crystal structure can be explained using the ideas [7-9], where authors assume that at such small value energy of the magnetic field (B~1T) existence of magnetic sensitivity effects can be related not to the magnetic field energy but to the electronic spins which are managed by magnetic field. Using the results obtained in [7-9] one can conclude that spin conversion in the nanoclusters as complexes of point defects (CPD) is responsible for the change of diamagnetic crystal properties under act of weak magnetic field. Weakening and break of chemical bonds is the final result of spin conversion in the nanoclusters on the base of point defects, existing in real crystal structure.

Induced the magnetic field decomposition of the Si-O complexes and "vacancy-O" ones promote the migration of the free vacancies to the surface and forming the more relief surface. Due to such mechanism an amount of chemically active centers increases. Further growth of oxide film can be induced by, in our view, the molecular oxygen taken from surrounding atmosphere, diffused to the Si/SiO_2 interface, as well as the atomic one, created after disintegration of the indicated complexes. Significant growth of the oxide layer thickness is the main reason of the mechanical stress changes. The removal of such stress according to [10], take place due to migration of silicon atoms to the Si/SiO2 interface. Laser ellipsometry analysis of the oxide thickness for three investigated states confirmed the results, obtained by means of XPS method.

4. Conclusion

Observed in the present work effects, induced by MT, can be attributed to influence of the weak magnetic field on the stability of the chemical bonds in nanoclusters formed by structural defects. This mechanism provides modification of the structure and properties of Si single crystal.

References

[1] Molotskii M and Fleurov V 1995 Magnetic effects in electroplasticity of metals, *Phys.Rev.* **52** 15829-34
[2] Makara V, Steblenko L, Klochenko Yu, Naumenko S, Patran O , Kravchenko V and Dranenko O 2005 Magnetic – field - induced modification of properties of silicon lattice defects, *Solid State Phenomena*, **108**, 339-344
[3] Golovin Yu I, Morgunow R B and Dmitrievskii A A 2000 Influence of a weak magnetic field on spin-dependent relaxation of structural defects in diamagnetic crystals, *Materials Science and Engineering* A, **288** (2) 261-265
[4] Molotskii M and Fleurov V 1997 Spin effects in plasticity, *Phys.Rev.Lett.* 78 2779-82
[5] Golovin Yu I, Morgunov R B, Baskakov A A, Badylevich M V and Shurmak S Z 1999 Effect of a magnetic field on the plasticity and photo and electroluminescence of ZnS single crystals, *JETP Letters*, **69** 127-132
[6] Levin M N and Zon B A 1999 Magnetic field induced generation of A-like centers in Cz-Si crystals, *Physics Letters* A **260** 386-390
[7] Molotskii M I 2000 Theoretical basis for electro and magnetoplasticity, *Materials Science and Engineering* A 248-258
[8] Levin M I, Semenov V N and Naumov A V 1997 Pulsed magnetic field processing of silicon substrates priror to thermal spray film deposition, *Technical Physics Letters* **27** 89-95
[9] Salikhov K M, Molin Yu N, Sagdeev R Z and Buchachenko A L 1984 Spin polarization and magnetic field effects in radical reactions, ed. Molin Yu N (Elsevier, Amsterdam)
[10] Seah M P 2005 Ultrathin SiO_2 on Si. VI Evaluation of uncertainties in thickness using XPS, *Surf. Interface Anal.* 37 300-309

Journal of Physics: Conference Series **289** (2011) 012027 doi:10.1088/1742-6596/289/1/012027

Changes of magnetic properties of $Co_{78}Si_9B_{13}$ metallic glass in the crystallization process

E Jakubczyk [1], L Krajczyk [2], Z Stępień [1]

[1]Institute of Physics, Jan Długosz University, Al. Armii Krajowej 13/15, 42-200 Częstochowa, Poland
[2]W. Trzebiatowski-Institute of Low Temperature and Structure Research, Polish Academy of Sciences, Okolna 2, 50-950 Wrocław, Poland

e-mail: e.jakubczyk@ajd.czest.pl

Abstract. The crystallization process of $Co_{78}Si_9B_{13}$ metallic glass starts at 648 K and as a result of this the ε–Co(Si) phase with needle morphology is formed. The second stage of the crystallization follows after annealing at 773 K and as a consequence of this the main phases: α-Co(Si) and $(Co,Si)_3B$, $(Co,Si)_2B$ are created. Crystallites of the α-Co(Si) phase have layer morphology. The coercivity increases after the both stages of the crystallization.

1.Introduction

As-quenched metallic glasses are in the metastable state. This metastability is understandable, because the solid state is obtained as a result of rapid solidification. Therefore, the as-quenched metallic glasses possess the structure characteristic of the liquid with short range ordering (SRO).The as-received metallic glasses are in a higher energetic state because atoms during the quick transition from the liquid to the solid state do not take up an equilibrium position, in which they would have lower energy. So, the metallic glasses undergo structural changes first through the structural relaxations of type chemical and/or topological short range ordering (CSRO, TSRO) and next medium (MRO) and finally to long range ordering (crystallization) [1,2]. Under the influence of different stimulating factors (e.g. annealing at different temperatures or isothermal annealing during different time intervals and annealing by contribution of additional conditions as magnetic field, and stress) they evolve with different rate to the stable polycrystalline state [3-6]. Depending on their composition the process of the crystallization proceedes through one or many stages and various types of transformation are possible: primary, eutectic, polymorphous or perytectic [7,8]. The process of structural changes is accompanied by changes in the physical properties. The magnetic properties (coercivity, saturation magnetization, remanence, Curie temperature) are particularly sensitive to the structural changes [9]. Considering the application of the ferromagnetic metallic glasses, it is important problem to specify the conditions in which the amorphous structure is satisfactorily stable in a given time and temperature range.

In this paper are presented the results of the structural changes of the $Co_{78}Si_9B_{13}$ metallic glass on the magnetic properties, which are important applicational parameters, i.e. saturation magnetization, remanence and coercivity. Progressive process of crystallization was stimulated by a controlled annealing. By means of X-ray diffraction and transmission electron microscopy (TEM) the sequence of the creation of the crystalline phases from the amorphous matrix was stated.

2. Experimental

The $Co_{78}Si_9B_{13}$ metallic glass was prepared by the melt spinning technique in Faculty of Materials Science and Engineering of Warsaw University of Technology in Poland.

X-ray diffraction was done at the room temperature for the non-annealed as well as isochronally (4 h) annealed samples at various temperatures (573 K - 823 K) and for the isothermally (648 K) annealed samples during different time intervals (10^4 s - 2· 10^4 s). The annealing was always made for the as-received samples in an inert argon atmosphere. The X-ray studies were performed using a DRON-2.0 diffractometer with a horizontal goniometer of GUR-5 type. The X-ray tube had a molybdenium target $(\lambda_{K\alpha} = 0.71069 \cdot 10^{-10} m)$ and a graphite monochromator in the primary beam.

The microstructure and phase composition were studied by transmission electron microscopy (TEM) using JEOL-JEM 3010 microscope with EDS detector. Thin foils for transmission electron microscopy were prepared by ion polishing.

Partially the investigations of the microstructure and phase composition were studied also by the high resolution transmission electron microscopy (HRTEM) using Philips CM 20 Super Twin microscope. For these studies fine powder sample was prepared by grinding the sample in mortar and dispersing in methanol with ultrasonic agitation. A droplet of suspension was deposited on a microscope grid covered with holey carbon film.

The magnetic properties were determined from the measured hysteresis loops at room temperature using a vibration sample magnetometer (VSM-Lake Shore) operating in a magnetic field up to 2 T. From the hysteresis loops for non-annealed samples as well as annealed at 648 K and 823K the values of magnetic parameters like coercivity, remanence and saturation magnetization were determined.

3.Results and discussion

The values of the magnetic parameters: coercivity $_JH_c$, remanence M_r and specific saturation magnetization M_s [10,11] for non-annealed samples as well as annealed at 648 K and 823 K were obtained from the hysteresis loops (Fig.1) and are listed in table 1.

Table 1.
The magnetic parameters: coercivity $_JH_c$, remanence M_r and specific saturation magnetization M_s for the $Co_{78}Si_9B_{13}$ alloy .

sample	$_JH_c$ [kA/m]	M_s [Am²/kg]	M_r [Am²/kg]
non-annealed	0,15	104,14	0,48
annealed at 648 K	7,34	82,88	6,99
annealed at 823 K	40,43	90,23	19,92

The increase of the coercivity after annealing at 648 K and 823 K correspond to Herzer diagram for the nanocrystallites , according to which: $H_c \propto D^6$ (D – average grain diameter) [12].

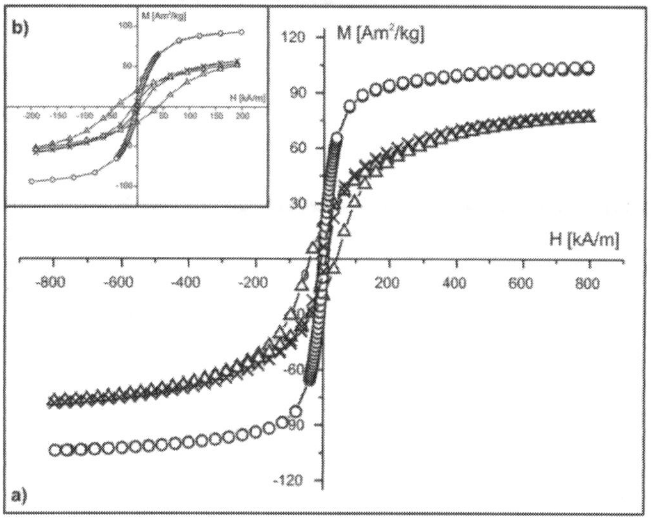

Figure 1. The hysteresis loops for $Co_{78}Si_9B_{13}$ metallic glass:
o – non-annealed,
x – annealed at 648 K for 4 h,
Δ – annealed at 823 K for 4 h.

The obtained values of coercivity and the average grain diameters show the following dependences: $({}_{J}H_{c,823K} / {}_{J}H_{c,648K})^{1/6} = 1,33$ and $\overline{D}_{823K} / \overline{D}_{648K} \approx 1,31$. It agrees with Herzer diagram [12]. The increase of coercivity (49 and 270 times bigger after the first and second stages of crystallization, respectively, than in the non-annealed state) is caused by the characteristic morphology of the created crystallites during crystallization, the needle one after the first stage (Fig.2) and the layer one after the second stage (Fig.3). The pictures of the microstructure obtained by the electron microscopy for the samples after the annealing at the 648 K and 823 K show different morphology of the crystallites. These are crystallites with needle morphology (Fig.2) and a numerous group with layer morphology (Fig.3). The second ones are polycrystallites of fused crystallites from the early stage of crystallization and the ones growing in the second stage of crystallization.

In order to analyse of the structural changes and identify the crystalline phases formed in the transformation of $Co_{78}Si_9B_{13}$ metallic glass from the parent amorphous matrix to the crystalline state besides TEM study the X-ray investigations were performed. The X-ray diffraction made for the non-annealed as well as annealed samples (Fig.4a). The X-ray diffraction performed for the samples annealed at different temperatures (573 K - 823 K) during 4 h allowed to observe the occuring structural changes. It was stated that after the annealing at the temperature 648 K the metallic glass starts the primary crystallization, and the crystallizing phase is the hexagonal ε–Co(Si) [13].This was proved by the X-ray diffraction patterns (Fig.4b) obtained for the samples annealed isothermally at this temperature in different times (10^4 s - $2\cdot10^4$ s).

The phase identyfication proved the existence of the crystalline metallic phase of type α-Co(Si) as well as the boron phases: $(Co,Si)_2B$ $(Co,Si)_3B$ and a small part of the metallic phase of type ε–Co(Si) after the final annealing (823 K) [13, 14]. The creation of the metallic phase α-Co(Si) and boron phases is a result of the eutectic crystallization, which occurs after the annealing at the temperature 773K. The trace amount of the phase ε–Co(Si) can be explained by the polymorphous reaction, which results in transformation of the phase ε–Co(Si) into α-Co(Si) phase. This transformation correlates with the phase diagram Co-Si, which shows that 693 K is the temperature of transformation:
ε–Co(Si) → α-Co(Si) [15].

The typical microstructure of the $Co_{78}Si_9B_{13}$ alloy after annealing at 823 K is presented in the Fig.3. and this characteristic image proves that the alloy possesses of the full crystallization. However, the investigations by means of the high resolution transmission electron microscope (HRTEM) point that in the volume of the alloy the rest of the amorphous phase exist (Fig.5).

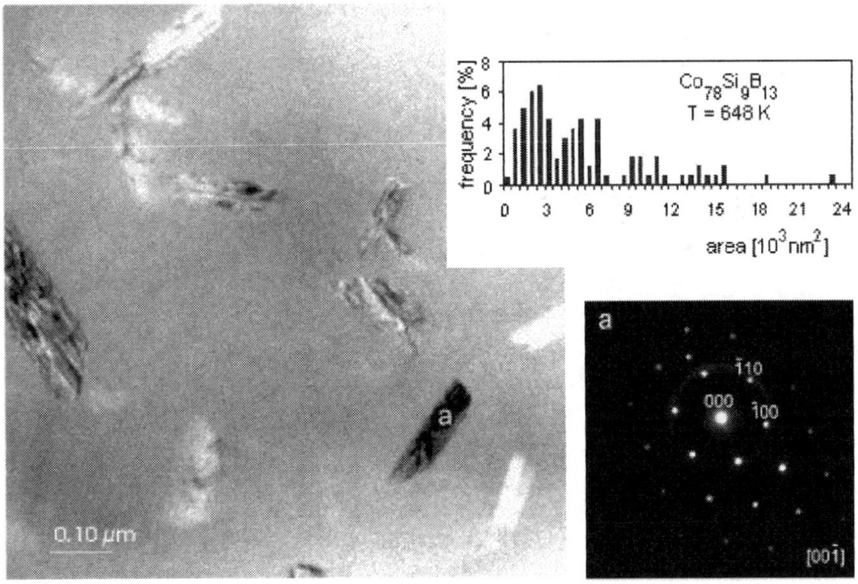

Figure 2. TEM micrograph for the sample of the $Co_{78}Si_9B_{13}$ alloy annealed at 648 K for 4 h and grain size distribution, a – the electron diffraction pattern for ε–Co(Si).

Figure 3. TEM micrograph for the sample of the $Co_{78}Si_9B_{13}$ alloy annealed at 823 K for 4 h and grain size distribution, a – the electron diffraction pattern for Co_2Si.

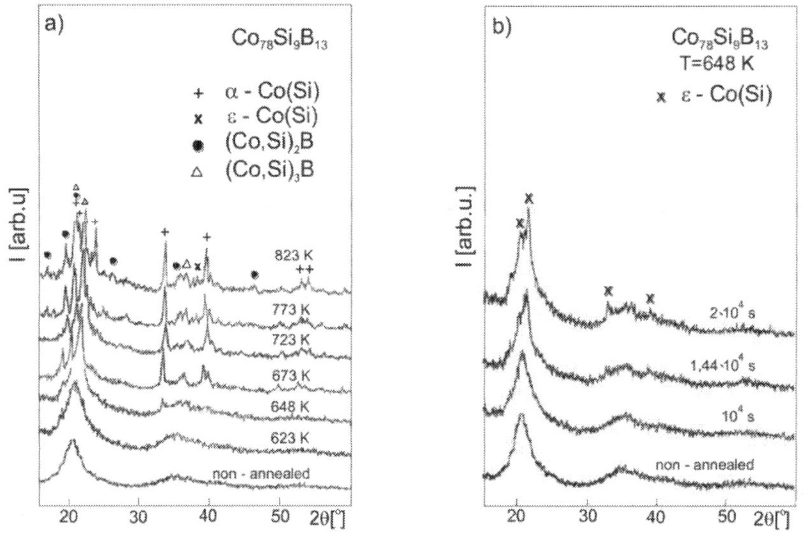

Figure 4. X-ray diffraction patterns for samples of the $Co_{78}Si_9B_{13}$ alloy annealed: at different temperatures (a), during different time intervals at 648 K (b).

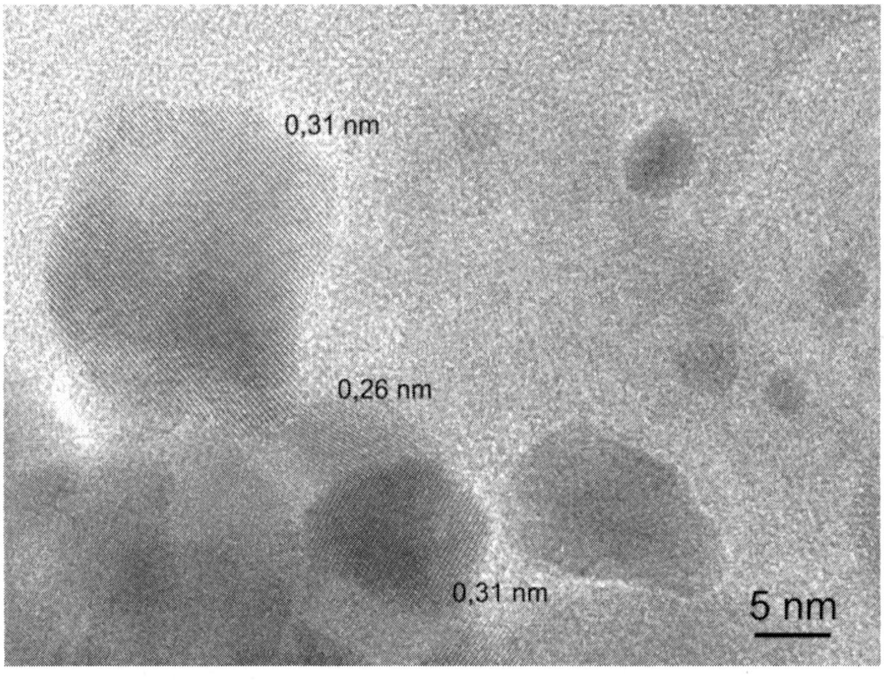

Figure 5. HRTEM image for the sample of the $Co_{78}Si_9B_{13}$ alloy annealed at 823 K for 4 h.

In the Fig.5. in the amorphous background are the visible nanocrystallites between 2,5 and 25 nm sizes. The phase analysis of them was carried out by FFT (Fast Fourier Transform) patterns obtained from the HRTEM image of the nanocrystallites proves that these are CoB compounds [16].

4. Conclusions

- The crystallization process of $Co_{78}Si_9B_{13}$ metallic glass proceeds through two main stages at 648 K and 773 K.
- The hexagonal metallic phase ε-Co(Si) formed during the first stage of crystallization transforms through the polymorphous reaction into regular phase α-Co(Si).
- Both stages of the crystallization are accompanied by different morphologies of the crystallites: needle and layer in the first and the second stages of crystallization, respectively.
- The characteristic morphology of the crystallites is responsible for the increase of coercivity after the first and the second stages of crystallization.
-HRTEM study for the samples of the alloy annealed at 823 K for 4 h proves that in the volume of the alloy the rest of the amorphous phase exists.

Acknowledgements

The autors wish to thank for experimental support Professor J. Lelątko in TEM measurements and dr K. Pawlik in VSM measurements.

References

[1] Van den Beukel A and Radelaar S,1983, On the kinetics of structural relaxation in metallic glasses , *Acta Metal.*, 31, 419-427.
[2] Van den Beukel A, 1993, On the kinetics of structural relaxation in metallic glasses, *Key Eng. Mater.*, 81-83, 3-16.
[3] Kolano R, Kuźmiński M, Kawior W and Wójcik N., 1994,, Induced transverse magnetic anisotropy and domain structure in Co-based amorphous ribbons, *J. Magn. Magn. Mater.*, **133**, 321-324.
[4] Jakubczyk E and Kolano R., 2000, Induced transverse magnetic anisotropy in Co-based amorphous alloys, *Visnyk Lviv Univ.*,Ser. Physic, **33**, 233-239.
[5] Ślawska-Waniewska A, Siemko A and Lachowicz H,1991, Some aspects of simultaneous stress- and field-annealing of metallic glasses, *J. Magn. Magn. Mater.*, **101**, 40-42.
[6] Chanrda Shekar N V and Govinda Rajan K, 2001, Kinetics of pressure induced structural phase transitions- A review, *Bull. Mater. Sci.*, **24**, 1-21.
[7] Köster U,1993 , Phase transformations in rapidly solidified alloys, *Key Eng. Mater.*, **81-83**,647-662.
[8] Köster U and Herold U, 1981,*Crystallization of the metallic glasses*, (Topics in Applied Physics,vol.46: *Glassy Metals I*, ed. Güntherodt H.J, Beck H, Springer Verlag)
[9] McHenry M E, Willard M A and Laughlin D E, 1999, Amorphous and nanocrystalline materials for applications as soft magnets, *Progress in Materials Science*, **44,** 291- 433.
[10] Zeisberger M, Dutz S, Müller R., Hergt R, Matoussevitch N and Bönnemann H, Metallic cobalt nanoparticles for heating applications, 2007, *J. Magn. Magn. Mater.*, **311,** 224- 227.
[11] Vojtaník P, Andrejco R, Varga R, Kováč J, Csach K and Lovas A, 2004, Magnetic and structural properties of amorphous Co-Cr-Si-B alloys, **54**, D113-D116.
[12] Herzer G, 1990, Grain size dependence of coercivity and permeability in nanocrystalline ferromagnets, *IEEE Transactions on Magnetics*, **26**, 1397-1402.
[13] Taylor A, Floyd R W, 1950, Precision measurements of lattice parameters of non-cubic crystals, *Acta Crystallographica*, **3**, 285-289.
[14] Havinga E E, Damsma H and Hokkeling P,1972, Compounds and pseudo- binary alloys with the $CuAl_2$(C16)-type structure., I. Preparation and X-ray results, *Journal of the Less-Common*

Metals, **27**, 69-186.
[15] Massalski T, Binary Alloy Phase Diagrams, American Society for Metals, Metals Park,Ohio 44073.
[16] Gianoglio C, Quadrini E, 1978, Cobalt boride compounds, *Metallurgia Italiana*, **70**, 122-126.

Journal of Physics: Conference Series **289** (2011) 012028 doi:10.1088/1742-6596/289/1/012028

Luminescence centers in $Y_3Al_5O_{12}$:La single crystals

Y Zorenko, T Zorenko, T Voznyak

Laboratory of Optoelectronic Materials, Department of Electronics, Ivan Franko National University of Lviv, 107 Gen. Tarnavskyj Str., 79017 Lviv, Ukraine

E-mail: zorenko@electronics.wups.lviv.ua

Abstract The nature of luminescence centers created by La^{3+} isoelectronic impurities (*IIs*) in $Y_3Al_5O_{12}$ (YAG) crystals has been considered. The La^{3+} dopant in YAG crystals gives rise to intensive UV luminescence in the complex band peaked at 4.28 eV at 300 K with a decay time of the main component of 575 ns. It has been shown that the complex UV emission band in YAG:La crystals in the 10-300 K range is a superposition of the luminescence of excitons localized near La dopants (LE(La) centers), the recombination luminescence of La^{3+} ions in the positions of Y^{3+} cations (La_Y centers) and the low-intensive recombination luminescence of Y_{Al} antisite defects in the bands peaked at 4.65-4.62, 4.29-4.25 and 3.86-3.84 eV, respectively. In the visible range the emission spectra of YAG:La crystals also consist of the luminescence of F^+ and F-centers in the bands peaked at 3.1 and 2.69 eV with a decay time of the main components of 2.3 ns and in the µs-ms range, respectively. It has been shown that the luminescence of F^+ and F-centers in YAG:La crystals can be excited by emission of LE(La_Y) and La_Y centers.

1. Introduction

Isoelectronic impurities (*IIs*) with respect to core cation (for example, La^{3+} and Y^{3+}, Lu^{3+} or Sc^{3+} and Al^{3+}) can effective localize low energy excitation (electrons and holes or excitons) at the expense of the non-Coulomb potential arising at substitution of core cation by *IIs* [1, 2]. For this reason *IIs* are being widely used for creation of luminescence materials based on Al_2O_3-Y_2O_3-Lu_2O_3 oxides of different structural types with radiation in the UV range (250-350 nm) [1, 3, 5-10]. One can also refer to analogues of *IIs* the antisite defects (AD) of A_B type (A^{3+} cations in sites of B^{3+} cations) which are dominant type of defects in single crystals of complex oxides with the $A_3B_5O_{12}$ garnet and ABO_3 perovskite structure at their high-temperature (1800-2000 °C) crystallization from melt by the Czochralski and Bridgman methods [1, 2, 11-15].

Among the whole set of ions of III group, the Sc^{3+} and La^{3+} *IIs* possess the highest efficiency of radiative recombination in the Al_2O_3-Y_2O_3-Lu_2O_3 oxides of different structural types [2]. In this connection the luminescence of these two impurities has been studied in single crystals and single crystalline films of Al_2O_3:Sc sapphire [1, 3] and $Y_3Al_5O_{12}$:Sc [1-4], $Lu_3Al_5O_{12}$:Sc [6-9] and $Lu_3Al_5O_{12}$:La [9] garnets, as well as in single crystalline films of $YAlO_3$:Sc and $YAlO_3$:La perovskites [10]. At the same time, the luminescence of La^{3+} *II* in crystals and films of such widespread garnet as $Y_3Al_5O_{12}$ (YAG) has not been earlier investigated at all. Therefore, the aim of this work was detailed study of the nature of the luminescence formed by La^{3+} *IIs* in YAG crystals under excitation by synchrotron radiation (SR).

2. Samples and experimental technique

YAG:La crystals were grown by the Czochralski method in Ar +1.5 % O_2 atmosphere from the charge of 4N purity placed in Ir-crucible in the Institute of General Physics of RAS. The concentration of La_2O_3 impurity was 0.6 mole % in the melt. The La concentration in the crystal was 0.085 at. %; thus the segregation coefficient of La dopant amounted to 0.14. It should be noticed that large (1.165 Å) La^{3+} ions are localized in YAG crystals exclusively in the dodecahedral sites of Y^{3+} cations (1.01 Å)

Journal of Physics: Conference Series **289** (2011) 012028 doi:10.1088/1742-6596/289/1/012028

forming La_Y centers [16]. At the same time, Y^{3+} cations besides the dodecahedral sites of garnet lattice can also substitute for the octahedral sites of Al^{3+} cations (0.53 Å) forming Y_{Al} AD (0.90 Å) [1, 2, 11-15]. This process is substantially stimulated by entering La^{3+} large ions in the regular sites of Y^{3+} cations what results in the so-called effect of "displacement" Y^{3+} cations in the octahedral sites of Al^{3+} cations with smaller dimensions [11].

Investigation of the luminescence of LuAG:La crystals at 10 and 300 K was performed under excitation by SR with an energy 3-25 eV at Superlumi experimental station in HASYLAB at DESY (Hamburg, Germany). The emission and excitation spectra were measured in both the integral regime and the 1.2-6 ns, 150-200 ns time intervals (fast and slow components, respectively) in the limits of SR pulse with a repetition time of 200 ns. The decay kinetics of the luminescence was measured in the time ranges 0-200 ns at 10 K. The excitation spectra were corrected for the spectral transmittance of Al-grating and the intensity of SR beam; but emission spectra were not corrected.

3. Experimental results and discussion

The La^{3+} dopant causes in YAG crystals the intensive UV luminescence in the complex band peaked at 4.28 eV at 300 K (Fig.1 a). By analogy with studying LuAG:La single crystalline films [9], for analysis of the structure of this band in YAG:La crystals we have used the luminescent time-resolved spectroscopy under excitation by SR in the range of fundamental absorption of this garnet.

The the luminescence spectra of YAG:La crystals in the UV range at 10 K (Fig.1a) and 300 K (Fig.1b) under excitation by SR the YAG exciton range (7.6-6.45 eV) and in the range of the interband transitions (13.75 eV) contain superposition of two intensive bands with E_{max}=4.65-4.62 and 4.29-4.24 eV, which by analogy with LuAG:La films [9] can be related to the emission of the excitons localized near La_Y ions (LE(La) centers) and the recombination luminescence of La_Y centers, respectively, as well as the low-intensity emission band of Y_{Al} AD with E_{max}=3.885-3.84 eV [17]. The luminescence of La_Y center arises at the radiative recombination of electrons, preliminary localized at La_Y centers, with holes in valence band [9], whereas the luminescence of Y_{Al} AD is caused by the recombination of holes preliminary localized at these centers with electron of conduction band [1, 2]. In both these cases there occur formation and radiation relaxation of excitons bound on La_Y and Y_{Al} centers [9, 17]. The above mentioned components of UV luminescence of YAG:La crystals can be spectrally resolved only at 10 K (Fig.1b), whereas at 300 K certain structure of complex band of the UV luminescence can be observed only at the analysis of difference in emission spectra of YAG:La crystals obtained at different energies of excitation (Fig.1a, curve 3).

In the longer wavelength range the emission bands with E_{Max}=3.1 and 2.69 eV is also observed in the luminescence spectra of YAG:La crystals (Fig.1a) which are caused by the luminescence of F^+ and F-centers (an oxygen vacancy with trapped one or two electrons, respectively) [18].

The excitation spectra of different UV luminescence centers in YAG:La crystals in the exciton range at 10 K (Fig.2a) and especially at 300 K (Fig.2b) are complex superposition of excitation bands of the luminescence of LE(La), La_Y and Y_{Al} centers. Meanwhile, in the excitation spectra of the luminescence of YAG:La crystals at 10 K at registration of the luminescence in the vicinity of the corresponding emission bands we could spectrally extract two excitation bands of the LE(La) centers luminescence with E_{max}= 6.95 and 7.64 eV (Fig.2, curve 1) and La_Y centers with E_{max}= 6.645 and 7.33 eV (Fig.2, curve 2). These bands correspond to two excited states of excitons (most probably singlet and triplet states) bound with the corresponding luminescence centers. Due to small intensity of the emission of Y_{Al} AD at 10 K, the dominant band of the luminescence of these centers was observed in the form of a bump with E_{max}=6.83 eV in the excitation spectrum of the luminescence of LuAG:La crystals monitored at 3.81 eV at 300 K (Fig.2a, curve 3). However, due to large overlapping of the excitation spectra of different emission centers in the exciton range the position of high-energy excitation bands of the Y_{Al} AD centers luminescence could not be determined.

The excitation spectra of the F^+ and F-centers luminescence in YAG:La crystals are shown in Fig. 2c. The excitation spectrum of the F^+-center luminescence (curve 1) contains the low-intensity band with E_{max}=4.165 eV and the dominant band at 4.9 eV in the YAG transparency range, which correspond to the radiative 1B→1A transition in these centers [18], as well as two intensive bands in the exciton

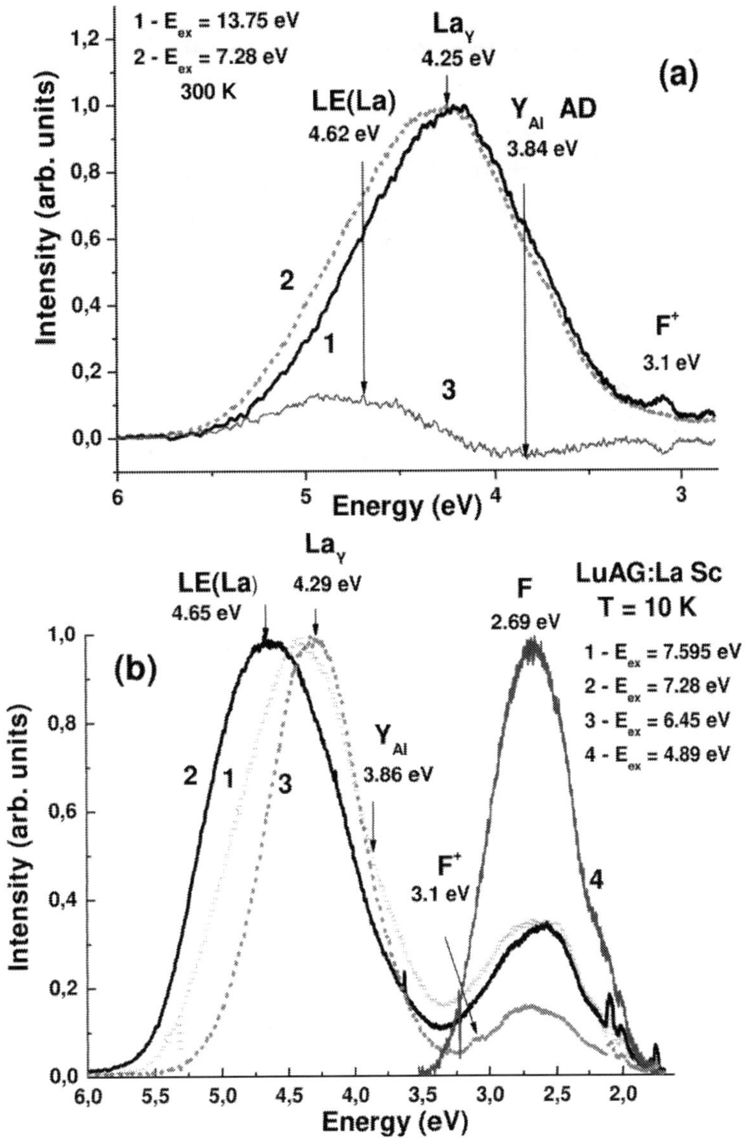

Figure 1. Luminescence spectra (normalized) of LuAG:La crystals at 300 K (a) and 10 K (b) under excitation by SR in the range of interband transition (13.75 eV) (1a), in the exciton range at 7.595 eV (1b), 7.28 eV (2a, 2b) and 6.45 eV(3b) as well as in the range of F-center absorption (4.9 eV) (4b). Curve 3a shows difference between emission spectra 2a and 1a.

range with E_{max}=6.57 and 6.93 eV. The excitation spectrum of the F-centers luminescence in the YAG transparency range besides of the intensive band with E_{max}=4.9 eV contains also the spectrally-resolved band with E_{max}=5.37 eV corresponding to the radiative $^3P \rightarrow {}^1S$ transitions in these centers. The excitation spectrum of the F-center luminescence in the exciton range contains two intensive bands with E_{max}=6.57 and 6.93 eV and the dominant band with E_{max}=7.3 eV. Most probably, the position of the last excitation band corresponds to the energy of formation of exciton bound with F-center. On the whole, the positions of maxima of the excitation band of F^+ and F-center luminescence in YAG:La crystals well correlate with the analogous spectral characteristics of these centers in Al_2O_3 [19] and $YAlO_3$ [18] crystals with taking into account the difference in the forbidden gap of these oxides.

It should be noted that large overlapping of excitation spectra of the F^+ and F-centers luminescence (Fig.2c) can be an evidence of the mutual photo-conversion of $F^+ \leftrightarrow F$ center in YAG crystals which has been earlier observed in Al_2O_3 [19] and $YAlO_3$ [18] crystals. It is also worth to note that the excitation bands of the F^+ and F-center luminescence in YAG:La crystals in the 4-5 eV range

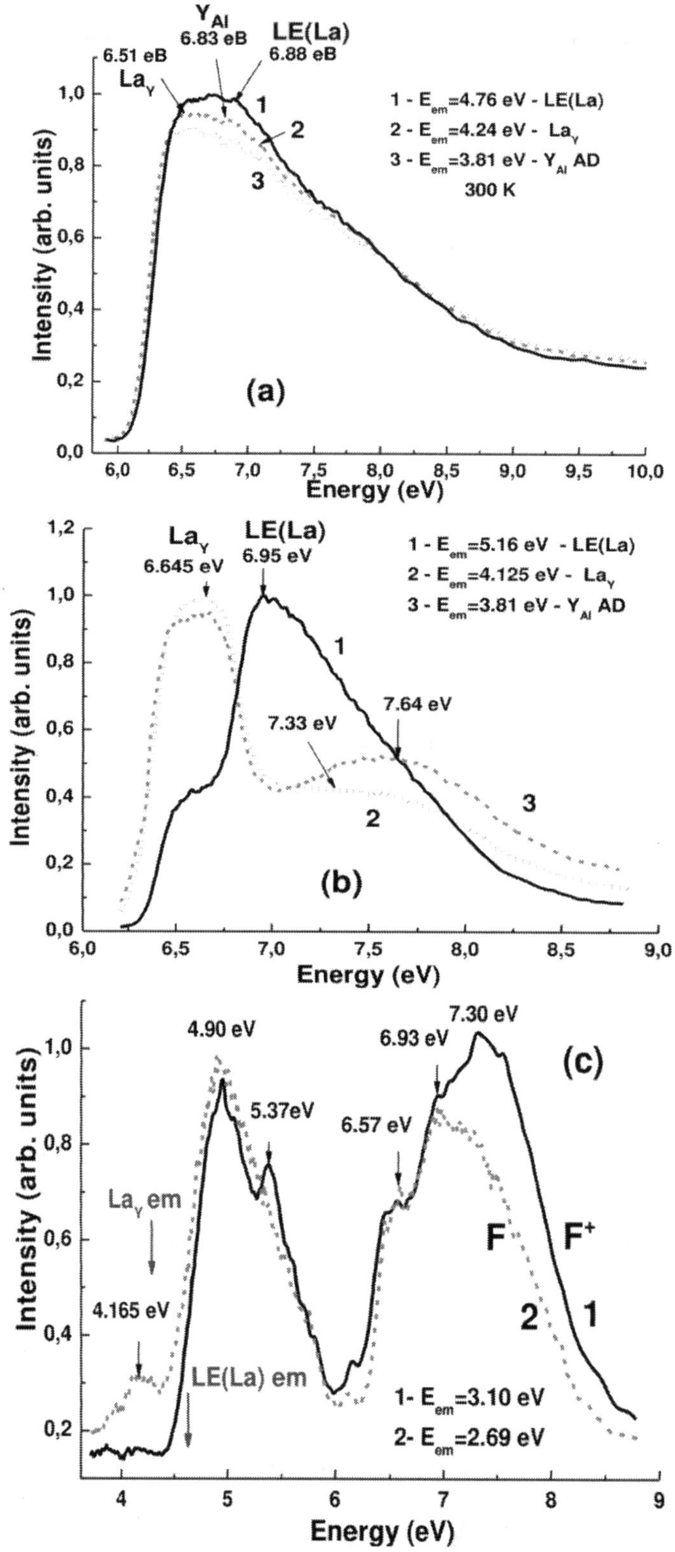

Figure 2. Excitation spectra (normalized) of LE(La) (1a,1b), La_Y (2a, 2b), Y_{Al} (3a, 3b), F^+ (1c) and F-centers (2c) luminescence monitored at 5.15 eV (1b) and 4.76 eV (1a), 4,24 eV (2a) and 4.125 eV (2b), 3.81 eV (3a, 3b), 3.1 eV (1 c) and 2.69 eV (2c), respectively, at 300 K (a) and 10 K (b, c) in LuAG:La crystals. The maxima of emission bands of LE(La) and La_Y centers are indicated by arrows in Fig.2c.

overlap to a great extent with the emission bands of LE(La) and La_Y centers (Fig.2c). This results in excitation of the F^+ and F-center luminescence by the emission of the centers formed by La^{3+} *II*. This fact is supported by the closeness of maxima of excitation bands of the F^+, F-centers luminescence and luminescence bands of LE(La), La_Y centers in the 6.5-7.0 eV range (Fig.2c). Similar effect of the excitation of F^+- and F-center luminescence by the emission of Sc^{3+} *II* has been earlier observed by us in Al_2O_3:Sc crystals and caused large deterioration of selectivity of the emission of this phosphor under long-duration excitation by electron beam [3].

The decay kinetics of different components of the UV luminescence of YAG:La crystals is shown in Fig.3a. All the decay curves have the similar shape. Generally, the decay curves of the UV luminescence of YAG:La crystals can be presented as a superposition of three components caused by the transition from the excited states of the LE(La) centers, and excitons bound with the La_Y ions and Y_{Al} ADs. At the same time, using the appropriate energies of excitation and registration of luminescence at the maxima of excitation and emission bands of different centers, it is possible to separate at 10 K the dominant emission decay of LE(La), La_Y and Y_{Al} centers (Fig.3a). The parameters of two-component fitting of the corresponding decay curves are summarized in Table 1.

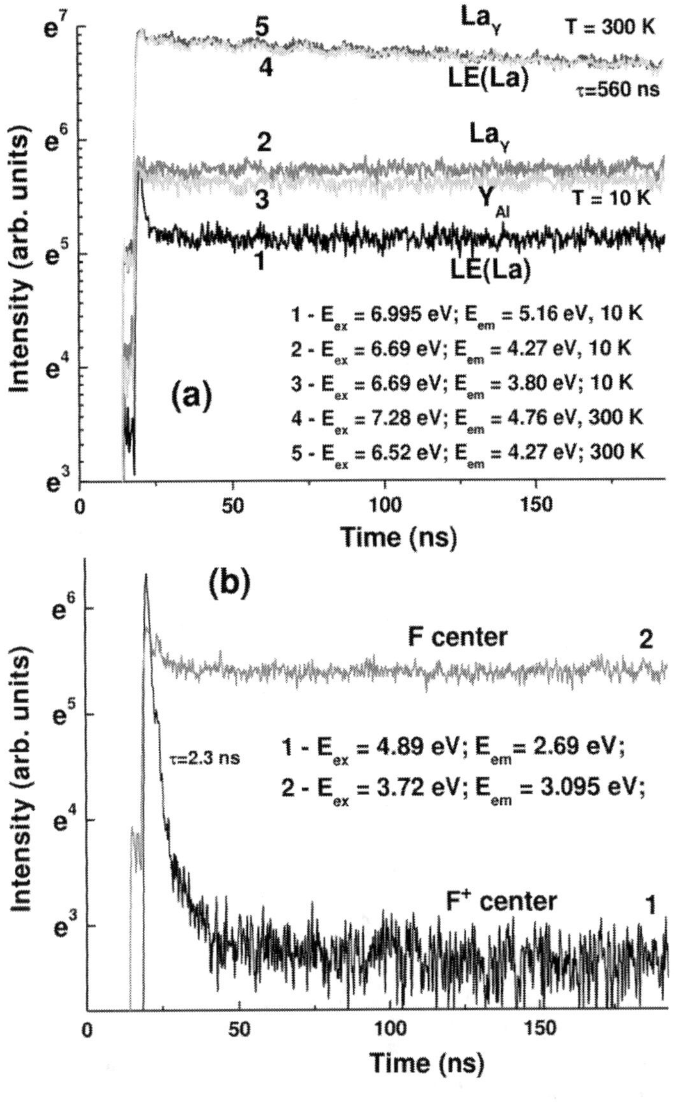

Figure 3. Decay kinetics of luminescence of LE(1a, 4a), La_Y (2a, 5a), Y_{Al} AD (3a), F^+ (1b) and F-centers (2b) in LuAG:La crystals at 10 K (1a-3a, 1a, 2b) and 300 K (4a, 5a) under excitation by SR with different wavelengths in characteristic points of excitation spectra in Fig.2.

Table 1. The parameters of two-component fitting of the decay curves corresponding to luminescence of different centers in YAG:La crystal at 300 K and 10 K

Type of center	τ_1, ns	A_1	τ_2, ns	A_2
1 – LE(La), 300 K	1.3	166	6560	139
2 - La_Y, 10 K	4.4	12	9.9×10^6	246
3 - Y_{Al}, 10 K	4.5	28.6	11×10^6	219
4 – LE(La), 300 K	0.6	115.6	550	377
5 - La_Y , 300 K	0.65	49.28	575	473
1 - F^+ - center, 10 K	2.3	525.5	488	15
2 - F – center, 10 K	4.6	111	12970	181

As can be seen from Table 1, the complex decay kinetics of all components of the UV luminescence of YAG:La crystal with the characteristic decay times of components in the ns and hundred ns ranges suggests the radiative relaxation from the *singlet* and *triplet* esxcited states of different types of the excitons related to the above mentioned centers. This result well matches to the data for the decay of different emission components of the intrinsic luminescence in single crystals of $A_3B_5O_{12}$ (A=Y, Lu, La, Gd; B=Al, Ga) garnets [17, 20].

At the room temperature range the decay kinetics of LE(La), La_Y and Y_{Al} centers are strongly accelerated (Fig.3a). Specifically, at 300 K the main slow component of the La_Y center luminescence decays with a time of about 575 ns (Fig.3a, curve 4).

The decay kinetics of F^+- and F-centers luminescence under excitation in the respective intrinsic transition of these centers, are shown in Fig.3b. Due to allowed character of the $^1B\rightarrow^1A$ singlet-singlet transition in F^+-centers [18, 19], the main component of the luminescence of these centers decays with a time of 2.3 ns (Fig.3b, curve 1). This value well correlates with the decay time of the F^+-centers luminescence in $YAlO_3$ crystals [18]. Due to forbidden character of the $^3P\rightarrow^1S$ triplet-singlet transition in F-centers [18, 19], the main component of the luminescence of these centers decays in the ms range (Fig.3b, curve 2). Meanwhile, due to short time interval for the luminescence decay measurements (200 ns), the decay time of this component can not be evaluated anyway in our experiment.

4. Conclusions

The nature of luminescence centers created by the La^{3+} isoelectronic impurities in $Y_3Al_5O_{12}$ (YAG) single crystal grown from melt with La^{3+} content of 0.085 at. % has been considered. The UV emission of YAG:La crystal in the 10-300 K range is a superposition of the luminescence of the excitons localized near La^{3+} dopant in the positions of Y^{3+} cations (LE(La_Y) centers), the recombination luminescence of La_Y centers as well as the low-intensive recombination luminescence of Y_{Al} antisite defects in the bands with maxima at 4.65-4.62, 4.29-4.24 and 3.86-3.84 eV, respectively.

The energy structure of the mentioned emission centers in YAG:La crystal is determined. The presence of two peaks in the excitation spectra of the LE(La) center luminescence (6.95 and 7.64 eV) and La_Y centers luminescence (6.645 and 7.33 eV), as well as the complex decay kinetics of their emission with the decay time of components in the ns and hundred ns ranges, respectively, suggest the radiative relaxation from the *singlet* and *triplet* state of excitons related to these centers in YAG:La crystals. The above mentioned results are consistent, on the whole with the results of studying the nature of the luminescence of isoelectronic impurities and antisite defects as special analogues of such dopants in single crystals of Al- and Ga-garnets [9, 10, 17, 20]. This indicates the universality of a model approach for the radiative relaxation of low-energy excitation in such a class of luminescent materials based on garnets compounds [2, 17, 20].

In the visible range the emission spectra of YAG:La crystal also consist of the luminescence of F^+ and F-centers in the bands peaked at 3.1 and 2.69 eV with a decay time of the main components of 2.3 ns and in the μs-ms range, respectively. The overlapping the excitation band of the F^+ and F-center luminescence in the 4.5-6 eV and 6.2-7.3 eV ranges presupposes the mutual photo-conversion of these centers. The excitation spectra of the F^+ and F-center luminescence in YAG:La crystals in the 4-5 eV range overlap with the emission band of LE(La) та La_Y centers what causes excitation of the F^+ and F-center luminescence by emission of the centers created by La^{3+} impurity.

Acknowledgement
The authors thank Dr. N. Ryskin from the Institute of General Physics of RAS for LuAG:La crystals. The authors also express their gratitude to Prof. Dr. G. Zimmerer and Dr. G. Stryganyk from II Institute of Experimental Physics University of Hamburg for their assistance during our experiments in HASYLAB at DESY. The work was performed in the framework of HASYLAB II-05-105 research project.

References

[1] Valbis Ya A 1985 Isoelectrinic impurity centers and luminescence of ionic crystals (Extended Abstract of Doctoral Dissertation, Institut of Physics of Solid State, Salaspils, Latvia).
[2] Zorenko Yu 2005 *Physica Status Solidi* C **2** 375.
[3] Zorenko Yu V 1997 *Ukrainian Physical Journal* **42** 559.
[4] Rachko Z A and Yansons Ya L 1987 Optics and Spectroscopy **63** 63.
[5] Valbis Ya A, Volzhenskaya L G, Dubov Yu G and Zorenko Yu V 1987 *Optics and Spectroscopy* **63** 624.
[6] Ryskin N N, Dorenbos P, van Eijk C W E and Batygov S Kh 1994 *J. Phys.: Condens. Matter* **6** 10423
[7] Zorenko Yu V, Konstankevich I V, Gorbenko V I and Yurchishin P I 2002 *J. Applied Spectroscopy* **69** 665.
[8] Zorenko Yu, Gorbenko V, Voloshinovskii A, Stryganyuk G, Nedilko S, Degoda V and Chykova O, 2005 *Phys. Stat. Sol.*(c) **2** 105.
[9] Zorenko Yu V 2006 *Optics and Spectroscopy* **100** 572.
[10] Zorenko Yu, Gorbenko V, Voznyak T, Zorenko T, Nikl M and Nejezchleb K 2008 *Journal of Luminescence* **128** 595.
[11] Ashurov M, Voronko Yu, Osiko V and Solol A 1977 *Phys. Stat. Sol.*(a) **42** 101.
[12] Lupei V, Lupei A, Tiseanu C, Georgescu S, Stoicescu C and Nanan P M 1995 *Phys. Rev.* **B 51,** 8.
[13] Lupei A., Stoicescu C.and Lupei V. 1997 *J. Cryst. Growth* **177** 207.
[14] Bo Liu, Mu Gu, Xiaolin Liu, Shiming Huang and Chen Ni 2009 *Applied Physics Letter* **94** 121910.
[15] Kuklja M M 2000 *J. Phys.: Condens. Matter* **12** 2953.
[16] Zorenko Y, Konstankevych I, Gorbenko V and Zorenko T 2002 *Mol. Phys. Rep.* **36** 127.
[17] Zorenko Yu, Voloshinovskii A, Savchyn V, Vozniak T, Nikl M, Nejezchleb K, Mikhailin V, Kolobanov V and Spassky D 2007 *Phys. Stat. Sol.* (b) **244** 2180.
[18] Zorenko Yu V, Voloshinovskii A S and Konstankevych I V 2004 *Optics and Spectroscopy* **96** 532.
[19] Surdo A I, Kortov V S and Pustovarov V A 2001 *Radiation Measurements* **33** 587.
[20] Zorenko Yu, Voloshynovskii A, Vistovsky V, Grinberg M, Kornylo A, Łukasiewicz T and Świrkowicz M 2007 *Phys. Stat. Sol.* (b) **244** 3271.

Development of novel UV emitting single crystalline film scintillators

Yu Zorenko[1], V Gorbenko[1], V Savchyn[1], T Voznyak[1], M Nikl[2], J A Mares[2], T Martin[3], P –A Douissard[3]

[1] Laboratory of Opoelectronic Materials (LOM), Electronics Department of Ivan Franko National University of Lviv, 79017 Lviv, Ukraine
[2]Institute of Physics of ASCR, 162 53 Prague, Czech Republik
[3]ESRF, Instrument Support Group, 6 rue Jules Horoeitz, 38043 Grenoble, France

E-mail: zorenko@electronics.wups.lviv.ua

Abstract. The work is dedicated to development of new types of UV –emitting scintillators based on single crystalline films (*SCF*) of aluminimum perovskites and garnets grown by the liquid phase epitaxy (LPE) method. The development of the following three types of UV *SCF* scintillators is considered in this work: i) Ce-doped *SCF* of Y-Lu-Al-perovskites with Ce^{3+} emission in the 360-370 nm range with a decay time of 16-17 ns; ii) Pr-doped *SCF* of Y-Lu-Al garnets with Pr^{3+} emission in the 300-400 nm range with a decay time of 13-17 ns; iii) La^{3+} and Sc^{3+} doped *SCF* of Y-Lu-Al-garnets, emitting in the 290-400 nm range due to formation of the $La_{Y,Lu}$, $Sc_{Y,Lu}$ and Sc_{Al} centers with decay time of 250-575 ns. The results of testing the several novel UV-emitting *SCFs* scintillators for visualization of X-ray images at ESFR are presented. It is shown that the UV emission of the LuAG:Sc, LuAG:La and LuAG:Pr *SCFs* is efficient enough for conversion of X-ray to the UV light and that these scintillators can be used for improvement of the resolution of imaging detectors in synchrotron radiation applications.

1. Introduction

During the last 30 years the liquid phase epitaxy (LPE) shows oneself as a beneficial method for the development of new types of luminescent materials based on *single crystalline films (SCF)* of oxide compounds (garnets, perovskites, sapphire, silicates, tungstates)[1-3]. The fields of application of such *SCF* are now extended to α- and β- scintillators, screens for visualization of X-ray images, cathodoluminescent screens as light sources for scanning optical microscopes, laser media and luminescent converters of LED radiation [1-9].

In several previous works [2,4,10,11] we have demonstrated the creation by LPE method of the *SCF* scintillators based on $Y_3Al_5O_{12}$:Ce (YAG:Ce) and $Lu_3Al_5O_{12}$ (LuAG:Ce) garnets emitting in visible (450-750 nm) range with a decay time of 70 and 50 ns respectively. We have also shown [12-16] that these *SCF* scintillators in comparison with bulk *single crystal (SC)* analogues are characterized by faster Ce^{3+} emission decay under high energy excitation, substantially less (by 4.5-8 times) content of slow emission components and higher energy resolution due to the absence in them of Y_{Al} and Lu_{Al} *antisite defects (ADs)* and low concentration of vacancies as luminescence and trapping centers.

In this work, we focus on the development of emitting in the UV range *SCF* scintillators based on garnet and perovskites compounds. The reasons for the development of such scintillators are as follows:

1.Shift of the emission spectra of *SCF* scintillators into the UV range with respect to recently developed *SCFs* of LuAG:Ce anf YAG:Ce garnets, emitting in visible range, in principle can result in the faster emission decay, larger light yield (LY) and higher energy resolution [4,15]. Specifically, we plan to extend the class of novel SCF scintillators which effectively emit in UV ranges, on Ce-doped perovskites (YAP:Ce and LuAP:Ce) and Pr^{3+}-doped garnets (LuAG:Pr and YAG:Pr).

2. Microimaging techniques with synchrotron or X-ray radiation for applications in microtomography and industry [5-7] demands fast, on-lne X-ray image detectors with spatial resolution in the μm or sub-μm range. For this task, a X-ray image detectors based on YAG:Ce SCF, microscope optics, a low-noise CCD camera, operated at X-ray energies of 10-50keV have been recently developed [5,6]. The spatial resolution of X-ray images of 0.8 μm FWHM is achieved . Future increase of the spatial resolution of detector requires creation of the *SCF* scintillating screens emitting the UV light. Registration in the UV range could increase the spatial resolution of a detector according to the formula:$0.61*\lambda/NA$, where λ is the emission wavelength, NA is the numerical aperture of the optics.

In this work we have studied the scintillation properties of several types of UV emitting *SCF* scintillators prepared by LPE method. For visualization of X-ray images in the UV light we also tested at ESRF[3] the novel *SCF* scintillators based on LuAG garnet and LuAP perovskites compounds which have significantly higher density (ρ=6.67 and 8.34 g/cm^3) and effective atomic number (Z_{eff}=59 and 62) [6,7,10,11] as compared to commonly used YAG (ρ=4.6 g/cm^3, Z_{eff}=29) [4,5]. As mentioned above, for creation of the intense UV light, the Pr^{3+} as well as La3+ and Sc^{3+} ions can be used as activators in LuAG *SCF* and Ce^{3+} ions in LuAP *SCF* respectively.

2. Growth of *SCF* scintillators and experimental technique

The series of Pr, La and Sc-doped YAG and LuAG and Ce-doped YAP and LuAP *SCF* scintillators were grown in LOM Lviv University by the LPE methods onto YAG and YAP substrates, respectively, with super-cooled melt – solution (MS) based on the PbO-B_2O_3 flux at relatively low temperatures (950-1100 °C) as compared to Czochralski – grown *SC* (~2000°C). The thickness of the *SCF* scintillators was varied in the 4.8-45 μm range. It is important to note, that we did not use any additional doping to reduce the significant differences in the lattice constants of the LuAG- and LuAP- based SCF and the YAG and YAP substrates which are equal to about 1.0 and 1.4% respectively [18-20].

The lower growth temperature of *SCF* results in absence of Y_{AL} and Lu_{AL} *ADs* and decrease of concentration of other type of defects in *SCF* in comparison with bulk *SC* analogues [13,17]. On the other hand, flux components can be introduced in the *SCF* that might have detrimental influence on their emission and scintillation properties [21-24]. Usually, the *SCFs* prepared from the PbO –based flux, can contain lead ions preferably in the Pb^{2+} charge state. Since the PbO – based melt dissolves the Pt-crucible, Pt^{4+} ions can also be introduced in the garnet or perovskites lattices. As a consequence, various locally non- compensated lattice defects can be created, which may result in the decrease of the scintillation efficiency and light yield of *SCF* scintillators (LY) [2,4,20,24].

Therefore, the LY of Pr-, La- and Sc-doped YAG and LuAG *SCF* and Ce-doped YAP and LuAP *SCF* scintillators in the UV range can strongly depend on the concentration in them of activators as well as Pb^{2+} and Pt^{4+} dopants. In principle, the suitable LY of the mentioned *SCF* can be achieved by means of optimization of the activator content in MS [2,10,21,22,24]. Specifically, due to significant difference in the segregation coefficient of La^{3+}, Sc^{3+} and Pr^{3+} ions at the crystallization of YAG and LuAG *SCF* onto YAG substrates [20,21] and Ce^{3+} ions at crystallization of YAP and LuAP *SCF* onto YAP substrates [20,24], the concentrations of Sc_2O_3, La_2O_3, Pr_2O_3 and CeO_2 activated oxides in MS was varied in the 4.6-6.7,1-8.2, 3.1-8.6 and 2-20 mole % ranges, respectively. At the same time, the concentration of different impurities in the *SCF* depends not only on the content of activated oxides in MS, but is also strongly influenced by the *SCF* growth temperature. Of the most importance is that the concentrations of dopants as well as Pb^{2+} and Pt^{4+} flux impurities increase with decreasing the growth temperature and vice versa [21,22,24]. For these reasons we have used relatively high growth temperatures above 950°C for *SCF* preparation.

Journal of Physics: Conference Series **289** (2011) 012029 doi:10.1088/1742-6596/289/1/012029

The content of different dopants in *SCF* was determined using JEOL JXA-8612 MX electron microscope and presented in table 1 for different series of *SCF* scintillators.

Table1. Relative LY of the best samples in series of UV-emitting SCF scintillators in comparison with standard bulk *SC* analogues under excitation by α-particles of Pu^{239} sources (5.15MeV). The concentration of Y_{Al} and Lu_{Al} *ADs* estimated by XRD analysis is presented for undoped YAG and LuAG *SC*

Dopant	Scintillators	Activator content, at. %	Maximum of emission band, nm	Decay time of main emission component, ns	LY, % Am^{241}(3 μs)/ Pu^{239}(0.5 μs)
Ce	YAP:Ce *SCF* (PbO flux)	0.08	366	13.1	36.1/25.5
	YAP:Ce *SCF* (BaO flux)	0.053	373	16.1	- /23.8
	$Y_{0.4}Lu_{0.6}AP$:Ce *SCF* (PbO flux)	~0.04	363	16.0	- /10.2
	LuAP:Ce *SCF* (PbO flux)	0.013	358	16.9	- /7.65
	YAP:Ce *SC*	~0.1	366	16.2	100/100
	$Lu_{0.3}Y_{0.7}AG$:Ce *SC*	~0.1	375	17.1	83.2/56.3
Pr	YAG:Pr *SCF*	0.085	323	13	33.7/21.6
	LuAG:Pr *SCF*	0.66	305	17	29.8/24.9
	YAG:Pr *SC*	0.19	327	17.6	75.4/60.8
	LuAG:Pr *SC*	0.31	308	18.6	59.3/55.6
La	YAG:La *SCF*	0.045	297		13.6/6.9
	$Lu_{2.74}Y_{0.25}Al_5O_{12}$:La *SCF*	0.04	282	170	15.9/11.0
	YAG:La *SC*	0.085	298	575	- /21.0
	undoped YAG *SC*	0.19*	292; 332	350; 1490	19.8/10.9
	undoped LuAG *SC*	0.575*	296; 334	440, 3020	- /18.8
Sc	$Y_{2.96}Sc_{0.475}Al_{4.555}O_{12}$ *SCF*	2.38	322	~750	38.7/26.7
	$Y_{2.84}Lu_{0.14}Sc_{0.37}Al_{4.65}O_{12}$ *SCF*	1.85	314	~650	57.4/48.2
	$Y_{2.76}Lu_{0.18}Sc_{0.715}Ga_{0.095}Al_{4.25}O_{12}$ *SCF*	3.60	330	~800	28.0/17.8
	$Lu_{2.67}Sc_{0.585}Al_{4.745}O_{12}$ *SCF*	2.93	312	245; 390	27.6/21.8
	$Y_{2.995}Sc_{0.077}Al_{4.864}O_{12}$ *SC*	0.39	314	~580	69.3/45.9
	$Lu_{2.773}Sc_{0.022}Al_{5.205}O_{12}$ *SC*	0.11	290	1330	31.4/21.8

For estimation of the scintillation efficiency in the UV range of the developed *SCF* scintillators with garnet and perovskites structure the relative LY measurements of all corresponding series *SCF* samples were performed in comparison with standard samples of YAG:Pr, LuAG:Pr, YAG:La, YAG:Sc, LuAG:Sc,YAP:Ce and $Lu_{0.3}Y_{0.7}AG$:Ce bulk *SC* scintillators under excitation by α-particles of Pu^{239} (5.15 MeV) and Am^{241} (5.4857 MeV) sources with a penetration length in studied materials of about 10-12 μm. For LY measurements we used the detectors based on FEU-110 and hybrid DEP PPO 475B PMT types which have the maximum sensitivity in the 250-375 and 200-400 nm ranges and mulitichannel single – photon counting system working within a time interval of 0.5 and 3 μs, respectively. The results of YL measurements are presented in Table 1.

For the correct characterization of the luminescent properties of the tested series of UV-emitting *SCF* we used measurements of their cathodoluminescence (CL) spectra under pulsed e-beam excitation (a pulse duration of 2 μs and a frequency of 30-3 Hz) with energy of electrons of 9keV and a beam current of 100 μA. Emission spectra were corrected for the spectral dependence of the detection part consisting of DMR-4 monochromator and FEU-106 photomultiplier (PMT).

The decay kinetics of luminescence were measured at 300K in the time interval 0-200 ns under excitation *SCF* and *SC* scintillators by synchrotron radiation (SR) with a pulse duration of 0.126 ns at Superlumi experimental station in HASYLAB,DESY.

Journal of Physics: Conference Series **289** (2011) 012029 doi:10.1088/1742-6596/289/1/012029

3. Luminescence spectra and LY of *SCF* scintillators

3.1 Ce-doped perovskites

The CL spectra of YAP:Ce and LuAP:Ce SCF (Fig.1, curve 1 and 2) present intensive emission band in the UV (320-450 nm) range related to the $5d^1 \rightarrow 4f(^2F_{5/2,7/2})$ transitions of Ce^{3+} ions [19,20,24]. It is important to note that the short-wavelength part of the Ce^{3+} emission spectra in YAP:Ce and LuAP:Ce *SCF*, growth from PbO – based flux, can be strongly distorted by the UV emission band of Pb^{2+} ions peaked at 340 nm [25] (Fig.1). This caused a notable short-wavelength shift of the emission spectrum of YAP:Ce *SCF*, grown from the PbO-based flux, with respect to the luminescence spectrum of the same *SCF*, grown from the lead-free flux. Other low-intensity complex emission band in YAP:C and LuAP:Ce *SCF*, related to the luminescence of excitons localized around of Pb^{2+} ions, is located in the visible range with maximum approximately at 590 nm (Fig.1).

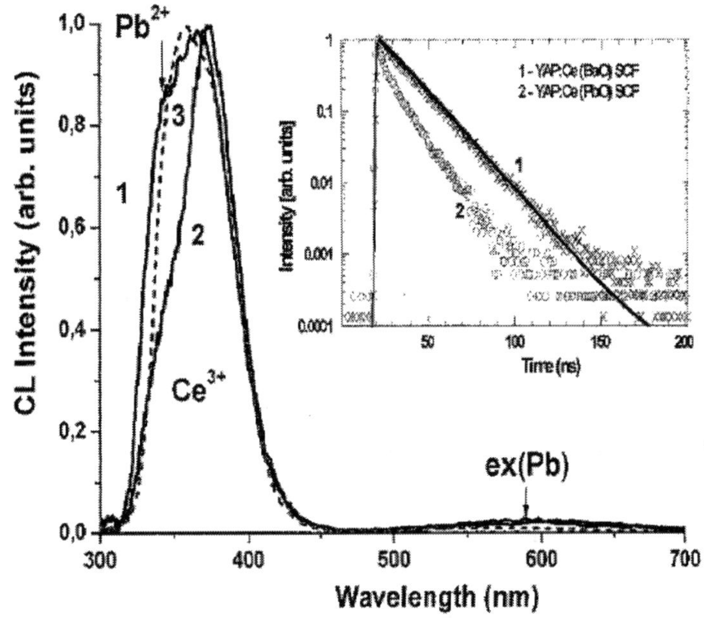

Figure 1. Normalized CL spectra of YAP:Ce (1,2) and LuAP:Ce (3) *SCF*; insert – normalized luminescence decay kinetics of YAP:Ce *SCF*, grown from the BaO- (1) and PbO-based (2) flux under excitation in Ce^{3+} absorption band at 300 nm; T=300K

As can be seen from Table 1, the LY of YAP:Ce and especially of LuAP:Ce *SCF* scintillators , grown even at the optimal CeO_2 content in the MS of 10-20 mol%, is significantly lower (by 3.9 anf 7.6 times, respectively) then that of their YAP:Ce and (Y-Lu)AP:Ce *SC* analogues. The main reason for such low LY of YAP:Ce and LuAP:Ce *SCF* scintillators is the large incorporation of Pb^{2+} ions in *SCF* of perovskites in comparison with SCF of garnets [23] due to larger volume of cub-octahedral position in perovskites lattice with respect to the dodecahedral position of garnet lattice, where Pb^{2+} ions are localized. We also note that the LY of (Y-Lu)AP:Ce *SCF* systematically decrease with increasing the Lu content up to a value of 7.65% for LuAP:Ce *SCF* as compared to that of YAP:Ce *SC*. This effect can be explained by the large incorporation of Pb^{2+} ions in LuAP-based *SCF* with respect to YAP *SCF* due to preferable Lu-Pb pair incorporation in comparison with Y-Pb pair in the case of *SCF* crystallization of both types of perovskites onto YAP substrates [20].

We have also observed the visible acceleration of decay kinetics of the Ce^{3+} luminescence in YAP:Ce *SCF*, grown from PbO-based flux (Fig.1, insert, curve 2). The decay curve for this YAP:Ce *SCF* is strongly non-exponential and can be quantitatively characterized by an average decay time of 13.1 ns

(Table 1). Such acceleration of the decay kinetics of the Ce^{3+} luminescence in YAP:Ce *SCF* scintillators grown from PbO-based flux, can be caused by the energy transfer from Ce^{3+} ions to Pb^{2+} related centers, which have the excitation bands in the 350-370 nm range, well overlapped with Ce^{3+} emission band [26]. In opposite to this *SCF*, the decay kinetics of the Ce^{3+} luminescence in YAP:Ce *SCF*, grown from BaO – based flux, is strongly exponential (insert of Fig.1, curve 2) with a decay time of 16 ns, which is very close to the decay time of the Ce^{3+} emission in YAP:Ce *SC* (Table 1).

From calculation of the integral of normalized decay curves for YAP:Ce *SCF*, grown from the BaO- and PbO-based fluxes, which is equal to 100 and 46%, respectively, we can conclude that in the last *SCF* scintillators the losses of more than 50% excitation energy take place due to energy transfer away from Ce^{3+} $5d^1$ excited state to Pb^{2+} - based centers [24]. Such huge loss can partly explain such low LY of YAP:Ce and LuAG:Ce *SCF* scintillators, grown from PbO – based flux (Table 1).

3.2 Pr^{3+} doped garnets

The CL spectra of YAG:Pr and LuAG:Pr show intensive and fast emission in the 290-450 nm range with main maxima at 305 and 322 nm, respectively, caused by the $5d^14f^1 \rightarrow 4f^2(^3H_4,^3H_5,^3H_6,^3F_{3(4)})$ transition of Pr^{3+} ions. The sharp-line emission bands peaked at 487, 501 and 563 nm as 609-620, 638 and 663 nm in the visible range, are related to the transition from the $^3P_0$ and $^1D_2$ levels of the $4f^2$ shell to the $^3H_{6,5,4}$ levels of the ground state of Pr^{3+} ions (Fig.2,curve 2). The content of slow visible emission is significantly lower in LuAG:Pr *SCF* than in YAG:Pr *SCF* (Fig.2).

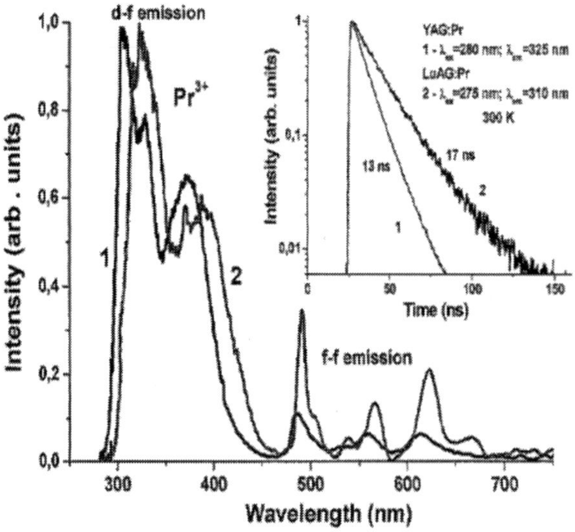

Figure 2. CL spectra and decay kinetics (insert) of Pr^{3+} luminescence in YAG:Pr (1) and LuAG:Ce (2) *SCF* at 300K

As can be seen from Table 1, the LY of YAG: Pr and LuAG:Pr *SCF* scintillators is significantly lower (by 2.2-2.8 times) then that of their *SC* analogues. This result contradicts with the results of our previous investigations of the LY of Ce-doped YAG and LuAG SCF scintillatores, emitting in the visible (515-530 nm) range, with their SC analogues [13].Specifically, for these compounds the comparable LY of *SCF* and *SC* scintillators occurs [13]. Taking into account the absence of Y_{Al} and Lu_{Al} *AD* and low concentration of vacancy – type defects in YAG and LuAG *SCF* [13,14], the main reason for lower LY of the Pr^{3+} - doped *SCF* scintillatores in comparison with *SC* of these oxides is the strong quenching influence of Pb^{2+} flux related impurity. Thus, based on the results presented in Table 1, we can conclude that significantly larger influence of Pb^{2+} flux dopant on the UV luminescence of Pr^{3+} ions takes place in YAG:Pr and LuAG:Pr *SCF* in comparison with the influence of this dopant on the Ce^{3+} luminescence n the visible range in Ce-doped *SCF* of these garnets [13,23]. The mechanism of $Pb^{2+} \rightarrow Pr^{3+}$ energy transfer is considered in detail in [26].

3.3 La doped garnets

The CL spectra of YAG:La SC, YAG:La and LuAG:La *SCF* are shown in Fig.3a. The intensive complex emission bands in the 250-450 nm range are caused by La^{3+} isoelectonic impurities. Namely, these bands in YAG:La and LuAG:La *SCF* presents superposition of two close-lying bands peaked at 298, 305 nm and 267,282 nm (Fig.3a,curves 2 and 3 respectively). The short-wavelength bands are related to the luminescence of exciton localized around $La_{La,Y}$ centers (La^{3+} ions in the dodecahedral sites of Y^{3+} or Lu^{3+} cations), when the long-wavelength bands are caused by the recombination luminescence of La_Y and La_{Lu} centers. In YAG:La and LuAG:La *SCFs*, grown from PbO-based flux, the long-wavelength wngs of complex UV emission bands, caused by La dopant, are also overlapped with the UV emission band of Pb2+ ions peaked at 350-360 nm [23] (Fig.3a, curve 2 and 3). It is worth to note that the nature of the emission centers formed by La^{3+} dopant in YAG and LuAG host is similar in the whole to the nature of the centers responsible for the relatively intensive intrinsic UV luminescence of YAG and LuAG bulk *SCs* (Fig.3 b, curves 1 and 2, respectively). Specifically, this band in YAG and LuAG *SCs* presents superposition of the luminescence of excitation localized around Y_{Al} and Lu_{Al} *ADs* (Y^{3+} and Lu^{3+} cations in the octahedral sites of Al^{3+} cations) in the band peaked at 292 and 296 nm and the recombination luminescence of Y_{Al} and Lu_{Al} *ADs* in the bands peaked at 332 and 334 nm, respectively [13].

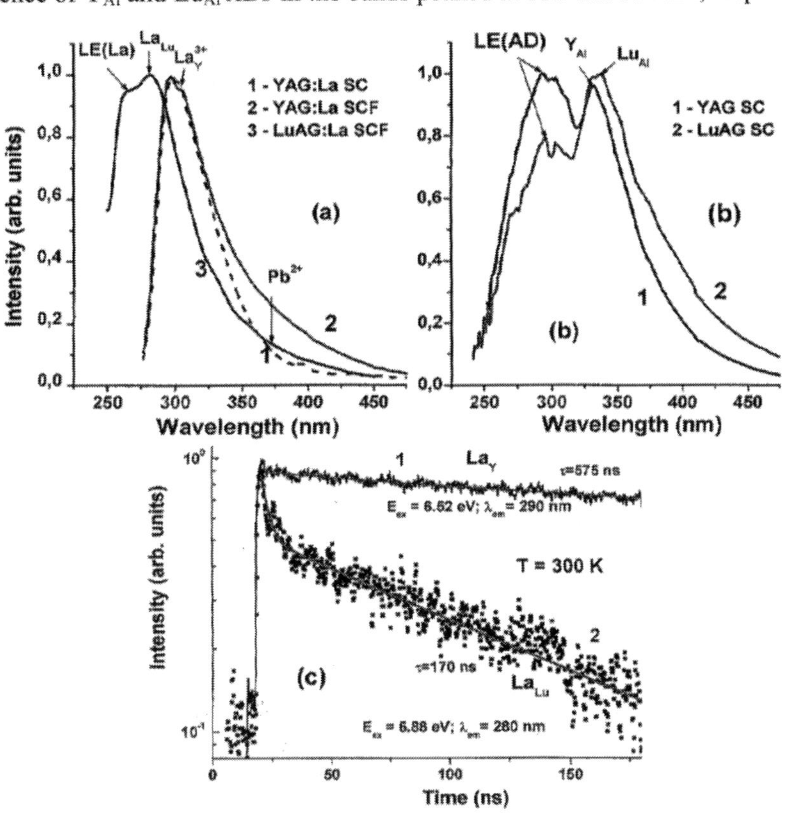

Figure 3.(a)-CL spectra of YAG:La *SC*(1), YAG:La *SCF*(2) and LuAG:La *SCF*(3) at 300K; (b)-CL spectra of undoped YAG (1) and LuAG (2) *SC* at 300K;(c)-decay kinetics of luminescence of La_Y (1) and La_{Lu} (2) centers at 290 and 285 nm in YAG:La and LuAG:La *SCF* at 300K under excitation by SR at 6.52 and 6.88 eV, respectively.

The decay kinetics of luminescence of La_Y and La_{Lu} centers in YAG:La and LuAG:La *SCF*, under excitation in the exciton range with an energy of 6.52 and 6.88 eV, respectively, is shown in Fig.3c. The decay times of main component of La_Y and La_{Lu} centers luminescence are 560 and 170 ns, respectively.

The LY of YAG:La and LuAG:La *SCF* and *SC* scintillators is not so much influenced by Pb^{2+} flux impurity but strongly depend on the La content in garnet host. As can be seen from Table 1, the LY of YAG:La and LuAG:La *SCF* scintillators with La content of 0.04-0.05 at.% is by 2-3 times lower in comparison with YAG:La (0.085at.%) *SC*. Also the LY of YAG:La and LuAG:La *SCF* is significantly (by 2-4 times) lower than that of Sc-doped YLuAG and LuAG SCF. Such a low LY of La-doped *SCF* is mainly caused by very low (0.005) segregation coefficient of La^{3+} ions at the crystallization of YAG:La and LuAG:La *SCF* in comparison with that (0.14) for YAG:La *SC* analogues [21]. As a result, the rather small (0.04-0.045 at.%) concentration of La^{3+} ions can be achieved in LPE-grown YAG:La and LuAG:La *SCF* (Table 1) even at large (above 8 mole %) content of La_2O_2 dopant in MS in comparison with an optimum value of 0.4-1.85 at.% for Sc-doped *SCF* analogues of these garnets.

3.4 Sc-doped garnets

In contrast to La^{3+} ions, Sc^{3+} isoelectronic impurity in YAG and LuAG host has relatively high (0.8-0.55 and 0.2-0.4, respectively) segregation coefficients [21,22]. This allows readily achieving the optimum values of Sc doping for both *SC* and *SCF* of these garnets in range 0.4 – 2.4 at.% at which the highest LY scintillators are observed (Table 1). The incorporation of Sc^{3+} ions in YAG and LuAG *SC* and *SCF* at total scandium content x<0.3 formula units (f.u) results in the Sc^{3+} ions localization both in the dodecahedral {c}- and octahedral (a)-sites in ratio of about 40:60% [27] with formation $Sc_{Y,Lu}$ and Sc_{Al} centers, respectively. At higher content the Sc^{3+} ions occupy predominantly the (a)-sites of garnet lattice (insert in Fig.4) [27]. At the same time, the relation between the $Sc_{Y,Lu}$ and Sc_{Al} centers in YAG and LuAG host can be changed by substitution of the {c}- and (a) – sites of garnet lattice by other co-dopants with smaller or larger dimensions. Specifically, due to partial substitution of $^{Y3+}$ cations by the Lu^{3+} ions, or Al^{3+} cations by the Ga^{3+} ions, the content of $Sc_{Y,Lu}$ center in $(YLu)_3Al_5O_{12}$ and $(YLu)_3Al_5GaO_{12}$ *SCF* can strongly decrease up to 7-8% (Table 1).

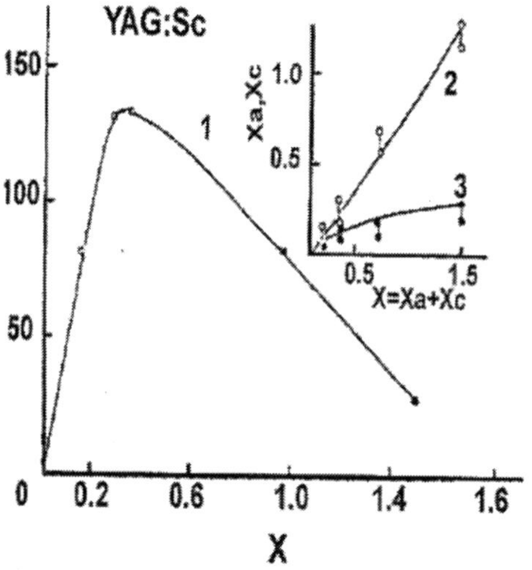

Figure 4. Dependence of integral intensity of CL (1) and concentration of Sc_{Al} (2) and Sc_Y (3) centers on total scandium content in YAG:Sc *SCF*.

Thus, in Sc-doped garnets the formation of two types of emission centers ($Sc_{Y,Lu}$ and Sc_{Al}) takes place which can contribute to the intensive UV luminescence of these compounds and compete in the processes of energy transfer [27,28]. Specifically, above mentioned distribution of Sc^{3+} ions over the {c}- and (a) – sites explains the complex dependence of the positions of the maximum emission bands of YAG:Sc and LuAG:Sc *SCF* on the Sc content (Fig.4). At relatively small (x=0.05-0.15 f.u.) Sc concentration, the

Journal of Physics: Conference Series **289** (2011) 012029 doi:10.1088/1742-6596/289/1/012029

emission of Sc_Y and Sc_{Lu} centers in the bands at 313 and 290 nm, respectively, dominates in the luminescence of YAG:Sc and LuAG:Sc *SCF* (Fig.5, a and b, curve 1). The bands at 353 and 335 nm, related to the emission of Sc_{Al} centers in YAG and LuAG *SCF*, respectively, become dominant at larger (x=0.15-0.4 f.u.) scandium concentration (Fig.5a and 5b, curve 2) which correlate with substantial increase of the LY of these *SCF* (Fig.4, Table 1). Maximum of the LY of YAG:Sc and LuAG:Sc *SCF* is reached at x=0.3-0.4 f.u. (Fig.4). At highest Sc content the concentration quenching of Sc^{3+} related centers occurs.

The luminescence decay kinetics of the $Sc_{Y,Lu}$ and Sc_{Al} centers at 300K is shown in Fig.5c on example of LuAG:Sc *SCF*. Under excitation with energy of 6.88 eV in exciton range the main components of the Sc_{Lu} center luminescence in LuAG:Sc decay with a time of 245 ns (Fig.6, curve 1). The decay time of the dominant component of the Sc_{Al} center luminescence at 300K (Fig.5c, curve 2) is equal to 390 ns.

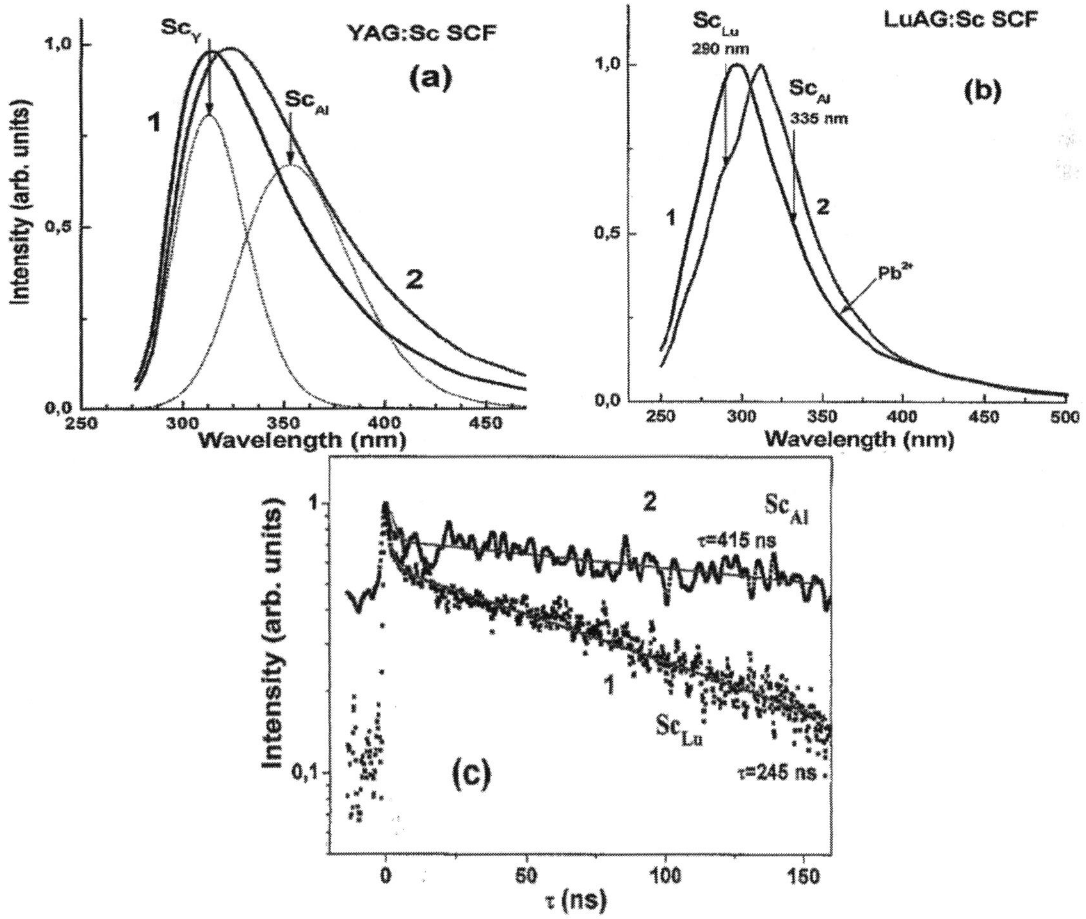

Figure 5 CL spectra of YAG:Sc (a) and LuAG:Sc (b) *SCF* with different Sc content at 300K. The content of Sc_2O_3 oxide in MS was 4.6 and 6.7 mole % (curves 1) and 10.4 and 9.8 mole % (curves 2), respectively. The decomposition of spectrum 2 in Fig.5a is given by the dashed lines. (c)-decay kinetics of the luminescence of Sc_{Lu} centers at 280 nm (1) and Sc_{Al} centers at 335 nm (2) in LuAG:Sc *SCF* under excitation by SR with an energy of 6.88 ev at 300K.

The LY of YAG:Sc and LuAG:Sc *SCF* and *SC* scintillators is strongly depend on the total Sc content and distribution of Sc^{3+} dopant over the {c}- and (a)-positions in garnet host. Specifically, the concentration of $Sc_{Y,Lu}$ and Sc_{Al} centers in YAG *SCF* can be significantly changed by substitution of the {c}- and (a)-sites of garnet lattice by Lu^{3+} and Ga^{3+} ions with smaller or larger ionic radii, respectively, than that in Y^{3+} and Al^{3+} cations. Such substitution is strongly reflected on the LY of *SCF* scintillators. Specifically, the Lu^{3+} co-doping leads to increasing the LY of YAG:Sc *SCF*, whereas the Ga^{3+} co-doping decrease of the LY of these scintillators (Table 1). As can be seen from Table 1, the LY of (YLu)AG:Sc and LuAG:Sc *SCF* scintillators at the optimal Sc^{3+} content 0.4-0.6 f.u. is comparable with their *SC* analogues and reaches values about 50% of YAP:Ce and (LuY)AP:Ce standard crystal. This result shows that influence of Pb^{2+} dopant on the UV luminescence of Sc^{3+} - based centers is not so much significant as that on the luminescence of Ce^{3+} ions in *SCF* of perovskites and the luminescence of Pr^{3+} ions in *SCF* of garnets. It is most important that this allows to use PbO-based flux for producing *SCF* scintillators with high LY emitting in UV range.

4. X-ray excited luminescence spectra of SCF

The emission spectra in 300-750 nm range of the tested LuAG:Pr/YAG, LuAG:La/YAG, LuAG:Sc/YAG and LuAP:Ce/YAP epitaxial structures with the SCF thickness of 0.8, 1.7,5.0 and 2.3 μm, respectively, under X-rays excitation in the direction perpendicular to *SCF* surface, were measured with Oriel 77233 grating (1200l/mm, blazed at 500 nm) monochromator and Hamamatsu R4632 PMT assembly. The X-rays were operated with a Cu anode (25kV, 40mA) and a 25μm copper filter for selection of the dominant 8keV line of copper for preferable excitation of *SCF* scintillators. At the same time, the partial excitation of YAG and YAP substrates by the 8 keV X-ray is also expected due to low (1.7-8 μm) thickness of *SCF*. All spectra were corrected for the grating wavelength response and the photomultiplier quantum efficiency. The results are presented in Fig.6.

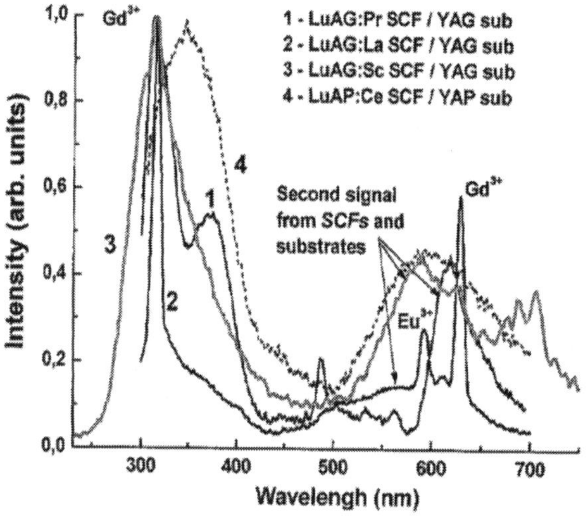

Figure 6. X-ray excited ($Cu_{K\alpha}$) luminescence spectra of LuAG:Pr/YAG (1), LuAG:La/YAG (2), LuAG:Sc/YAG (3) and LuAP:Ce/YAP (4) epitaxial structures at 300K.

The X-rays excited emission spectra of LuAG:Pr/YAG, LuAG:Sc/YAG and LuAP:Ce/YAG epitaxial structures in UV range (Fig.6, curves 2 and 3) with a *SCF* thickness of 8, 5 and 2.3 μm, respectively, are very close to the CL spectra of their *SCF* components (Fig.1,curve 1, Fig.2, curve 1 and Fig.3b, curve 2,respectively). At the same time, the X-rays excited emission spectra of LuAG:La/YAG epitaxial structure in UV range due to small (1.7 μm) *SCF* thickness present superposition of the luminescence of LuAG:La *SCF* and YAG substrate (Fig.2, curves 2 and 3, respectively). It is necessary to note that apart from the visible emission of Pr^{3+} and Pb^{2+} ions, all epitaxial structures also reemit the part of UV light in the visible range (Fig.6, curves 1-4). Specifically, the sharp line bands peaked at 592, 610 and 627 nm in the emissions spectra of LuAG:La/YAG epitaxial structure are caused by Eu^{3+} ions emission and the second order of the UV emission of Gd^{3+} ions as trace impurities in YAG substrate. Therefore, for detection of X-ray image only in UV light, the use of a band-pass filter is strongly required, in order to cut the visible emission of scintillators.

We also estimate the conversion efficiency of the tested *SCF* scintillators into the UV and visible ranges under X-ray excitation. The conversion efficiency is measured with an imaging system equipped with microscope objective 4x and CCD camera. The X-rays from copper anode, operated at 20kV, 50mA, were additionally separated by 25μm copper filter for receiving 8keV line of copper. The *SCF* are exposed for 30 s and an average value of the intensity is taken on the final image. The image is corrected from detection system response to wavelength and *SCF* thickness (X-ray absorption). All the results are compared to the conversion efficiency of the reference YAG:Ce *SCF* scintillator. The characteristics of tested *SCF* samples under X-ray excitation are summarized in Table 2.

Table 2 Characteristics of LuAG:Sc/YAG, LuAG:La/YAG, LuAG:Pr/YAG and LuAP:Ce/YAP epitaxialstructures under excitatiom by 8keV X-ray (CuKα) excitation. Conversion efficiency of LuAG:La/YAG structure was measured without correction on the part of UV emission below 320 nm, which was cut by the objective used.

***SCF* scintillators**	**LuAG:La**	**LuAG:Sc**	**LuAG:Pr**	**LuAP:Ce**
Thickness, μm	1.7	5.0	8	2.3
Main emission lines (nm)	280, 315	313.5	315, 375	360
Conversion light yield, % with respect to YAG:Ce *SCF*	95 % (λ >320 nm)		100 % (λ>320 nm)	20 %
Afterglow @20ms (2s exposure to X-rays)	0.05 %		0.29 %	0.14 %
Afterglow @1000ms (2s exposure to X-rays)	0.001 %		0.007 %	0.07 %

5.Afterglow

The afterglow (delayed luminescence) is measured after X-rays excitation (Mo-anode) with an exposure time of 2 s (typical exposure time at ESRF is 0.1-10 s). The setup is made of a PMT2020Q and a SR400 gated photon counter (Stanford Research Instrument) working in the counting mode. The time resolution of the setup is 4 ms. The results of measurements of afterglow of LuAG:La, LuAG:Pr and LuAG:Ce *SCF* scintillators in comparison with recently developed [5,6] LuAG:Eu and YAG:Ce *SCF* are presented in Fig.7 and Table 2.

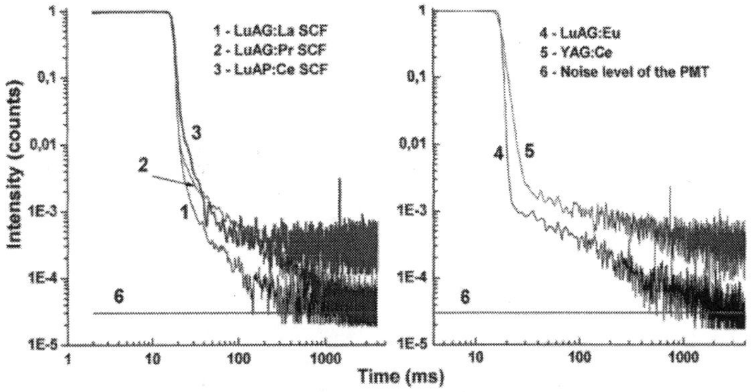

Figure 7. Afterglow of LuAG:La(1),LuAG:Pr(3) and LuAP:Ce *SCF* scintillators in comparison with recently developed [5,6]LuAG:Eu and YAG:Ce *SCF* scintillators following a 2s X-ray pulse exposition

As can be seen from Fig.7, the LuAG:La and LuAG:Pr SCF scintillators shows three decay decades in the 12 and 44 ms time intervals and four decay decades in the 30 and 100 ms time intervals, respectively (curves 1 and 2). The LuAP:Ce CF scintillators shows at least 3 decay decades in the 30 ms time interval. For comparison, recently developed [5,6] LuAG:Eu and YAG:Ce *SCF* shows three decay decades in the 5 and 55 ms time intervals, respectively (Fig.7, curve 4 and 5). At the same time, four decay decades for LuAG:Eu *SCF* was obtained in the significantly largest (384 ms) time intervals. Unfortunately, the dynamic of the measurement does not allow to obtain the temporal response from LuAP:Ce and YAG:Ce *SCF* samples behind 4 decades (see Fig.7, curve 3 and 5, respectively).

6. Visualisation of X-Ray images with UV – emitting SCF scintillators

6.1 Flat field X-rays images

Flat field images of the surface of tested LuAG:La, LuAG:Pr and LuAG:Ce *SCF* scintillators were obtained using no-filtered X-ray excitation (Cu anode operated at 25 keV, 45 mA). The Sensicam (PO) CCD camera was used as a recording device. The objective magnification was 4x. For LuAG:Pr and LuAG:La *SCF* only a part of their emission spectra contributes to the images because using the glass objective cuts the emission with the wavelength below 320 nm. The results are presented in Fig.8.

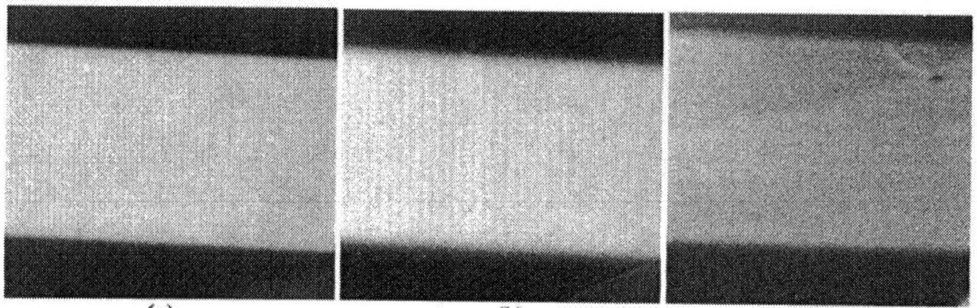

(a) (b) (c)

Figure 8. Flat-field image of LuAG:La (a), LuAG:Pr (b) and LuAP:Ce (c) *SCF*. Some defects on the surface of the *SCF* scintillators can be seen

6.2 Imaging in UV light

We also tested the LuAG:Pr and LuAG:La *SCF* scintillators at ESFR for visualization of image of X-ray radiated standard objects (resolution target) in the UV light. Our experimental system included the UV objective 15x, UV mirror, bandpass filter peaking at 320 nm and Sensicam QE CCE camera (from PCO). We could acquire images in the time less than 5 s with under excitation by X-ray quanta with an energy of 20 keV and a flux of $2 \cdot 10^{10}$ photons/sec both using the total emission spectrum (UV + visible) of LuAG:Pr or LuAG:La *SCF* scintillators (Fig.9a) or narrow part of the UV emission bands around 320 nm (Fig.9b).

The first image of resolution target was obtained with a spatial resolution of about 1.5 μm using only of the UV part of light of LuAG:La *SCF* scintillators (Fig.9b). The image resolution is comparable with quality of the image obtained with using total spectrum of LuAG:La and LuAG:Pr *SCF* (Fig.9a). These first results of testing are really encouraging. The UV emission of LuAG:La and LuAG:Pr *SCF* scintillators is efficient enough to be used for detection of the UV light following X-ray to UV conversion. That can be an interesting aspect for improving the resolution of high – resolution imaging detectors used in SR applications.

Nevertheless, we believe that the resolution of image in the UV light, presented in Fig.9b, can be significantly improved in future. The main problem of decreasing the resolution of the detector is connected with the large part of the UV light, emitting from the substrate, which then reemits in the visible range (Fig.6). Therefore, the ratio of the UV light coming from *SCF* and substrate must be substantially increased for the studied compounds by the way of increase of the total LY of *SCF* scintillators. We plan to test other UV-emitting *SCF* scintillators with relatively higher LY, specifically LuAG:Sc, LuYAG:Sc *SCF* (Table 1) as well as LuAG:Bi *SCF* [16] for visualization of X-ray image in the UV light. We also expect that a increase of the LY of LuAG:Pr and especially LuAP:Ce *SCF* scintillators can be achieved by means of using of **lead-free fluxes** for their crystallization by LPE method. Namely, we obtain the first good results on YAP:Ce *SCF* crystallization from $BaO\text{-}B_2O_3\text{-}BaF_2$ flux [24] with comparable LY in comparison with the same SCF grown from the PbO-based flux with strongly exponential decay kinetic of Ce^{+3} emission (Table 1).

7. Conclusion

Development of three types of UV –emitting single crystalline film (*SCF*) scintillators grown by liquid phase epitaxy methods (LPE) is considered in this work:

1) Ce^{3+} doped *SCF* of Y-Lu-Al perovskites, where the fast and intensive f-d luminescence of Ce^{3+} ions with a decay time of 16-17 ns occurs in the 360-370 nm range;
2) Pr^{3+} doped *SCF* of Y-Lu-Al garnets, where the fast and intensive f-d luminescence of Pr^{3+} ions in the 300-400 nm range with decay time 13-17 ns is realized;
3) *SCF* of Y-Lu-Al-garnet compounds doped with La^{3+} or Sc^{3+} isoelectronic impurities emitting in the 290-400 nm range due to formation of $La_{Y,Lu}$ or $Sc_{Y,Lu}$ and Sc_{Al} emission centers with the decay time of the luminescence in several hundred ns range.

(a)

(b)

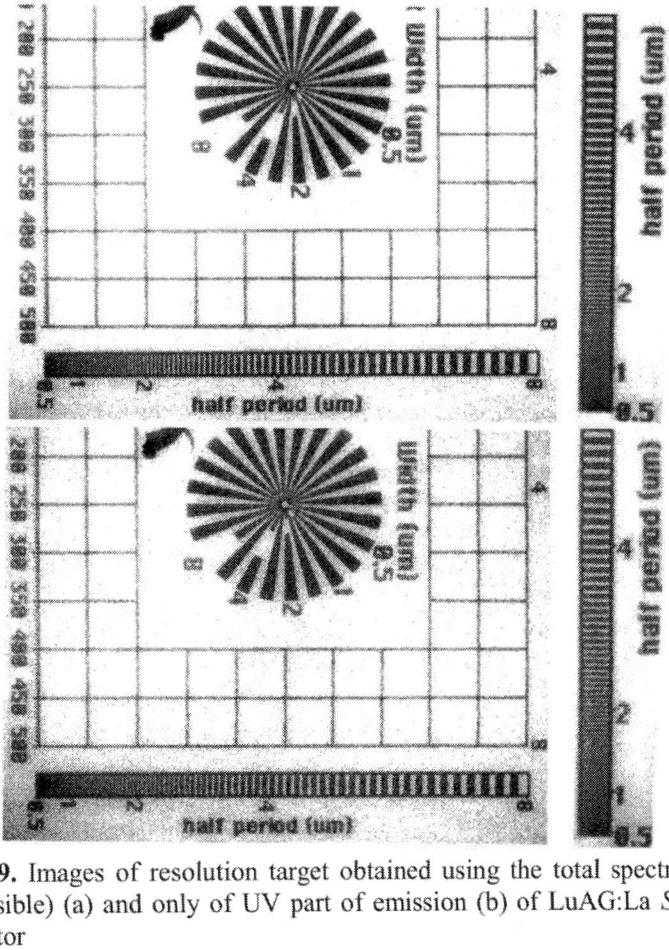

Figure 9. Images of resolution target obtained using the total spectrum (UV+visible) (a) and only of UV part of emission (b) of LuAG:La *SCF* scintillator

We have tested several novel UV-emitting LuAG:Pr, LuAG:La and LuAP:Ce SCF scintillators for visualization of X-ray images at ESFR. The UV emission of the LuAG:Sc,LuAG:La and LuAG:Pr *SCFs* is efficient enough for conversion of X-ray to UV light. This can be an interesting alternative to increase the resolution of imaging detectors used in synchrotron radiation applications. The first image with a spatial resolution of about 1.5 μm of resolution target was obtained using only of the UV part of light of the LuAG:La *SCF* scintillators.

The main problem in the development of these types of scintillators by LPE method from the traditional $PbO-B_2O_3$ flux is connected with the significantly larger influence of Pb^{2+} flux dopant on the UV luminescence of Pr^{3+} ions in *SCF* of garnets and Ce^{3+} ions in *SCF* of perovskites than in the case of recently developed YAG:Ce and LuAG:Ce *SCF* scintillators emitting in the visible range. This is the main reason for lower (by 2-3 times) LY of Pr – doped YAG and LuAG *SCF* and significantly lower (by 4-7.6 times) LY of Ce-doped YAP and LuAP *SCF* scintillators in comparison with their single crystals analogues. At the same time, the influence of Pb^{2+} dopant on the UV luminescence of La^{3+} and especially Sc^{3+} - based centers is not so significant than on the luminescence of Ce^{3+} and Pr^{3+} ions in *SCF* of perovskites and garnets. This allows using PbO-base flux for producing *SCF* scintillators emitting in the UV range with high LY.

The possible ways for improvement of figure-of-merit of UV-emitting *SCF* scintillators are discussed. Specifically, future development of UV scintillators based on Ce^{3+} -doped *SCF* of perovskites, and Pr^{3+}, La^{3+} and Sc^{3+} doped *SCF* of garnets strongly demands the use of alternative lead-free fluxes for their crystallization. In this work, we present the set of results related to using the $BaO-B_2O_3-BaF_2$ flux for crystallization of UV –emitting YAP:Ce *SCF* perovskites. We show, that the advantageous properties of YAP:Ce *SCF* scintillators, grown from the BaO – based flux, with respect to those grown from the traditional PbO – based flux, are connected with removing the additional channels of dissipation of excitation energy on the luminescence and trapping charge carriers at Pb – based centers.

Acknowledgement
This research was supported by Czech Science Foundation (project no. 202/08/0893) and Ministry of Education and Science of Ukraine (projects No DZ/296-2008 and no SL-28F).

References

[1] Robertson J M, van Tol M V 1984 Cathodoluminescent garnet layers *Thin Solid Films* **114**(1-2) 220-40

[2] Zorenko Yu V, Novosad S S, Pashkovskii M V, Lyskovich A B, Savitskii V G, Batenchuk M M, Malyutenkov P S, Patsagan N I, Nazar I V and Gorbenko V I 1990 Epitaxial structures of garnets as scintillators detectors of ionizing radiation *Journal of Applied Spectroscopy* **52**(6) 645-49

[3] Ferrand B, Chambaz M and Couchaud M 1999 Liquide phase epitaxy. A versalite technique for the development of miniature optical components in single crystal dielectric media. *Optical Materials* **11** 101-14

[4] Globus M, Grinyov B, Ratner M, Tarasov V, Lyubinskiy V, Vyadi Y, Ananenko A, Zorenko Y, Gorbenko V and Konstankevych I 2004 New type of scintillation detectors for biological , medical and radiation monitoring applications. *IEEE Transact. on Nucl.Sci.* **51**(3) 1297-303

[5] Koch A, Raven C,Spanne P and Snigirev A 1998 X-ray imaging with sub-micrometer resolution employing transparent luminescent screens. *J.Opt.Soc.Am.* **A** 15 1940-51

[6] Koch A, Peyrin F, Heurtier P, Chambaz B, Ferrand B, Ludwig W and Couchaud M 1999 An X –ray camera for computed microtomography of biological samples with micrometer resolution using $Lu_3Al_5O_{12}$ and $Y_3Al_5O_{12}$ scintilators 1999 *Proc.SPIE* **3659** 170-9

[7] Martin T and Koch A 2006 Recent development in X-ray imaging with micrometer spatial resolution *Journal of Synchrotron Radiation* **13** 180-94

[8] Zorenko Yu, Batenchuk M, Pashkovsky M, Konstankevych I, Gorbenko V, Yurchushyn P, Martynova V and Duzyj T 1998 Single crystalline film sensors for cathode-ray tubes: possibilities of application, peculiarities and light parameters, *Proc.SPIE* **3359** 261-64

[9] Hrytskiv Z D, Zorenko Y, Gorbenko V,Peden A D and Shkliarskyi V I 2007 Single crystalline film screens for cathode-ray tubes: New life of television scanning optical microscopy. *Radiation Measurements* **42** (4-5) 933-36

[10] Zorenko Yu V, Konstankevich I V, Gorbenko V I and Yurchishin P I 2002 *Journal of Applied Spectroscopy* **69** 769-76

[11] Zorenko Yu, Gorbenko V, Konstankevich I, Grinev B and Globus M 2002 Scintillation properties of $Lu_3Al_5O_{12}$:Ce single crystalline films *NIM* A**486** 309-14

[12] Zorenko Yu, Gorbenko V, Konstankevych I, Voloshinovskii A, Stryganyuk G, Mikhailin V, Kolobanov V and Spassky D 2005 Single – crystalline films of Ce-doped YAG and LuAG phosphors: advantages over bulk crystals analogues *Journal of Luminescence* **114** (4) 85-94

[13] Zorenko Y, Gorbenko V,Mihokova E, Nikl M,Nejezchled K,Vedda A,Kolobanov V and Spassky D Single crystalline film sicintilators based on Ce-and Pr-doped aluminium garnets 2007 *Radiation Measurements* **42** 521-27

[14] Nikl M, Mihokova E, Pejchal J, Vedda A, Zorenko Yu and Nejezchled K 2005 The antisite Lu_{Al} defect – related trap in $Lu_3Al_5O_{12}$:Ce single crystal *Phys.Stat.Sol.* b **242** R119-21

[15] Prsusa P,Cechak T,Mares J A, Nikl M, Zorenko Yu V, Gorbenko V I, Tous J and Blazek K 2008 The α-particle excited scintillation response of the liquid phase epitaxy grown LuAG:Ce thin films *Applied Physics Letters* **2**(1) 1-3

[16] Zorenko Yu, Voloshinovskii A,Savchyn V, Vozniak T,Nikl M, Nejezchled K, Mikhailin V, Kolobanov V and Spassky D 2007 Exciton and antisite defect – related luminescence in $Lu_3Al_5O_{12}$ and $Y_3Al_5O_{12}$ garnets *Phys.Stat. Sol.* b **244** 2180-89

[17] Nikl M, Pejchal J, Yoshikawa A, Fukuda T, Krasnikov A, Vedda A, and Nejezchleb K 2006 Pr^{3+} doped novel single crystal scintillators ed.Getkin A and Grinyov B *SCINT 2005 Proc.of the 8 Inter.Conf. on inorganic scintillators and their use in science and industrial applications*, Alishta, Crimea Ukraine, Kharkiv 89-94

[18] Zorenko Y and Gorbenko V 2007 Growth peculiarities of the $R_3Al_5O_{12}$ (R=Lu,Yb,Tb,Eu-Y) single crystalline film phosphors by Liquid Phase Epitaxy *Radiation Measurements* **42** 907-10

[19] Zorenko Y, Gorbenko V, Konstankevych I, Voznjak T, Savchyn V, Nikl M, Mares J A, Nejezchleb K, Mikhailin V, Kolobanov V and Spassky D 2007 Peculiarities of luminescence and scintillation properties of YAP:Ce and LuPA:Ce single crystals and single crystalline films *Radiation Measurements* **42** 528-32

[20] Zorenko Y, Gorbenko V, Batentschuk M and Winnacker A 2009 Growth and luminescent properties of single crystalline films of $RAlO_3$ (R=Lu,Lu-Y,Y,Tb) perovskites *Phys.Stat.Sol.* accepted

[21] Zorenko Yu, Konstankevych I, Gorbenko V and Zorenko T 2002 Growth and characteristics of crystallophosphors on the base of single crystalline films of LuAG and YAG garnets 2002 *Molecular Physics Reports* **36** 127-32

[22] Zorenko Yu, Gorbenko V and Konstankevych I 2003 Peculiarities of growth of phosphors based on the single crystalline films of garnets by liquid phase epitaxy method *Surface* **5** C83-90

[23] Babin V,Gorbenko V,Makhov A,Mares J A, Nikl M, Zazubovich S and Zorenko Yu 2007 Luminescence characteristics of $Pb^{2}+\rightarrow Ce^{3+}$ energy transfer processes *Journal of Luminescence* **127**(2) 384-90

[24] Zorenko Y, Gorbenko V, Savchyn V, VoznyakT, Nikl M, Mares J A, Beitlerova A and Kucerkova R 2009 The luminescent and scintillation properties of YAP and YAP:Ce single crystalline films grown by liquid phase epitaxy from BaO-based flux *Phys.Stat.Sol.* a submitted;
Globus M and Grinyov B 2000 *Non – inorganic scintillators* (Akta, Kharkiv)

[25] Babin V, Bichevin V, Gorbenko V, Makhov A,Mhokova E,Nikl M, Vedda A, Zuzubovich S and Zorenko Yu 2009 Luminescence of dimmer lead centers in aluminium perovskites and garnets *Phys.Stat.Sol.* B **246**(6) 1318-26

[26] Gorbenko V,Krasnikov A, Nikl M, Zuzubovich S and Zorenko Yu 2009 Luminescence characteristics of LuAG:Pr and YAG:Pr crystalline films *Optical Materials* **31** 1805-07

[27] Valbis J A, Volzhenskaja L G,Dubov Yu G, Zorenko Yu V, Nazar I V and Patszagan N I 1987 Luminescence centers in single crystalline compounds of yttrium-aluminium garnet doped by scandium isoelectronic impurity *Optics and Spectroscopy* **63** 1058-63

[28] Zorenko Yu V 2006 Luminescence of La^{3+} and Sc^{3+} isoelectronic impurities in $Lu_3Al_5O_{12}$ single crystalline films *Optics and Spectroscopy* **100** 572-80

[29] Zorenko Y, Mares J A, Kucerkova R, Gorbenko V, Savchyn V,Voznyak T, Nikl M,Beitlerova A and Jurek K 2009 Optical, luminescence and scintillation characteristics of the Bi-doped LuAG and YAG single crystalline films *J.Phys*.D *Appl Phys*. **42** 7

Journal of Physics: Conference Series **289** (2011) 012030 doi:10.1088/1742-6596/289/1/012030

Structure prediction of Al_nO_m clusters

P Smok
Institute of Physics, AJD Częstochowa, Al. Armii Krajowej 13/15, PL-42200 Częstochowa, Poland

E-mail: p.smok@ajd.czest.pl

Abstract. Genetic algorithm simulations, using Buckingham potential to represent the anion-anion and cation-anion short-range interactions, were performed in order to predict the equilibrium positions of the Al and O ions in Al_nO_m clusters. In order to find the equilibrium structures of compounds a self-organizing genetic algorithm were constructed. The calculation were carried out for several clusters Al_nO_m, with different numbers of aluminium and oxygen atoms.

1. Introduction

Determining the global energy minimum of a molecule or a crystal is a serious problem. Depending on the goals and levels of investigation, the elementary object of consideration can be the overall of a crystal, a single molecule or an ensemble of molecules. The methods employed to determine the structure of the ground state and low energy of molecules can be categorized as either deterministic or stochastic. The deterministic methods, such as ab initio[1] and DFT[2] are generally considered to be most informative but are computationally to expensive to implement, especially for large clusters or an ensemble of molecules. Semi empirical methods employs potentials with parameters adjusted to experimental data. Such potentials can often include effects beyond the scope of the formalism[3]. Therefore results obtained by the means of these potentials are often used as a start point for DFT calculation. The stochastic methods such as Monte Carlo[4] or molecular dynamics simulated annealing[5,6] often encounter with difficulties[7]. They require high accuracy in using force fields, a proper accounting for interactions, and considerable computer time.

Genetic algorithms, proposed by Holland[8], are global optimization method which are inspired by Darivinian natural evolution. A conventional genetic algorithm can be described in four steps. a) It uses an encoding of the control variables, called individuals; b) Genetic algorithm searches from one population of possible solutions to another population; c) It uses only objective function information (not derivatives); d) GA bases on probabilistic rules(not deterministic). It is an intelligent and information efficient approach[9] to multi-variable, global optimization and has been successfully applied to a large number of complex problems[10-16].

Genetic algorithm developed and tested in this work starts from a randomly generated initial population. We use the Cartesian coordinates of the ions as a natural coding, and each cluster's fitness is computed based on its potential energy. After the fitness has been assigned to each cluster we change the position of each ion in the cluster. A new ion's position is accepted on condition that new cluster's fitness is better that the previous one. This procedure is repeated for many generations until the global minimum or at least a very good local minimum has been found.

In the following section of this paper we describe the above algorithm in detail. In section III the algorithm is applied to Al_nO_m clusters.

Journal of Physics: Conference Series **289** (2011) 012030 doi:10.1088/1742-6596/289/1/012030

2. Method

The model of molecular problem consists of N ions. The potential energy, V, of this system is given by the Buckingham potential:

$$V=\Sigma\ (q_i q_j/r_{ij}) + A_{ij}\exp(-r_{ij}/\rho_{ij}) - C_{ij}r^{-6} \quad (1)$$

Where the summation over i and j includes unique pairs of ions, and q_i is the charge of ion i, and r_{ij} denotes the distance between ions i and j. The first term represents the Coulomb interaction between an ion i of charge q_i and an ion j of charge q_j that are a distance r_{ij} apart and the remaining terms constitute the Buckingham potential representing the anion-anion and cation-anion short-range interactions. Although alternative functions can be used for the latter, the Buckingham form has proved to be particularly effective. The potential parameters, A_{ij}, ρ_{ij}, and C_{ij}, may be derived using standard empirical fitting techniques[17,18]. The sets of parameters A_{ij}, C_{ij}, and ρ_{ij} for each pair of ionic species are displayed in Table I.

Table 1. Values of the parameters used in Buckingham interactions[17,18].

Interaction	**A_{ij} (eV)**	**ρ (Å^{-1})**	**C_{ij} (eV Å^{-1})**
Al - Al	0.0	0.100	0.00
O - O	22764.0	0.149	27.88
Al - O	1460.3	0.299	0.00

A flow chart representing the operation of our cluster geometry optimization program is shown in Fig. 1. A starting generation consisting of a certain number N_{pop} (100 in our case) of clusters is generated. The Cartesian coordinates of each cluster ions are chosen randomly in a cubic box with length

$$L = (N{\cdot}\Omega)^{1/3} \quad (2)$$

where N is the number of ions per cluster, and Ω (800 Å in our case) is the volume per ion. In our algorithm we use very simple crossover and mutation operators. The coordinates of each ion are changed independently by a little distortion:

$$X_{i,i+1}=X_{i,n}\cdot(1+0,5\ DITS\cdot RND[-0,5;0,5]) \quad (3)$$

Where $DIST$ is a distortion parameter (0.125 in our case), $RND[-0.5,0.5]$ [19] is a random number from the range of -0.5 to 0.5, and i denotes x, y, z coordinates. After a coordinate of the ion has been changed the potential energy (Eq. 1) is calculated for each cluster, and the new position of the ion is accepted only on condition that the potential energy of the new ions configuration is smaller than the old one, otherwise the previous position of the ion is restored. This procedure is repeated until a stable configuration has been found or there is no change of ions position of the cluster for more than 1000 generation.

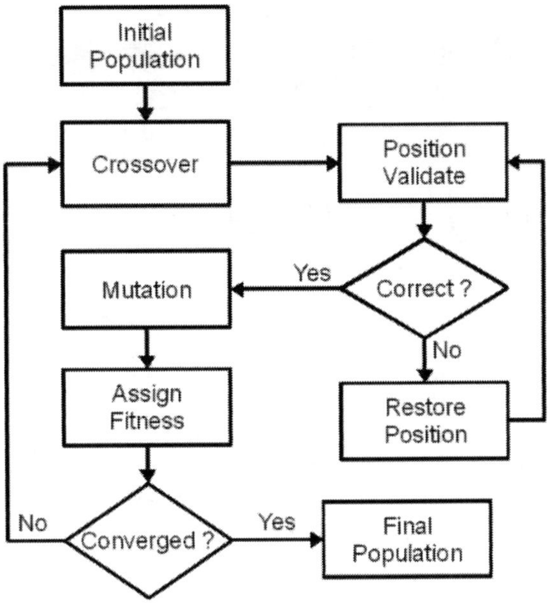

Figure 1. Schematic flow chart for the cluster geometry optimization GA program

To ensure that the whole space of consideration is taken into account we introduce a mutation operator. For each cluster we choose one ion for which a new position is generated randomly. The probability of the mutation we set to 0.0125, this step of the algorithm significantly slows the calculation. In order to control the process of forming the clusters we impose restrictions on the distance between ions and the positions of the ions. A minimal r_{min} distance between two ions we set to 0.5 Å and 1.5 Å, whenever a distance between two ions is smaller than r_{min} , a new position of the ion is generated. The second restriction reduces the space allocated to the cluster, every time the position of the ion is outside of the cubic box a new position is generated.

3. Results and discussion

We now present the simulation results for several populations composed of clusters with different numbers of aluminium and oxygen atoms. The results show a huge number of different configurations and therefore we had to divide clusters into two groups. The first group consists of clusters that have at least one symmetry element whereas the second group holds clusters without symmetry elements. Below we consider only clusters from the first group.

In order to present our results in clearly form we grouped all clusters according to the number of aluminium atoms. The lowest energies we observe for Al_1O_3 (-3.66 eV)(fig. 2c), Al_2O_4 (-7.72 eV)(fig. 4d.), Al_3O_6 (-11.93 eV)(fig. 5e.) and Al_4O_7 (-14.42 eV)(fig. 6f.). Figure 3 shows that, in all four groups, the binding energy increases when the number of oxygen atoms reach the critical value. The main cause of this fact is the strong repulsive interaction between oxygen ions. This stops clusters growth and leads to the splitting of the forming clusters into small fraction what significantly disturb the positions of the ions inside the clusters. The origin of this distortion can be found in nonbonding interaction between separated clusters. For these cases we do not observe clusters with symmetry elements.

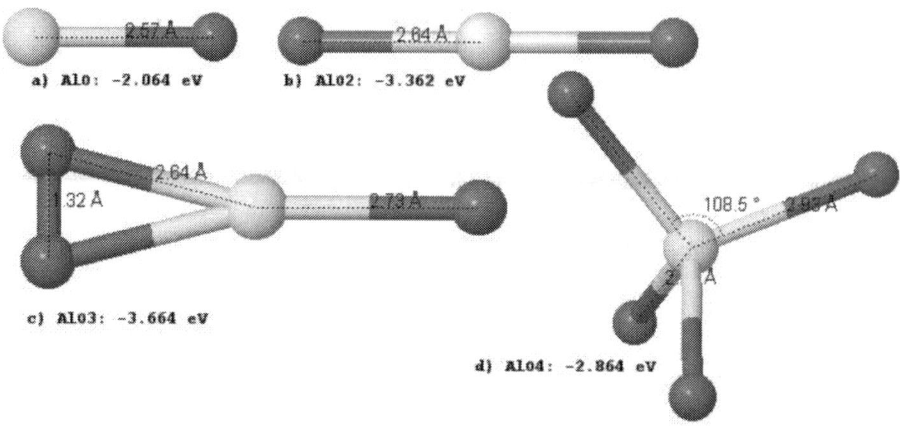

Figure. 2. Low energy clusters Al_1O_m. The numbers under the structures are binding energies in eV.

Further calculations showed simple solution of this problem. A strong repulsive interaction between oxygen ions can be reduced by forming O_2 groups(fig. 2c, fig. 4bcde, fig. 5ae, fig. 6abc) and O_3 groups. In some cases, especially for large number of atoms, O_3 groups take a shape of a line or an arc which are always composed into clusters. The energy binding of clusters containing O_3 groups are always higher than those with O_2 groups.

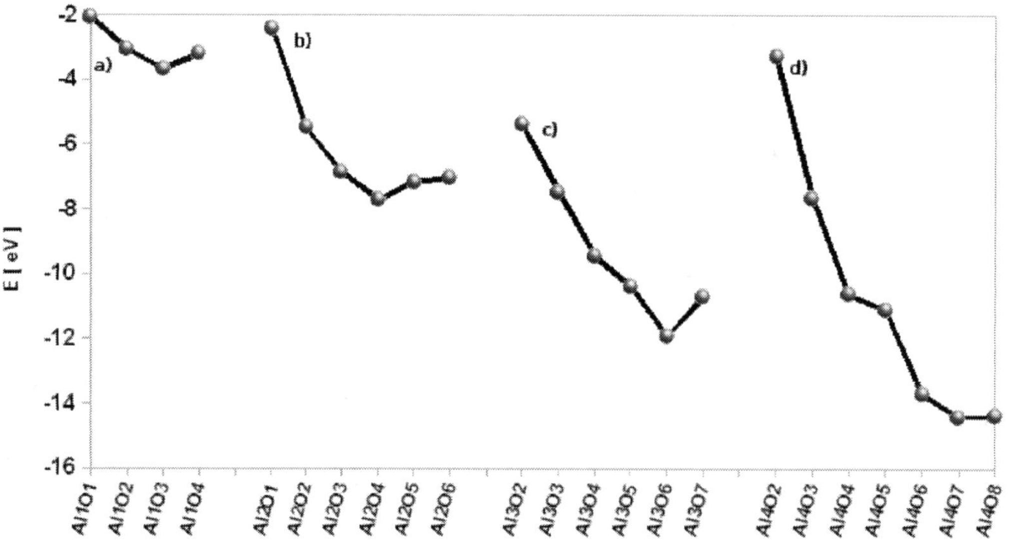

Figure 3. Binding energies for Al_nO_m clusters.

Journal of Physics: Conference Series **289** (2011) 012030 doi:10.1088/1742-6596/289/1/012030

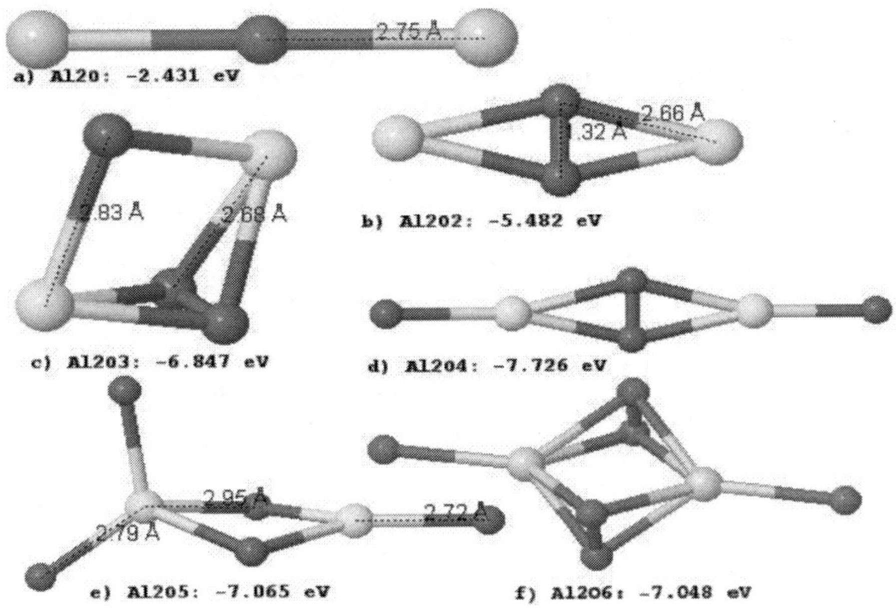

Figure 4. Low energy clusters Al_2O_m. The numbers under the structures are binding energies in eV.

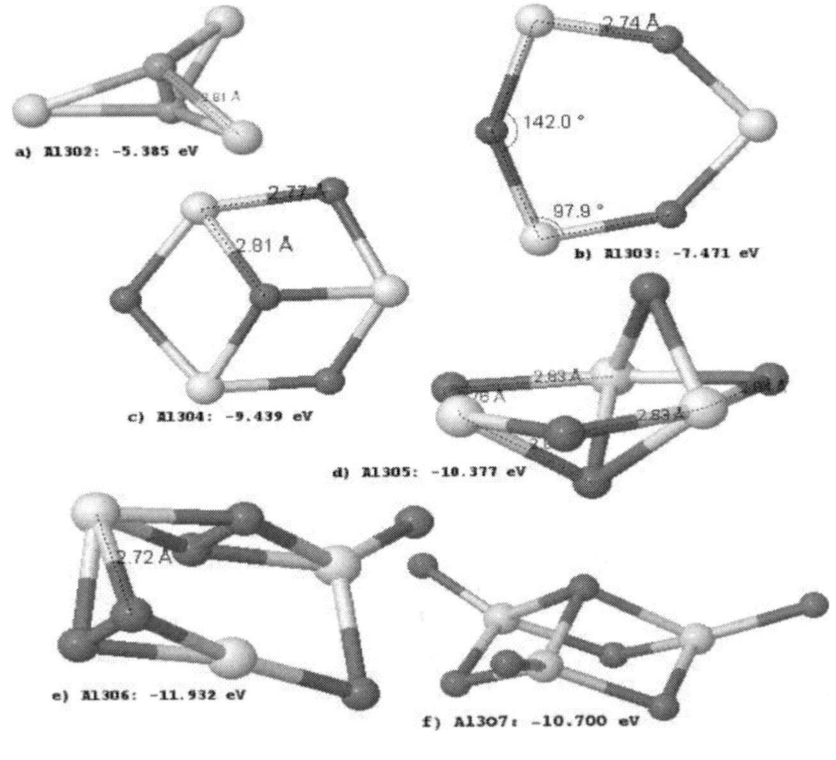

Figure 5. Low energy clusters Al_3O_m. The numbers under the structures are binding energies in eV.

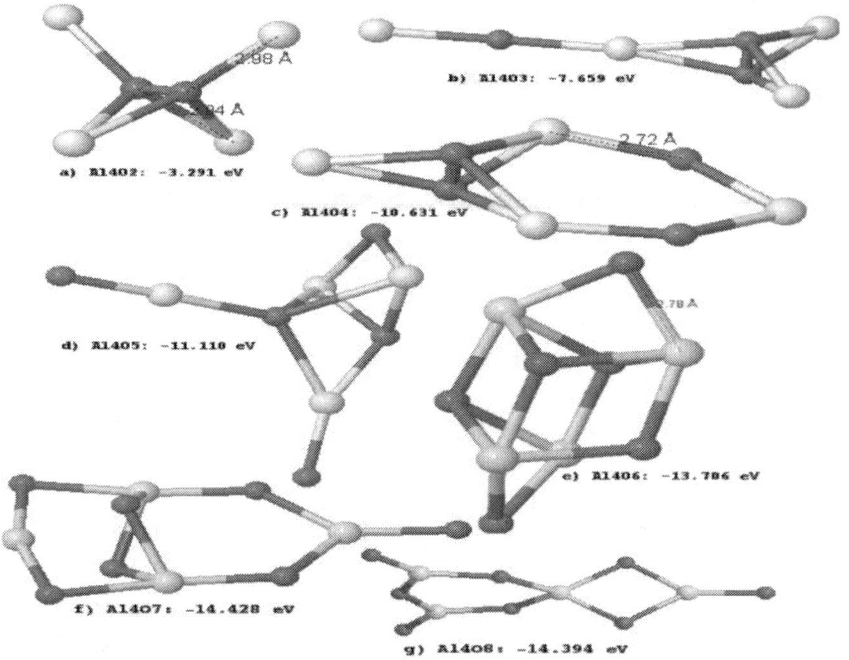

Figure 6. Low energy clusters Al_4O_m. The numbers under the structures are binding energies in eV.

The calculations showed that the clusters grow can be controlled by the r_{min} parameter. A small value (0.5 Å in our case) produces O_2 groups what leads to close clusters Fig. 4c, Fig. 4f, Fig. 5a, Fig. 5e, Fig. 6a, whereas a big value (1.5 Å in our case) produces loosely structured clusters Fig. 4e, Fig. 5b, 5c, 5d, 5f, Fig. 6d-f. Each O^{2-} oxygen ion is surrounded either by 2 or 3, 4 and 5 Al^{3+} ions. The average Al-O bondlength is 2.67 Å(fig. 7) and it is bigger than the Al-O distance observed in α-Al_2O_3[20,21]. The smallest (2.57 Å) Al-O distance we observe for AlO cluster whereas the highest value 2.98 Å we found for Al_4O_2. In contrary to Al-O the Al-Al distance distribution (fig. 7) has three peaks at 3.91 Å, 4.41 Å and 5.53 Å. The Al-Al distances are closely connected with the position of O^{2-} oxygen ions in the clusters. The first peak at 3.91 Å can be assigned to clusters with O^{2-} oxygen ions surrounded by three Al^{3+} ions(Al_3O_2, Al_3O_4, Al_3O_7, Al_4O_4, Al_4O_6), the third peak can be assigned with clusters containing linear Al_2O or Al_2O_2 conformation and the last one with cluster containing disordered Al_2O_2 conformation or Al_3O_3 rings. In consequence we do not observe direct Al-Al bonds. Tree sharp peaks can be seen for O-O distance distribution. The first peak at 1.32 Å is connected with direct O-O bonds, whereas the second one at 3.94 Å is strongly connected with clusters composed of O_2 conformations and O^{2-} oxygen ion(Al_2O_3, Al_2O_6, Al_3O_6). The last one at 5.13 Å can be assigned to cluster with Al^{3+} ions surrounded by three O^{2-} oxygen ions. We should noticed that the agreement between previous reported DFT and ab initio calculations[22-30] and our results is good. The difference in bondlength can be reduced by adjusting the ions charge.

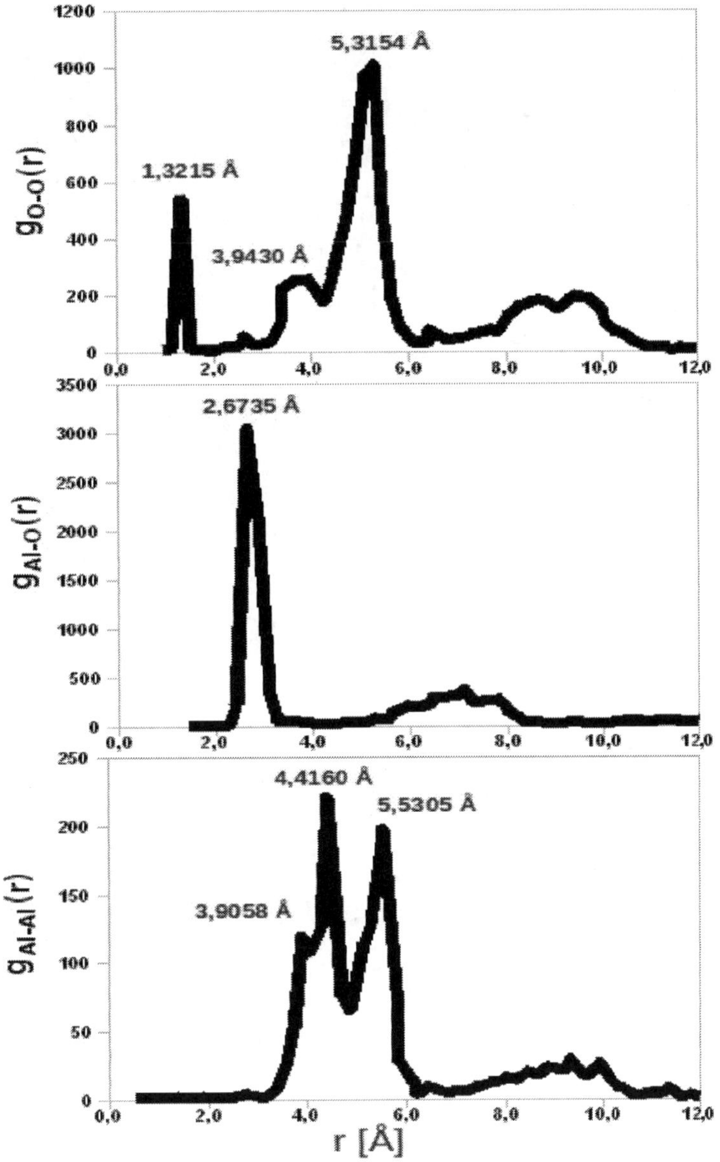

Figure 7. The distribution of interatomic Al-Al, Al-O and O-O distances.

4. Conclusions

The main aim of this paper was to report the method and describe the success of our approach in finding the equilibrium positions of Al and O ions in Al_nO_m clusters. We can control the process of forming Al_nO_m clusters either by the means of minimal distance between ions or the ions charges. Our algorithm is applicable to complicated chemical problems. There are no limits for number of ions or finite cluster. Comparison with DFT and ab initio results shows a convincing consistency with the cluster structures obtained in this work.

References

1. Hehre W J and Random L and Schleycher R and Pople J A 1986 *Ab Initio Molecular Theory* (Wiley: New York)
2. Ho K M and et al, 1998 *Nature*(London) **392** 582
3. Herman F and McLean A D and Nesbet R K 1973 (Plenum Press: New York)
4. Wille L T 1987 *Chem. Phys. Lett.* **133** 405
5. Farges J and de Feraudy M F and Raoult B and Torchet G 1986 *J. Chem. Phys.* **84** 3491
6. Garzon I L and Avalos Borja M 1989 *Phys. Rev.* **B40** 4749
7. Doy J P K and Wales D J 1997 *J Chem Soc Faraday Trans* **93** 4233
8. Holland J 1975 *Adaptation in Natural and Artifical Systems* (Ann Arbor: University of Michigan Press)
9. Hartke B 1993 *J. Phys. Chem.* **97** 9973
10. Dong X and Ming Wang G and Blaisten-Barojas E 2004 *Phys. Rev.* **B70** 205409
11. Chuang F-Ch and Wang C Z and Ho K H 2006 *Phys. Rev.* **B73** 125431
12. Takeuchi H 2006 J. *Chem. Inf. Model.* **46** 2066
13. Hamad S and Catlow C R A and Woodley S M and Largo S and Mejias A 2005 *J. Phys. Chem.* **B** **109** 15741
14. Oria O and Bazterra V E and Caputo M C and Facelli J C and Fuentealba P and Ferraro M B 2006 *Phys. Rev.* **A73** 053203
15. Tang M and Wang C Z and Lu W C and Ho K M 2006 *Phys. Rev.* **B74** 195413
16. Takeuchi H 2006 *J. Chem. Inf. Model.* **46** 2066
17. Catlow C R A and Stoneham A M 1983 *J. Phys.* **C16** 4321
18. Catlow C R A and James R and Mackrodt W C and Stewart 1982 R F, *Phys. Rev.* **B25** 1006
19. Matsumoto M and Nishimura T 1998 *ACM Transactions on Modeling and Computer Simulation* **8** 3-30
20. Gutierrez G and Johansson B 2002 *Phys. Rev.* **B65** 104202
21. Rambauty C and Jobicz H and Jaffrezicx H and Kohanoffk J and Fayeulley S 1998 *J. Phys. Condens. Matter* **10** 4221
22. Hongbin Wu and Li X and Wang X B and Ding C F and Wanga L S 1998 *J. Chem. Phys.* **109** 449
23. Alexey L Y and Zhidomirov G M 1999 *Catalysis Letters* **63** 91
24. Martinez A and Tenorio F J and Ortiz J V 2001 *J. Phys. Chem.* A**105** 11291
25. Adiga S P and Zapol P and Curtiss L A 2006 *Physical Review* **B74** 064204
26. Gowtham S and Lau K C and Deshpande M and Pandey R and Gianotto A K and Groenewold G S 2004 *J. Phys. Chem.* **A108** 5081
27. Martinez A and Tenorio F J and Ortiz J V 2003 *J. Phys. Chem.* **A107** 2589
28. Chang Ch and Patzer A B C and Sedlmayr E and Steinke T and Sulzle D 2000 *Chemical Physics Letters* **324** 108
29. Tenorio F J and Murray I and Martinez A and Klabunde K J and Ortiza J V 2004 *J. Chem. Phys.* **120** 1
30. Guevara-García A and Martinez A and Ortiz J V 2005 *J. Chem. Phys.* **122** 214309
31. Pinto H P and Nieminen R M and Elliott S D 2004 *Physical Review* **B70** 125402
32. Gowtham S and Lau K C and Deshpande M and Pandey R and Gianotto A K and Groenewold G S 2004 *J. Phys. Chem.* A **108** 5081

Journal of Physics: Conference Series **289** (2011) 012031 doi:10.1088/1742-6596/289/1/012031

Molecular dynamic investigation of atomic distribution in molten Al-Al_2Cu eutectic alloy

S Mudry[1], **I Shtablavyi** [1] **J. Rybicki** [2,3]

[1] Physics of Metals Department, Ivan Franko National University of Lviv, Kyrylo and Mephodii Str. 8, 79005 Lviv, Ukraine
[2] Faculty of Technical Physics and Applied Mathematics, Technical University of Gdansk, Narutowicza 11/12, 80-952 Gdansk, Poland
[3] Institute of Mechatronics, Nanotechnology and Vacuum Techniques, Koszalin University of Technology, Sniadeckich 2, 75-453 Koszalin, Poland

e-mail: sihor@ukr.net tel. +380-322-964594

Abstract. Structure of $Al_{0.83}Cu_{0.17}$ eutectic alloy has been studied in liquid state by means of X-ray diffraction and molecular dynamic methods. The total and partial structure factors are analyzed. These functions were used in order to calculate pair correlation functions and partial correlation functions, whose analysis allowed us to conclude that Al_2Cu-like atomic groups are distributed in Al-matrix.

1. Introduction

Growth of interest to Al-Cu binary alloys is motivated by opening of new spheres of their practical application although these materials are used during long time. Besides, Al-Cu binary is the base for synthesizing of such comparatively new (1984) materials as quasi crystals. In order to understand the mechanism of processes aimed the improvement of properties the complex studies of structure features including the liquid state are needed.

The structure of liquid Al-Cu has been investigated by diffraction methods and results are published in [1]. Analysis of these data allowed us to conclude that the structure of this alloy is far from random atomic distribution. Tendency to preferred interaction of unlike atoms is confirmed by structure sensitive physical properties measurements. Particularly, enthalpy of mixing shows the negative value with minimum corresponding to Al_2Cu stoichiometry. It was noted that chemical ordering in Al_2Cu persists upon melting but it is unclear whether this ordering influences the structure of another melts in this system.

In this work we represent the results on diffraction and molecular dynamic studies of $Al_{0,83}Cu_{0,17}$ eutectic melt.

2. Experimental

The diffraction studies were carried out using a high-temperature diffractometer with a special attachment that allows to investigate the solid and liquid samples at different temperatures up to 1800 K. Cu-K_α radiation monochromatized by means of LiF single crystal as a monochromator and Breg-Brentano focusing geometry were used. The scattered intensities as a function of the scattering angle were recorded within the range $1Å^{-1} < s < 7Å^{-1}$, with different angular step, which was equal to 0.05°

Journal of Physics: Conference Series **289** (2011) 012031 doi:10.1088/1742-6596/289/1/012031

within the region of principal peak and 0.5° at rest values of wave vectors. The measuring of scattered intensity was done with accuracy, better than 2%. In order to obtain the more accurate scattered intensities, the scan time was equal to 100 s. The diffracted intensity was recorded using a NaI(Tl) scintillator detector in conjunction with an amplification system. The sample was placed in a rounded cup of 20 mm diameter.

Scattered intensities as functions of scattering angels were recorded and corrected on absorption, anomalous dispersion and incoherent scattering [2].

The structure factors (SF) were obtained from angular dependences of scattered intensities. Pair correlation functions (PCF) were calculated from SF by means of Fourier-transformation. From this functions the main structure parameters – first and second peak positions k_1, k_2, r_1, r_2, number of neighbors – Z and first peak height $a(k_1)$ were determined.

In our work for MD simulations we have used nanoMD program, developed at the Gdansk University of Technology by M. Bialoskorski. This C++ program was created with typical non-equilibrium molecular-dynamic (NEMD) applications and has previously been used with success in various studies [3 - 5]. Our MD simulations were performed at constant volume regime. A usual time step is in order of 1fs, and in our case it was equil 2.5 fs. We have decided to choose a fourth-order Gear predictor-corrector algorithm as the numerical integrator. Nose, Nose-Hoover and Gaussian thermostats have also been implemented.

In this work we chose to use the Sutton-Chen (SC) potential [6, 7] which has a simple power law form and relatively long-range character.

The SC many-body potential has the form

$$U_{tot} = \sum_i U_i = \sum_i \epsilon \left[\sum_{j \neq i} \frac{1}{2} V(r_{ij}) - c\rho_i^{1/2} \right], \tag{1}$$

$$V(r_{ij}) = \left(\frac{\alpha_{ij}}{r_{ij}} \right)^n, \tag{2}$$

$$\rho_i = \sum_{j \neq i} \phi(r_{ij}) = \sum_{j \neq i} \left(\frac{\alpha_{ij}}{r_{ij}} \right)^m. \tag{3}$$

where r_{ij} is the distance between atom i and j. $V(r_{ij})$ is a repulsive pair potential between atoms i and j, accounting for the Pauli repulsion between the core electrons. The cohesion associated with atom i is captured in a local energy density ρ_i . ε sets the overall energy scale and c_i is a dimensionless parameter scaling the attractive term. a is a length parameter leading to a dimensionless form for V and r.

In calculating the structure, we included quantum corrections, leading to the quantum Sutton–Chen, or Q-SC force field.

For alloys we use the following combination rules to describe the interaction between different types of atoms:

$$\epsilon_{ij} = \sqrt{\epsilon_{ii}\epsilon_{jj}}, \tag{4}$$

$$n_{ij} = \frac{1}{2}(n_i + n_j), \tag{5}$$

$$m_{ij} = \frac{1}{2}(m_i + m_j), \tag{6}$$

$$\alpha_{ij} = \frac{1}{2}(\alpha_{ii} + \alpha_{jj}). \quad (7)$$

3. Results and discussion

Experimental structure factor for liquid $Al_{0,83}Cu_{0,17}$ eutectic alloy is interpreted comparing it with structure factors for liquid Al and Cu. As can bee seen (Fig.1)the significant disagreement between three SFs is observed. But it should be noted here that position of principal peak is significantly shifted in respect to corresponding peak of liquid aluminum. As can be seen from figure 1 and table 1 this parameter has a value 2.86 $Å^{-1}$ lying between corresponding values for liquid Al and Cu ($s_1^{Al} = 2.70 Å^{-1}$; $s_1^{Cu} = 3.00 Å^{-1}$) If suppose the formation of random atomic distribution in melt, this value should be significantly less and can be estimated using the additive law (s_1=2.75$Å^{-1}$). The same behavior is also observed in pair correlation function. Most probable interatomic distances change in similar way. These feature allowed us to conclude that atomic distribution in eutectic melt differs from one for atomic solution that was suggested.

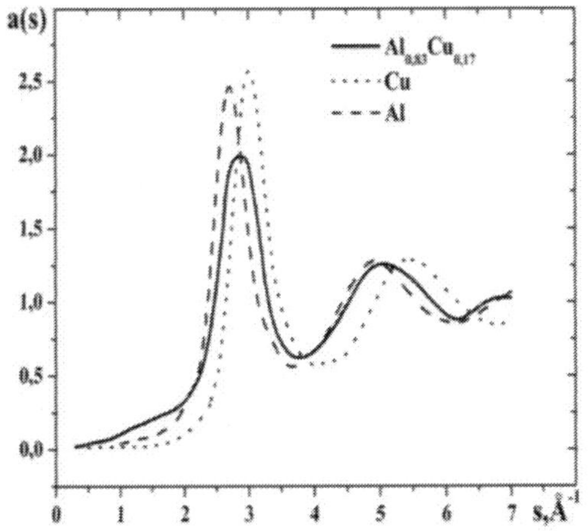

Figure 1. Structure factors for liquid $Al_{0,83}Cu_{0,17}$ eutectic alloy comparing with structure factors for liquid Al and Cu.

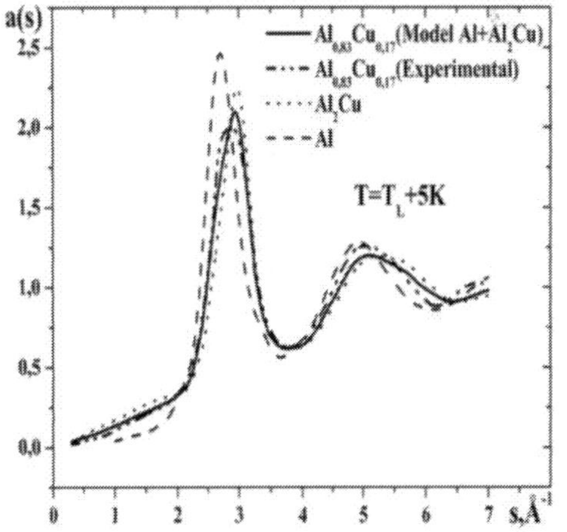

Figure 2. Experimental and model structure factors of liquid $Al_{0,83}Cu_{0,17}$ eutectic alloy.

Early we have analyzed the diffraction data for Al-Cu molten alloys with using of simple two models: random atomic distribution and quasieutectic structure [8]. It was shown that SF, calculated according to this model roughly agrees with experimental structure factor. At the same time it should be noted that some discrepancy is observed in values of principal peaks and lager disagreement reveals in main peaks region where the maximum existence display the presence of medium range order [9].

In order to avoid the discrepancies mentioned above in this work we use another model, which suppose an existence of Al_2Cu- clusters and Al-based structural units (Fig. 2). In case of this model the discrepancy between calculated and experimental values becomes less. Therefore one can suppose that Al_2Cu– nanogroups are arranged in matrix of Al.

In order to confirm this suggestion the structure of $Al_{0.83}Cu_{0.17}$ eutectic melt have been studied by means of molecular dynamic (MD) method. In our calculations the cubic cell containing 10000 atoms was chosen as initial structural unit. Volume of this cell was determined assuming that model density equals to experiment one. The calculated SF and experimental one show a good agreement (Fig. 3). Using a model atomic configuration a Faber-Ziman partial structure factors (PSF) and partial pair correlation function (PPCF) were calculated (Fig. 4).

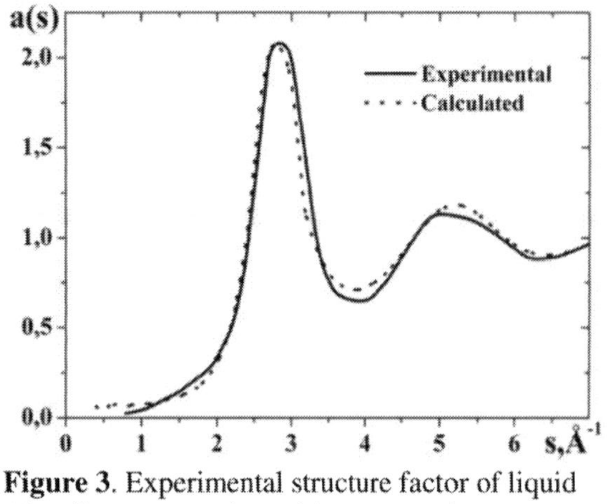

Figure 3. Experimental structure factor of liquid $Al_{0,83}Cu_{0,17}$ comparing with structure factor calculated by means of MD method.

PSFs and PPCFs related to Al-Cu and Cu-Cu distribution show the features in the wave vector interval 1-2 $Å^{-1}$. Such features were investigated in [7] and it was established that they are caused by substructure formation due to preferred interaction of unlike kind atoms. Partial structure factor related to Al-al atoms distribution is close to structure factor of pure liquid aluminum that is obvious because these atoms forms a matrix containing Al_2Cu-nanogroups. On the other hand the PSF describing partial Cu-Cu atomic distribution is in disagreement with SF for liquid copper indicating the lack of cu-like structure upon alloying.

At least the PSF related to Al-Cu interaction shows the averaged profile in respect to structure factors of Al and Cu, indicating the heterocoordinated atomic distribution.

Table 1. The main structure parameters of liquid $Al_{0,83}Cu_{0,17}$ eutectic alloy.

T,K	s_1,$Å^{-1}$	s_2,$Å^{-1}$	a(s_1)	Δs	r_1,Å	r_2,Å	Z
835	2,86	5,06	1,99	0,84	2,74	4,87	10,4
885	2,84	5,02	1,96	0,88	2,76	4,85	10,5
935	2,81	5,07	1,93	0,88	2,78	4,82	10,0
Al	2,70	4,94	2,46	0,68	2,82	5,30	11,5
Cu	3,00	5,44	2,57	0,66	2,57	4,80	11,3

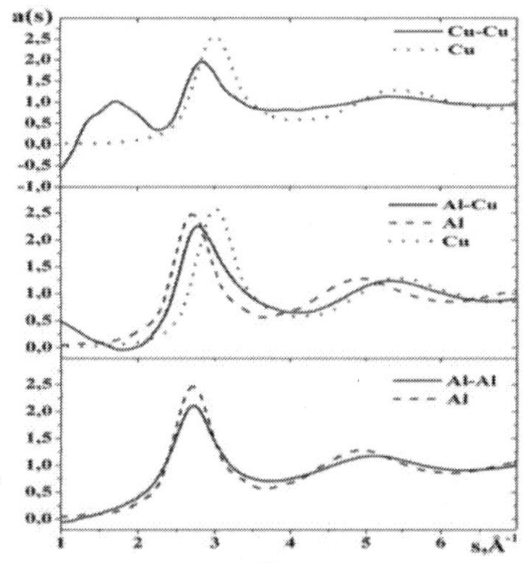

Figure 4a

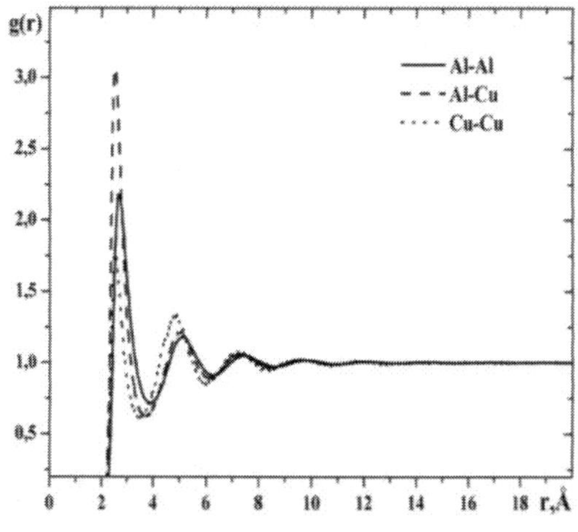

Figure 4b

Figure 4. Faber-Ziman partial structure factors (a) and partial pair correlation function (b).

4. Conclusions

The diffraction studying of $Al_{0.83}Cu_{0.17}$ eutectic alloy shows the inhomogeneous atomic distribution in liquid state at temperatures close to crystallization point. Molecular dynamic calculations confirm this kind structure.

Analysis of partial structure factors obtained by means of molecular dynamic shows the significant influence of Al_2Cu-like chemical ordering on atomic distribution in eutectic melt. It is shown that Al_2Cu-nanogroups are distributed in al-matrix.

References

[1] Brillo J, Bytchkov A, Egry I, Hennet L, Mathiak G, Pozdnyakova I, Price D L, Thiaudiere D and Zanghi D. 2006 Local structure in liquid binary Al–Cu and Al–Ni alloys *J. Non-Cryst.Sol.* **352** 4008-12

[2] Cromer D T and Waber J T 1965 Scattering factors compuyed from relativistic Dirac-Slater wave function *Acta Cryst* **18** 104-109.

[3] Dziedzic J, Rychcik-Leyk M and Rybicki J 2008 Degradation of a nano-cutting tool: An MD simulation *J. Non-Cryst. Sol.* **354**, 4309-15.

[4] Bialoskorski M and Rybicki J 2001 Mechanical Properties of Carbon Nanotubes: Simulation Program and Exemplary Results, Proc. of 8-th Workshop of PTSK Gdansk-Sobieszewo 8

[5] Bialoskorski, M, Rychcik-Leyk M, Rybicki J and Bregmanski G 2002 Molecular dynamics simulations of the ultraprecision cutting of metals, Proc. of the 9-th Workshop of PTSK Koszalin-Osieki 9

[6] Sutton A P and Chen J 1990 Long-range Finnis-Sinclair potentials *Philos. Mag. Le*tt **61** 139-146.

[7] Rafii-Tabar H Sutton A P 1991 Long-range Finnis-Sinclair potentials for f.c.c. metallic alloys *Philos. Mag. Lett* **63** 217-224.

[8] Shtablavyi I, Mudry S, Mykhaylyuk V and Rybicki J 2008 The structure of Al–Cu and Al–Si eutectic melts *J. Non-Cryst.Sol.* **354** 4469-74

[9] Elliott S R 1991 Medium-range structural order in covalent amorphous solids *Nature* **354** 445-452.

[10] Maret M Pomme T and Pasturel A 1990 Structure of liquid $Al_{80}Ni_{20}$ alloy *Phys.Rev*.B. **42** 1598-1604.

Numerical study of dependence of thermally stimulated currents in disordered solids at step heating regime on initial trap occupancy

W Tomaszewicz, P Grygiel

Department of Physics of Electronic Phenomena, Gdansk University of Technology, 11/12 Narutowicza Str., 80-233 Gdansk, Poland

E-mail: wtomasze@sunrise.pg.gda.pl

Abstract. In the present paper numerical investigation of thermally stimulated currents (TSCs) in disordered solids, measured at step heating regime, is performed. The utilized energy distribution of trapping states is the superposition of exponential and Gaussian distributions. The monomolecular carrier recombination regime is assumed. The resulting set of stiff differential equations governing TSCs is solved with the use of Gear's procedure. The calculations are made for several carrier generation and recombination rates. The TSC curves are next analyzed by the methods developed in previous papers and the calculated trap distributions are compared with the input distributions. It is concluded that the methods have satisfactory accuracy solely in the cases of week carrier retrapping or small trap occupancy.

1. Introduction

The measurements of thermally stimulated currents (TSCs) are widely applied technique of the determination of energy distribution of trapping states in disordered solids. The method consists in the photogeneration of excess carriers in the sample at low temperature and the subsequent sample heating, most frequently with a constant rate. The resulting dependence of TSC intensity on temperature is determined by the kinetics of carrier thermal release, retrapping and recombination.

In general, the interpretation of TSCs in disordered solids constitutes difficult and not fully resolved problem. This is due to uncertainity about the dominant mechanisms of carrier transport (multiple trapping (MT) or hopping) and recombination, as well as about the validity of approximations used for solving the corresponding equations. In recent years it has been established via numerical calculations that the MT model, assuming monomolecular carrier recombination, satisfactorily describes the TSCs measured in hydrogenated amorphous silicon (*a*-Si:H) [1, 2]. One of the factors complicating the TSC interpretation is the carrier retrapping, which results in gradual redistribution of captured carriers towards deeper traps. The reliable determination of energy distribution of trapping states is possible either in the case of sufficiently slow carrier retrapping [1, 2] or in the case of negligible trap occupancy [3]. The range of validity of the mentioned methods has been studied in [3] basing on the numerical calculations of TSCs for several values of carrier generation and recombination rates.

In some TSC measurements the complex heating schemes, consisting of several heating-cooling cycles, are also utilized. These measurements make possible to determine the charge released from the traps during sequential cycles as well as the corresponding trap depths (e.g., [4, 5]). However, the determination of the energetic trap distribution is subject to the same limitations as indicated above. One can state that the methods described in [4, 5] apply in the case of very slow carrier retrapping. The methods taking into account the carrier retrapping, developed in [6 - 8], are valid only in the case of small trap filling. In the present paper the numerical study of TSCs registered at step heating regime is performed and the range of applicability of these methods is investigated. The investigation is the continuation of that presented in [3], which concerns the TSCs at linear sample heating.

2. Formulas for determining trap distribution

In the step heating regime the sample temperature $T(t)$ is increased with the time t from initial temperature T_0 to higher and higher temperatures T_m (m – the cycle number) and then decreased to initial temperature, as shown in Fig. 1. The sample heating and cooling rates, $\beta_h = dT(t)/dt$ and $\beta_c = - dT(t)/dt$, are usually constant.

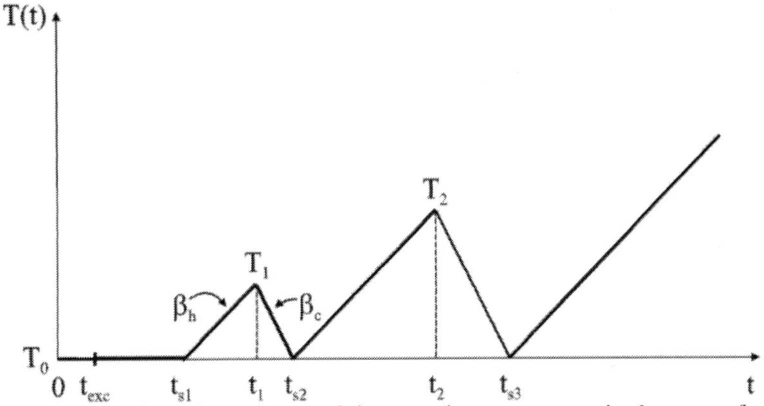

Figure. 1. Time dependence of the sample temperature in the case of step heating.

For complex heating schemes the initial rise of thermally stimulated conductivity, $\sigma(T)$, has approximately an activated character [4 - 8]. When the trapping states are continuously distributed in a wide energy region, the activation energy is approximately equal to demarcation energy ε_m, corresponding to the maximum temperature T_m of preceding heating cycle. The energy ε_m separates the traps with equilibrium ($\varepsilon < \varepsilon_m$) and non-equilibrium ($\varepsilon \geq \varepsilon_m$) occupancies. For the step heating regime the demarcation energy is given by the formula [6 - 8]

$$\varepsilon_m \approx k\left(c^* T_m - T^*\right), \quad (1)$$

Where

$$c^* = 0.967 \ln\left(46\,\mathrm{K} \cdot \nu / \beta\right), \quad T^* = 180\,\mathrm{K},$$

$$\beta = \beta_h \beta_c / (\beta_h + \beta_c)$$

and ν is the frequency factor. Therefore, the rise of TSC conductivity at the onset of (m+1)-th cycle is described by the formula

$$\sigma(T) \propto \exp\left(-\varepsilon_m / kT\right), \quad (2)$$

where k is the Boltzmann constant.

In the following it is assumed that the kinetics of carrier recombination has monomolecular character. When the carrier retrapping is negligible, the formula relating the trap density per energy unit, $N_t(\varepsilon)$, and the charge, ΔQ_m, collected during m-th heating-cooling cycle may be derived from that given in [9]. It has the form of

$$N_t\left(\varepsilon_m^*\right) \approx \frac{n_0}{\delta_f Q_\infty} \frac{\Delta Q_m}{\Delta \varepsilon_m}, \tag{3}$$

Where

$$\delta_f = 1 - \left(1 - \delta\right)\exp\left(-\tau_R C_t n_0\right),$$

$$\varepsilon_m^* = \left(\varepsilon_{m-1} + \varepsilon_m\right)/2, \tag{4}$$

$$\Delta\varepsilon_m = \varepsilon_m - \varepsilon_{m-1}.$$

In the equations n_0 is the initial density of carriers in the sample, δ_f is the final relative occupancy of traps having depth $\varepsilon \geq \varepsilon_m$, Q_∞ is the total collected charge corresponding to the whole area under TSC curve, $\delta = n_0/N_{tot}$ is the initial relative trap occupancy (with N_{tot} – the total trap density), τ_R is the mean time of carrier recombination and C_t is the carrier capture coefficient. It is seen that for the determination of absolute trap density, $N_t(\varepsilon)$, the value of initial carrier density n_0 and, in the case of final partial filling of the traps ($\delta_f < 1$), the values of δ and $\tau_R C_t n_0$ must be known.

When the initial trap occupancy is negligibly small, the relationship between $N_t(\varepsilon)$ and the charge Q_m, collected after m heating cycles has the following form [8]:

$$N_t\left(\varepsilon_m^*\right) \approx \frac{Q_\infty}{\tau_R C_t} \frac{\Delta Q_m^{-1}}{\Delta \varepsilon_m}, \tag{5}$$

where

$$\Delta Q_m^{-1} = Q_{m-1}^{-1} - Q_m^{-1}.$$

In this case the computation of absolute trap density requires the knowledge of $\tau_R C_t$ value. The above equation takes into account the carrier retrapping, contrary to Eq. (3). One can check that for small initial trap filling and negligible carrier retrapping, when $\delta \ll C_t n_0 \tau_R \ll 1$ and $\Delta Q_m \ll Q_m$ for arbitrary m, Eqs. (3) and (5) become identical.

3. Numerical method

The calculation method follows closely that of the paper [3], except for the sample heating scheme. It is assumed that the dominant contribution to the TSC is due to electrons. The kinetic equations for free electron density, $n(t)$, and trapped electron densities per energy unit, $n_t'(t,\varepsilon)$, are

$$\frac{\mathrm{d}}{\mathrm{d}t}\left[n(t) + n_t(t)\right] = G(t) - \frac{n(t)}{\tau_R}, \tag{6}$$

$$\frac{\mathrm{d}n_t'(t,\varepsilon)}{\mathrm{d}t} = C_t\left[N_t(\varepsilon) - n_t'(t,\varepsilon)\right]n(t) - \frac{n_t'(t,\varepsilon)}{\tau_r(\varepsilon,t)}, \tag{7}$$

$$n_t(t) = \int_0^{\varepsilon_t} n_t'(t,\varepsilon)\,\mathrm{d}\varepsilon, \tag{8}$$

with

$$\tau_r(\varepsilon, t) = \nu^{-1} \exp[\varepsilon / kT(t)].$$

Here, $G(t)$ is the carrier generation rate, $\tau_r(\varepsilon,t)$ is the mean carrier dwell-time in the traps of depth ε and ε_t is the maximum trap depth. The TSC conductivity is given by the formula,

$$\sigma_{\mathrm{TSC}}(t) = e\mu\, n(t), \tag{9}$$

where e is the elementary charge and μ is the free electron mobility.

As in [1 - 3], the model distribution $N_t(\varepsilon)$ of electron traps is the superposition of the exponential and Gaussian distributions. The chosen distribution corresponds approximately to that established in *a*-Si:H. The main calculation parameters are listed in Table 1. For given parameters the total density of traps $N_{tot} = 3\cdot 10^{19}$ cm^{-3} and the mean carrier capture time in empty traps $\tau_t = 1/C_t N_{tot} \approx 3.33\cdot 10^{-12}$ s.

Table 1
Parameters used in the numerical calculations

Parameter	Value
Trap density of the exponential distribution at the band edge, $N_{tE}(0)$	10^{21} cm^{-3} eV^{-1}
Characteristic energy of the exponential distribution, ε_{cE}	0.03 eV
Maximum trap density of the Gaussian distribution, $N_{tG}(\varepsilon_{mG})$	10^{16} cm^{-3} eV^{-1}
Centre of the Gaussian distribution, ε_{mG}	0.6 eV
Half-width of the Gaussian distribution, ε_{cG}	0.2 eV
Maximum trap depth, ε_t	0.9 eV
Carrier capture coefficient, C_t	10^{-8} cm^3 s^{-1}
Frequency factor, ν	10^{12} s^{-1}
Free electron mobility, μ	10 cm^2 V^{-1} s^{-1}

For the purpose of numerical calculations the model trap distribution is approximated by the 6000 discrete trap levels. The obtained set of stiff differential equations is solved with the use of modified Gear's subroutine DIFSUB [10]. In program the TSC conductivity $\sigma(t)$ is computed and next the relative collected charges $\Delta Q_m/Q_\infty$ and Q_m/Q_∞ are calculated.

It is assumed that the carrier generation rate is given by:

$$G(t) = G_0, \quad t < t_{exc},$$
$$G(t) = 0, \quad t \geq t_{exc}, \tag{10}$$

with the sample excitation time $t_{exc} = 10$ s. The maximum temperatures of sequential heating-cooling cycles (cf. Fig. 1) are determined by

$$T_m = T_0 + m\Delta T \tag{11}$$

with $T_0 = 20$ K and $\Delta T = 25$ K. The heating and cooling rates are $\beta_h = 0.05$ K/s and $\beta_c = 0.1$ K/s, the onset of first cycle is $t_{s1} = 100$ s, and the number of cycles equals to 12.

4. Results and discussion

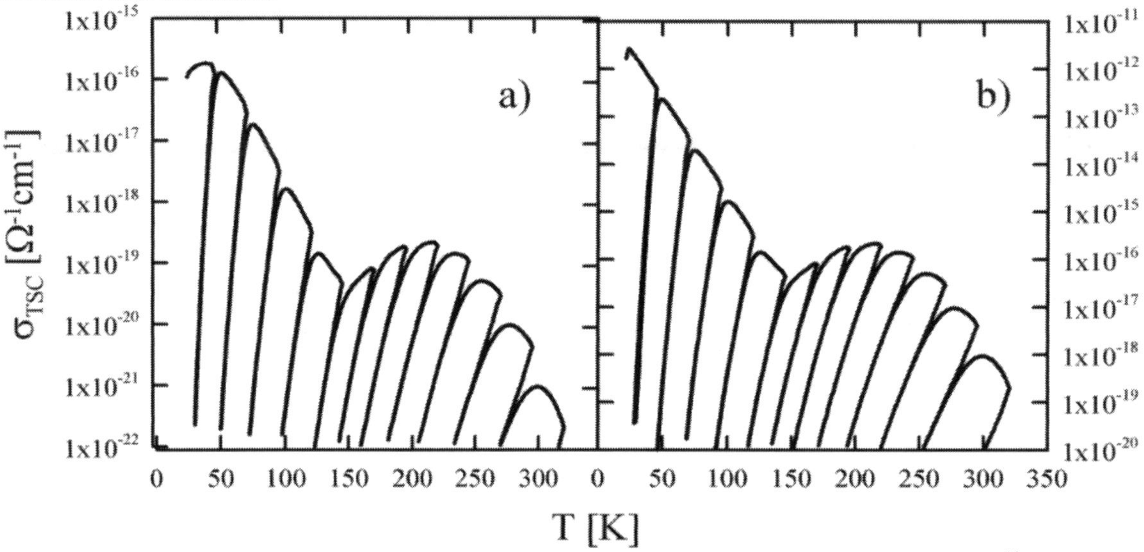

Figure 2. The TSC curves computed for low value of the carrier recombination time, $\tau_R = 10^{-10}$ s, and two values of the carrier generation rate, $G_0 = 10^{14}$ cm^{-3} s^{-1} (a) and 10^{18} cm^{-3} s^{-1} (b).

The calculated TSC curves corresponding to the values $\tau_R = 10^{-10}$, 10^{-8} and 10^{-6} s of the mean time of carrier recombination are presented, respectively, in Figs. 2, 3 and 4. In each figure two curves calculated for the values $G_0 = 10^{14}$ and 10^{18} cm^{-3} s^{-1} of the carrier generation rate are shown. The curves are marked by a) and b), respectively. Since $\tau_t \ll \tau_R$, the initial carrier density in the sample equals to $n_0 = G_0 t_{exc}$. Therefore, for given values of G_0 the initial trap filling amounts, respectively, to $\delta = 3.33 \cdot 10^{-5}$ and $3.33 \cdot 10^{-1}$. The curve envelopes are almost identical with the corresponding TSC curves computed in [3] for linear sample heating.

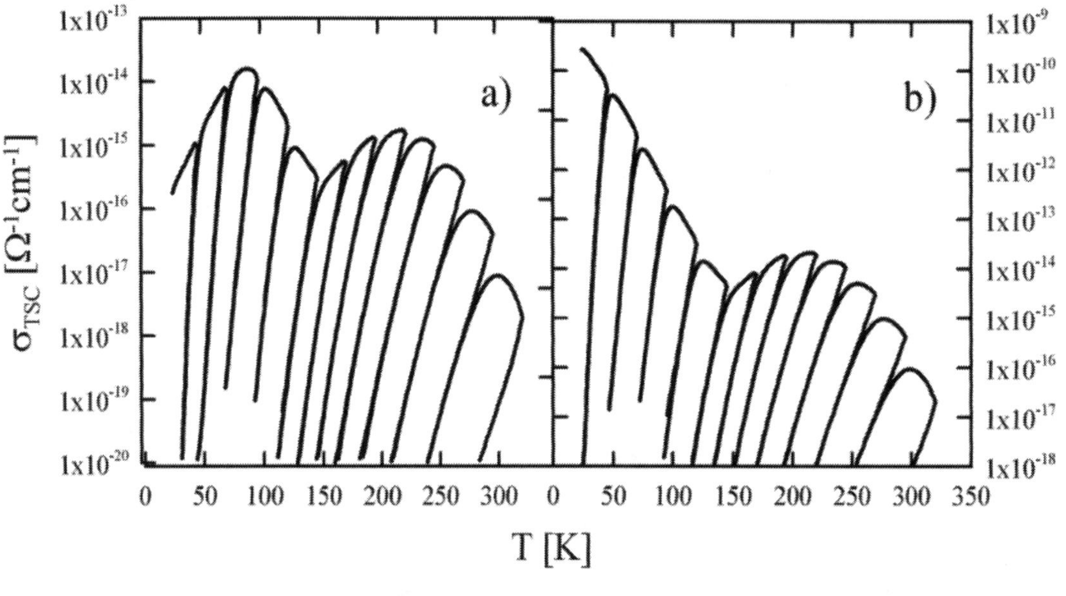

Figure 3. The TSC curves computed for intermediate value of the carrier recombination time, $\tau_R = 10^{-8}$ s, and two values of the carrier generation rate, $G_0 = 10^{14}$ cm^{-3} s^{-1} (a) and 10^{18} cm^{-3} s^{-1} (b).

Journal of Physics: Conference Series **289** (2011) 012032 doi:10.1088/1742-6596/289/1/012032

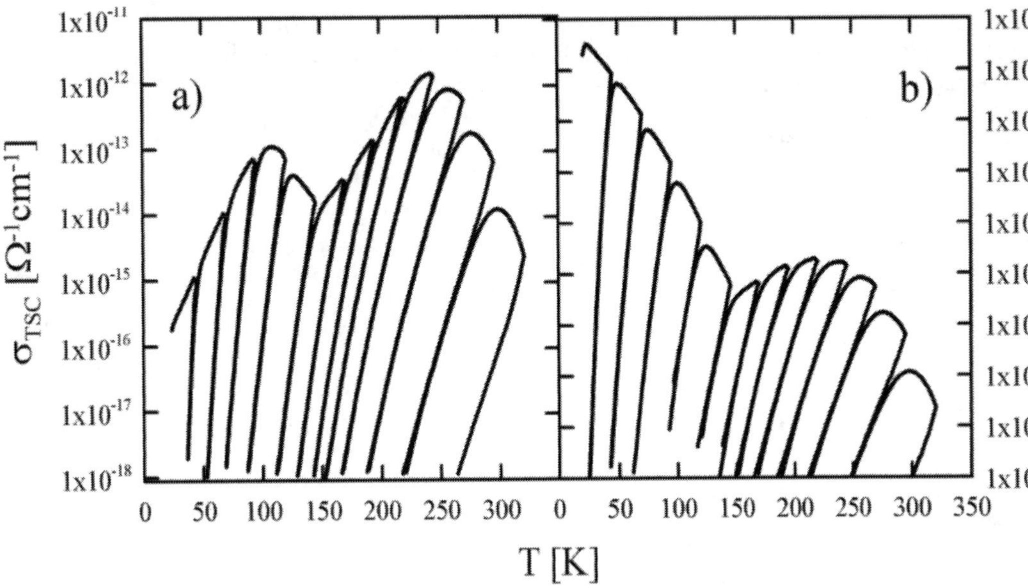

Figure 4. The TSC curves computed for high value of the carrier recombination time, $\tau_R = 10^{-6}$ s, and two values of the carrier generation rate, $G_0 = 10^{14}$ cm^{-3} s^{-1} (a) and 10^{18} cm^{-3} s^{-1} (b).

Figs. 2 - 4 illustrate the growing influence of carrier retrapping on the TSCs with the increase of carrier recombination time, τ_R. The shape of TSC curves in Figs. 2a and 2b corresponding to different values of G_0 is almost identical and their envelopes have the form similar to that of the trap distribution. This indicates that the carrier retrapping is practically negligible. The shape of TSC curves in Figs. 3a and 4a corresponding to lower generation rate G_0 significantly changes which shows the increasing influence of the carrier retrapping. Making use of Eq. (4) one can prove that the final trap occupancy is negligibly small for the curves from Figs. 2a and 3a and is almost complete for the curve from Fig. 4a. On the other hand, the form of TSC curves in Figs. 3b and 4b, which correspond to higher value of G_0, changes only slightly due to significant trap occupancy.

The obtained TSC curves are analyzed as follows. The activation energies ε_m of the initial TSC rise for sequential heating-cooling cycles are calculated from the plots of ln $\sigma_{TSC}(t)$ *versus* $1/T$. Since the initial rise of TSC is slightly faster than predicted by Eq. (2), the calculation of ε_m is somewhat arbitrary. The exemplary plot of the dependence of ε_m on the maximum temperature T_m is shown in Fig. 5. As is seen, there is rather good agreement with Eq. (1).

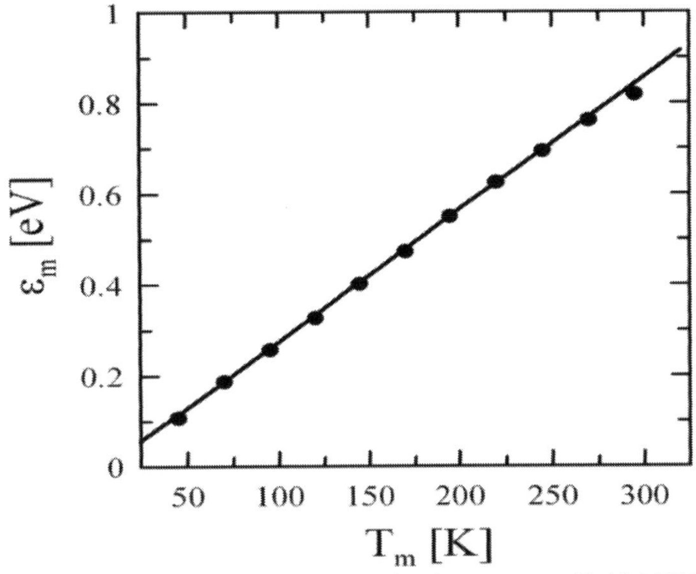

Figure 5. The dependence of the activation energy of initial TSC rise on the maximum temperature of previous heating cycle, obtained from the TSC curve in Fig. 2a (points) and calculated from Eq. (1) (solid line).

According to above discussion, for the TSC curves in Figs. 2a and 3a the energetic trap distribution may be determined from Eq. (5) and for the curves in Figs. 2b and 4b - from Eq. (3). The comparison of the calculated trap distributions (points) with the input trap distribution (continuous line) is presented, respectively, in Figs. 6a and 6b. For sufficiently small carrier recombination times the accuracy of reconstruction in both figures is quite satisfactory. However, the accuracy of reconstruction in Fig. 6b worsens with increasing recombination time of the carriers, i.e., with more intensive carrier retrapping. The present study shows that the correct interpretation of TSC measurements requires the knowledge of the extent of trap filling. This emphasizes the importance of the experimental investigations of TSC dependence on the carrier generation rate.

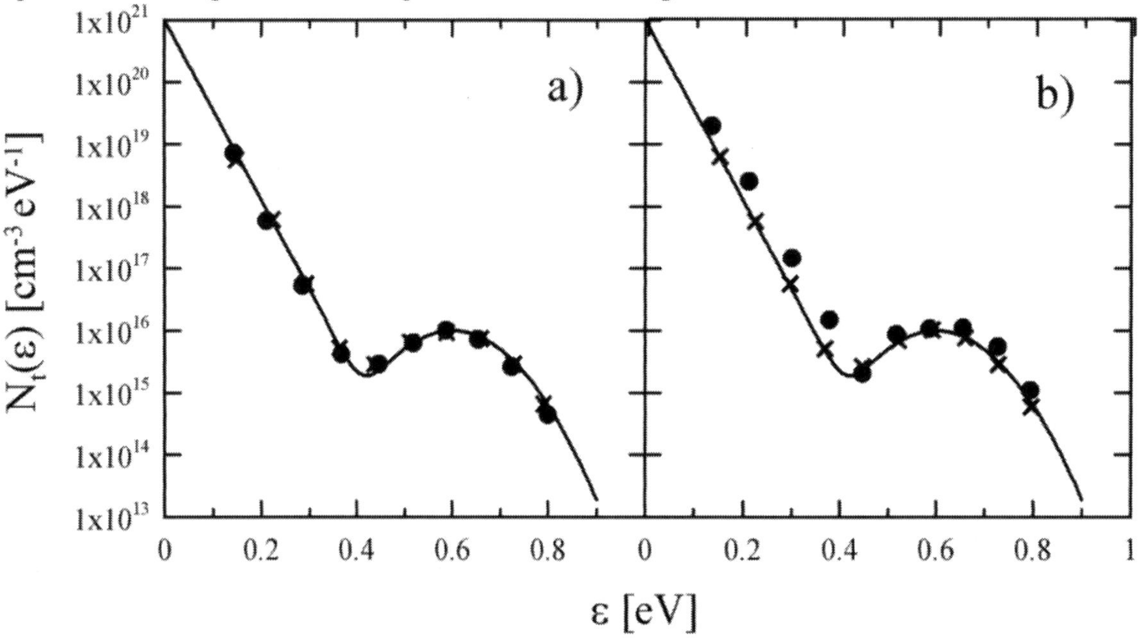

Figure 6. The output trap distributions (points) calculated from several TSCs given in Figs. 2 – 4, compared with the input distribution (solid line). The distributions in (a) indicated by (×) and (•) are obtained, respectively, from the curves in Figs. 2a and 3a making use of Eq. (5). The distributions in (b) marked by (×) and (•) are obtained using Eq. (3) from the curves in Figs. 2b and 4b, respectively.

It is worth to notice that very similar results were obtained in [3] for the linear sample heating. Although the TSC measurements at complex heating schemes are more complicated, they enable us to determine the dependence of demarcation energy on temperature in straightforward manner. On this basis one can distinguish between the cases of discrete trap levels and continuous trap distribution [4, 5].

5. Conclusion

We have demonstrated that the form of TSC curves in disordered solids strongly depends on the carrier generation and recombination rates. Subject to the monomolecular recombination kinetics, the reliable determination of the energetic trap distribution is possible either in the case of very week carrier retrapping or in the case of negligibly small trap occupancy.

References

[1] Schmidt J A Koropecki R R Arce R Dussan A and Buitrago R H 2004 Determination of the density of defect states by thermally stimulated conductivity studied from numerical simulation *J. Non-Cryst. Solids* **338-340** 322-5

[2] Main C Souffi N Reynolds S and Brüggemann R 2006 Numerical modeling of thermally-stimulated currents for the density-of-states determination in thin-film semiconductors *J. Non-Cryst. Solids* **352**

1028-31
[3] Tomaszewicz W and Grygiel P 2007 Numerical study of the dependence of thermally stimulated currents in disordered solids on initial trap occupancy *J. Phys.: Conf. Ser.* **79** 012020
[4] Gorokhovatskii Yu and Bordovskii G 1991 *Thermally Activational Current Spectroscopy of High-resistance Semiconductors and Dielectrics* (Moscow: Nauka) (in Russian)
[5] Schmechel R and von Seggern H 2004 Electronic traps in organic transport layers *Phys. Stat. Solidi* (a) **201** 1215–1235
[6] Tomaszewicz W 2004 Thermally stimulated carrier transport due to stepwise sample heating *Macromolecular Symposia* **212** 363-368
[7] Tomaszewicz W 2006 Thermally stimulated currents in amorphous solids due to complex heating schemes *J. Non-Cryst. Solids* **352** 4214-4220
[8] Tomaszewicz W and Grygiel P 2007 Thermally stimulated currents in disordered solids at step heating *Visnik Lviv. Univ.* Ser. Fiz. **40** 219-231
[9] Tomaszewicz W and Grygiel P 2000 On thermally stimulated currents under dispersive transport regime - trap saturation effects *Synth. Metals* **109** 263-6
[10] Gear W C 1971 *Numerical Initial Value Problems in Ordinary Differential Equations* (New York: Englewood Cliffs)

Study of the nature of Cu(II) complexes in aqueous ammonium oxalate solutions by ultraviolet-visible spectroscopy

E Mielniczek-Brzóska[1], K Sangwal[2]

[1] Institute of Chemistry and Environment Protection, Jan Długosz University of Częstochowa, ul. Armii Krajowej 13/15, 42-200 Częstochowa, Poland
[2]Department of Applied Physics, Lublin University of Technology, ul. Nadbystrzycka 38, 20-618 Lublin, Poland

E-mail: e.mielniczek@ajd.czest.pl

Abstract: Experimental results of an investigation of aqueous ammonium oxalate solutions containing Cu(II) impurity by ultraviolet–visible spectroscopy are described and discussed from the standpoint of speciation of complexes. The results show that absorption of light by aqueous ammonium oxalate solutions containing Cu(II) impurity in the range $-5 < \ln(c_i/c) < 2.5$ of the ratio of concentrations c_i and c of impurity and solute, respectively, leads to decrease or increase in the intensity of bands of the ultraviolet–visible spectral regions, and these changes may be expressed by full width at half maximum, molar extinction coefficient, peak wavelength and oscillator strength. The changes are caused by the coordination of $C_2O_4^{2-}$ ligand with $Cu(H_2O)_6^{2+}$ aquocomplex, and are related with the impurity–solute concentration ratio c_i/c. The coordination of $C_2O_4^{2-}$ ligand with $Cu(H_2O)_6^{2+}$ aquocomplex in the range $0 < \ln(c_i/c) < 2.5$ leads to the formation of $Cu(C_2O_4)$ complex, but the coordination of the $C_2O_4^{2-}$ ligand with $Cu(C_2O_4)$ complex in the concentration ratio range $-5 < \ln(c_i/c) < 0$ results in the formation of predominantly $Cu(C_2O_4)_2^{2-}$ complex. The effect of successive coordination of the $C_2O_4^{2-}$ ligand is well-defined in the ultraviolet spectral region but poor in the visible region.

1. Introduction

Impurities present in solutions used for growth of crystals have pronounced effect on the their growth kinetics and morphology [1–3]. The effect of ionic impurities on the growth kinetics and morphology of ionic crystals is explained from the standpoint of adsorption of individual ions at kink sites in step ledges or on the surface terrace between neighboring ledges present on the crystal surfaces in contact with the solution [1–3]. Adsorption of ions at kinks and surface terrace of ionic crystals is believed to occur as a result of electrostatic and van der Waals interactions. Although one aspires to have individual ions to adsorb on the growing surface of an ionic crystal, it is difficult to achieve individual ions in real growth systems because of the formation of coordination complexes between impurity cations and solute anions [4–6]. Indeed knowledge of the structure and stability of complexes in growth solutions containing impurities is necessary for understanding the relationship between growth kinetics of crystals and segregation coefficient of impurities in them because both of them are intimately connected with the processes of adsorption of impurity particles at the crystal–medium interface.

Spectroscopic studies are frequently employed to characterize new compounds, to analyze ions, compounds or mixtures, to follow reaction paths and rates, to investigate the constitution and concentration of species in different media, and to obtain information about the geometry of metal–ligand complexes and the type of coordination of ligands around the metal ions [7–12]. Among several diverse spectroscopic methods, ultraviolet–visible (UV–Vis) spectroscopy finds wide application in quantitative analysis of organic and inorganic compounds using dye reactions and determination of traces of impurities in mixtures of complexes and main components [7–9].

In this paper the experimental results of an investigation of dilute and concentrated aqueous ammonium oxalate solutions containing Cu(II) impurity obtained by ultraviolet-visible spectroscopy are described and discussed. The aim of the present paper is to follow the processes of formation of copper–oxalate complexes in relation to the concentration c of ammonium oxalate and the concentration c_i of Cu(II) and to establish the influence of these parameters on the chemical equilibria in the solutions. This paper is continuation of our earlier works [13,14], where some preliminary results on the investigation of copper–oxalate complexes were reported.

The structure of the paper is as follows: Section 2 briefly describes the experimental procedure used, while Section 3 gives the basic concepts used for the analysis of the experimental results described and discussed in the following sections. Analysis of chemical equilibria occurring in pure aqueous solutions and the results of spectrophotometric investigation in the ultraviolet and visible regions in relation to the structure of complex species present in aqueous solutions are discussed in Section 4. Finally, the main conclusions of the study are enumerated in Section 5.

2. Experimental

For preparing solutions, analytical grade ammonium oxalate monohydrate, $(NH_4)_2C_2O_4 \cdot H_2O$ (with the following impurities: 0.005% SO_4, 0.001% Cl, 0.001% Pb and 0.0005% Fe), copper chloride dihydrate, $CuCl_2 \cdot 2H_2O$ (with the impurities: 0.005% SO_4, 0.001% Fe, 0.0001% As, 0.004% Pb, 0.001% Ni, 0.005% Na and 0.004% nitrogen), and demineralized water were used. $(NH_4)_2C_2O_4 \cdot H_2O$ and $CuCl_2 \cdot 2H_2O$ were produced by Polish Chemicals, Lublin (Poland) and BDH Laboratories, Poole (England), respectively.

Ammonium oxalate aqueous solutions containing Cu(II) impurity were investigated at 25°C by using UV–Vis spectroscopy. Optical absorption measurements were made for unsaturated ammonium oxalate (AO) solutions of concentrations c between 3.52×10^{-3} and 0.2815 mol/dm^{-3} containing different $CuCl_2$ concentrations c_i between 5×10^{-3} and 4×10^{-2} mol/dm^3. UV–Vis absorption spectra were recorded by employing Shimadzu UV2401 spectrophotometer in conjunction with a computer. The 3.9 version of the UVPC Personal Spectroscopy Software supplied by the manufacturer records the values of absorbance and wavelength λ of the peak.

Analysis of speciation of complexes in aqueous AO solutions without and with Cu(II) impurity was carried out by using computer program Hyss [15]. Some preliminary results on these complexes have been reported earlier [13,14]. HySS is a program that can be used to perform two distinct functions: titration simulation and calculation of concentrations of species. Details concerning the algorithm used in the program and its other features are given in ref. [14].

3. Some basic equations and concepts

Appearance of an absorption spectrum is associated with the absorption of photon as a result of excitation of electrons of well-defined types or electrons belonging to small groups of atoms. The energy of transition and the structure and position of bands in the electron spectrum depend on the nature and position of ligands and multiple bonds in the particle as well as their mutual interactions. The problem of the dependence of a spectrum on the constitution of a particle is complicated.

Possible spectral transitions based on the energy difference ΔE involved during a transition may be arranged in the order [7]: $\sigma \rightarrow \sigma^* > \sigma \rightarrow \pi^* > \pi \rightarrow \sigma^* > n \rightarrow \sigma^* > \pi \rightarrow \pi^* > n \rightarrow \pi^*$. Transitions between orbitals $\sigma \rightarrow \sigma^*$, associated with transitions from binding orbitals in the ground state to unbinding orbitals, require the highest energy, and the bands corresponding to these transitions lie in the ultraviolet region $\lambda \leq 200$ nm. The $n \rightarrow \sigma^*$ transitions are of similar energies and are due to transition of an electron from unbinding atomic orbital to a molecular orbital of a higher energy ($\lambda \approx$ 170–210 nm). The lowest energy corresponds to $n \rightarrow \pi^*$ transition and depends on the type of compound. Absorption bands corresponding to this transition lie in the near-ultraviolet or visible region. Intensity of these bands is, as a rule, much lower than the intensity of bands due to other transitions. The $n \rightarrow \pi^*$ transition occurs in molecules in which a heteroatom with a free electron pair is joined with another atom by a multiple bond [7].

Intensity and width of a band in the ultraviolet and visible regions are the main parameters that characterize an absorption spectrum. Contours of the absorption bands are frequently described by Lorentzian- and Gaussian-type functions because the real contours of the bands lie between the contours predicted by these functions. In the case of Gaussian function, absorbance is expressed by [7,9]

$$A = A_{max} \exp\left\{-B(\nu_0 - \nu)^2\right\}, \quad (1)$$

where ν_0 is the value of the wavelength ν at the maximum absorbance A_{max}, and the B parameter is connected with the full width $\Delta\nu_{1/2}$ at half maximum (FWHM; hereafter called half width) of the peak by the relation

$$B = \frac{4\ln 2}{\Delta\nu^2_{1/2}}. \quad (2)$$

The half width $\Delta\nu_{1/2}$ is measured as the width of the peak of the maximum at its half height.

The electronic band intensity, characterized by the probability of transition of electrons from one state to another, is frequently described by the oscillator strength f. The higher the oscillator strength, the higher is the probability of electronic transition. When the band has the shape of a Guassian–type curve, the oscillator strength is given by [7–9]

$$f = 4.60 \cdot 10^{-9} \varepsilon \Delta\nu_{1/2}, \quad (3)$$

where ε is the molar extinction coefficient at the maximum of the band with $dm^3 \cdot mol^{-1} \cdot cm^{-1}$ units and $\Delta\nu_{1/2}$ has cm^{-1} units. The parameter f is close to 1 for allowed electron transitions, while it has relatively small value between 10^{-3} and 10^{-4} in the case of forbidden transitions.

Molar extinction coefficient ε of a solution is given by [4,7]

$$\varepsilon = A/lc, \quad (4)$$

where l is the thickness of the cuvette (in our case, l = 1 cm), A is the absorbance of the ammonium oxalate solution and c is the ammonium oxalate concentration in the solution.

In the case of multicomponent solutions and mixtures, the resultant (or total) absorbance A_i is given by the additivity rule for absorbance of individual components. The total absorbance A_i is the sum of the absorbances of individual components 1, 2, ... j (i.e. A_1, A_2, A_3, .., A_j), and may be expressed in the form [7]

$$A_i(\lambda)/l = \sum \varepsilon_j(\lambda)c_j = \varepsilon_1(\lambda)c_1 + \varepsilon_2(\lambda)c_2 + \varepsilon_3(\lambda)c_3 + \ldots.., \quad (5)$$

where ε_j and c_j are the molar extinction coefficient and the concentration of component j, respectively.

4. Results and Discussion

4.1. Ammonium oxalate solutions in ultraviolet region

Pure ammonium oxalate solutions without Cu(II) impurity reveal characteristic absorption spectra in the ultraviolet region of light. Fig. 1 illustrates the observed $A(\lambda)$ absorption spectra of unsaturated solutions containing different concentrations of ammonium oxalate. From the observed $A(\lambda)$ spectra of ammonium oxalate solutions the values of molar extinction coefficient ε and half width $\Delta\nu_{1/2}$ were obtained. The plots of $\ln\varepsilon$ against λ are presented in Fig. 2a while the dependence of the values of $\Delta\nu_{1/2}$ on $\ln c$ is shown in Fig. 2b.

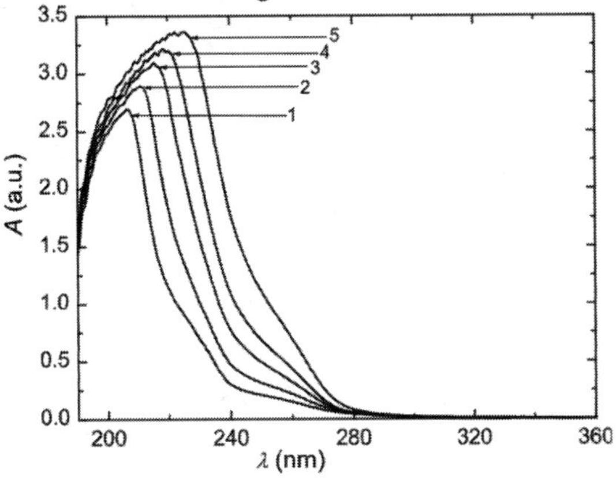

Figure 1. Observed $A(\lambda)$ absorption spectra of unsaturated solutions containing different concentrations of ammonium oxalate. Concentration of AO in solutions: (1) 0.00352, (2) 0.00704, (3) 0.01407, (4) 0.02111, (5) 0.03518 mol/dm^3.

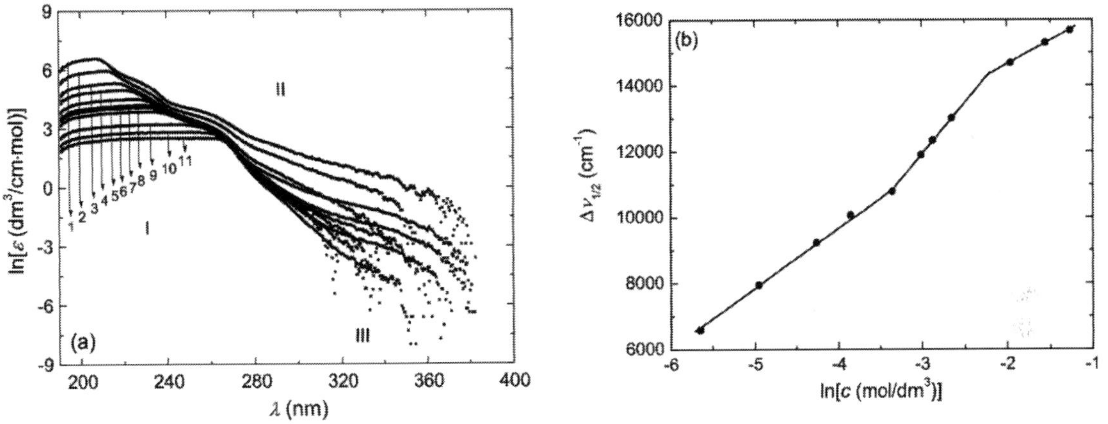

Figure 2. (a) Plots of $\ln\varepsilon$ on wavelength λ for unsaturated aqueous AO solutions. Concentration of AO in solutions: (1) 0.00352, (2) 0.00704, (3) 0.01407, (4) 0.02111, (5) 0.03518, (6) 0.04926, (7) 0.05629, (8) 0.07037, (9) 0.14074, (10) 0.2111, and (11) 0.28147 mol/dm^3, (b) Plot of half width $\Delta\nu_{1/2}$ on $\ln c$. Note three distinct slopes of $\Delta\nu_{1/2}(\ln c)$ plot.

It may be seen from Fig. 2a that the value of $\ln\varepsilon$ for 200 nm $< \lambda <$ 380 nm decreases with an increase in AO concentration c in the entire interval of wavelength λ. However, in this wavelength range, the plot of $\ln\varepsilon$ against λ reveals three distinct wavelength intervals (i.e. I, II, and III) characterized by linear dependence of $\ln\varepsilon$ on λ of different slopes. Fig. 2b of the plot of $\Delta\nu_{1/2}$ against $\ln c$ also shows three distinct intervals of AO concentrations c, corresponding to $-6 < \ln c < -3.4$, $-3.4 < \ln c < -2.3$ and $-2.3 < \ln c < -1$, respectively, in which the values of the slope of the linear $\Delta\nu_{1/2}(\ln c)$ dependence are different. The transitions in the value of the slopes occur at AO concentration c of 0.033 mol/dm^3 and 0.1 mol/dm^3, respectively.

It is known [12] that in the UV region dimethyl oxalate and several carboxylic acids and their esters give a characteristic band of low intensity in 210–260 nm region. For example, for the dimethyl oxalate absorption band, λ_{max} and ε_{max} are 246 nm and 51 dm^3·mol^{-1}·cm^{-1}, respectively. It is also known [16] that excitation of electrons in radicals containing carbonyl group C=O results in a band at around 290 nm, but the position of this type of absorption band depends on the nature of other parts of the radical. Consequently, for this group λ_{max} usually lies between 270 and 290 nm with molar extinction coefficient ε_{max} between 10 and 20 dm^3·mol^{-1}·cm^{-1}. In this case, the activity of C=O chromophore is associated with $n \rightarrow \pi^*$ transition and may be attributed to the free electron pair of the oxygen atom. It is also interesting to note [12] that in oxalic acid and 10^{-2} mol/l potassium oxalate solutions low-intensity bands are not observed. However, upon the addition of a strong mineral acid which suppresses the dissociation of oxalic acid, the shape of their absorption band becomes similar to that of dimethyl oxalate [12]. These observations suggest that an increase in the concentration of associated species $H_2C_2O_4$ in oxalic acid solutions can decrease the band intensity in the 210–260 nm region.

In aqueous AO solutions the following species are formed as a result of dissociation and hydrolysis reactions: $C_2O_4^{2-}$, $HC_2O_4^-$, $H_2C_2O_4$, $(NH_4)_2C_2O_4$, NH_4^+ and NH_4OH. However, the concentrations of the above species depends on the initial AO concentration in the solution. The initial AO concentration and the stability constants β of complex species forming in the solutions determine the solution pH. Analysis of speciation of complexes in AO solutions [13] by computer program Hyss [15] revealed that the concentration of species like $C_2O_4^{2-}$, $HC_2O_4^-$, $H_2C_2O_4$, NH_4^+, NH_4OH and $(NH_4)_2C_2O_4$ increases with an increase in AO concentration, but the relative concentrations of the dominating complexes also increase, resulting in an increase in solution pH. Speciation analysis also showed [13] that: (a) in the range of low AO concentration $c < 0.03$ mol/dm^3, the dominating species are: $C_2O_4^{2-}$, $HC_2O_4^-$, $H_2C_2O_4$, and NH_4^+; (b) at relatively high concentrations in the range $0.03 < c < 0.1$, the main species are: NH_4^+, $C_2O_4^{2-}$, $H_2C_2O_4$, $HC_2O_4^-$ and $(NH_4)_2C_2O_4$; while (c) in the range of very high concentrations $c > 0.1$, the complex species are: NH_4^+, $C_2O_4^{2-}$ and $(NH_4)_2C_2O_4$.

From the above discussion it may be concluded that, with an increase in the concentration of ammonium oxalate in the investigated solutions, the concentration of associated species such as $H_2C_2O_4$ and $(NH_4)_2C_2O_4$ also increases, which results in a decrease in the molar extinction coefficient ε_{max} in the 210–260 nm region (Fig. 2a). For ammonium oxalate solutions of concentration $0.004 < c \leq 0.3$ the molar extinction coefficient varies between 750 and 20 dm^3·mol^{-1}·cm^{-1} (i.e. $6.6 < \ln\varepsilon_{max} < 3$; cf. Fig. 2a) and decreases with increasing ammonium oxalate concentration. It should be noted that, irrespective of the ammonium oxalate concentration, practically for all investigated solutions the molar extinction coefficient ε_{max} lies between 10 and 20 dm^3·mol^{-1}·cm^{-1} in the 250–290 nm region and is associated with the activity of the C=O chromophore.

4.2. Aqueous ammonium oxalate solutions containing Cu(II) impurity – Speciation of copper complexes

Fig. 3a illustrates the data on the concentration of different copper complexes in AO aqueous solutions containing different Cu(II) concentrations in the form of the dependences of ln[species] on $\ln(c_i/c)$. From Fig. 3a it is interesting to note two different trends of the dependences of various complexes in the concentration ratio ranges: $-5 < \ln(c_i/c) < 0$, and $0 < \ln(c_i/c) < 2.5$. In the range $0 < \ln(c_i/c) < 2.5$, both $\ln[Cu(C_2O_4)_2^{2-}]$ and $\ln[Cu(C_2O_4)]$ decrease with an increase in $\ln(c_i/c)$ but the decrease is very pronounced in the case of $Cu(C_2O_4)_2^{2-}$ complex. In this concentration ratio range, $\ln[Cu(H_2O)_6^{2+}]$ increases with an increase in $\ln(c_i/c)$. In the concentration ratio range $-5 < \ln(c_i/c) < -1$, $\ln[Cu(H_2O)_6^{2+}]$ and $\ln[Cu(C_2O_4)]$ increase with an increase in $\ln(c_i/c)$, the increase being more in the case of $Cu(H_2O)_6^{2+}$ complex than in $Cu(C_2O_4)$. However, in this concentration ratio range, $\ln[Cu(C_2O_4)_2^{2-}]$ somewhat decreases with an increase in $\ln(c_i/c)$.

The dependences of ln[species] on $\ln(c_i/c)$ may be divided into three $\ln(c_i/c)$ ranges, as illustrated in Fig. 3b. In the concentration ratio ranges C and A, defined by $-5 < \ln(c_i/c) < -3$ and $1 < \ln(c_i/c) < 2.5$, respectively, for a given c_i the values of $\ln[Cu(H_2O)_6^{2+}]$, $\ln[Cu(C_2O_4)]$ and $\ln[Cu(C_2O_4)_2^{2-}]$ change practically linearly with increase in $\ln(c_i/c)$. However, in the concentration ratio range B, defined by $-3 < \ln(c_i/c) < 1$, for a given c_i the values of $\ln[Cu(H_2O)_6^{2+}]$ and $\ln[Cu(C_2O_4)]$ suddenly increase while those of $\ln[Cu(C_2O_4)_2^{2-}]$ decrease with increase in $\ln(c_i/c)$. The changes in $\ln[Cu(H_2O)_6^{2+}]$, $\ln[Cu(C_2O_4)]$ and $\ln[Cu(C_2O_4)_2^{2-}]$ are highly pronounced in the range $-1.5 < \ln(c_i/c) < 0.5$ i.e. 0.22 <

$c_i/c < 1.6$. These lower and upper limits of c_i/c represent the preferential formation of $Cu(C_2O_4)_2^{2-}$ and $Cu(H_2O)_6^{2+}$ complexes, respectively, while $c_i/c = 1$ corresponds to the preferential formation of the neutral $Cu(C_2O_4)$ complex.

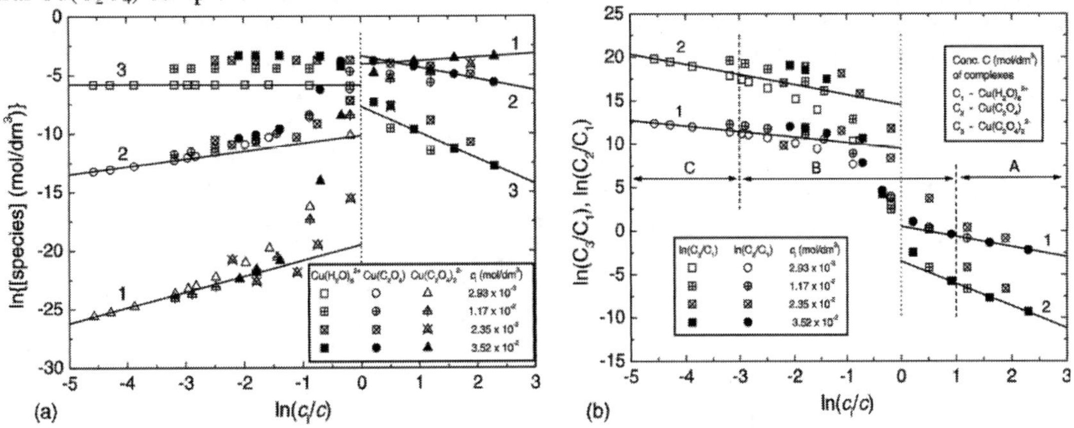

Figure 3. (a) Dependence of logarithm of concentrations of different complexes (denoted as ln[species]) on $\ln(c_i/c)$ for unsaturated aqueous AO solutions containing different Cu(II) concentration c_i. (b) Data of (a) shown as plots of $\ln\{[Cu(C_2O_4)]/[Cu(H_2O)_6^{2+}]\}$ and $\ln\{[Cu(C_2O_4)_2^{2-}]/[Cu(H_2O)_6^{2+}]\}$ against $\ln(c_i/c)$ for c_i. Plots of ln[species] against $\ln(c_i/c)$ may be divided into three $\ln(c_i/c)$ ranges A, B and C, as shown in (b).

The dependence of ln[species] on $\ln(c_i/c)$ in the concentration ratio ranges A and C may be described by the empirical relationship:

$$\ln[\text{species}] = \ln[\text{species}]_0 + Y \ln(c_i/c), \qquad (6)$$

where $\ln[\text{species}]_0$ is the extrapolated value of ln[species] when $\ln(c_i/c) = 0$ and Y is the slope of the plot of ln[species] against $\ln(c_i/c)$. The values of $\ln[\text{species}]_0$ and Y are listed in Table 1. The values of $\ln[\text{species}]_0$ somewhat depend on the Cu(II) concentration c_i, but these were not considered during the analysis of the data. In Fig. 3a, the linear plots are drawn in regions C and A, corresponding to $-5 < \ln(c_i/c) < -3$ and $1 < \ln(c_i/c) < 2.5$, for $c_i = 2.93 \times 10^{-3}$ mol/dm^3 and $c_i = 3.52 \times 10^{-2}$ mol/dm^3, respectively. These concentration ratio ranges correspond to the preferential formation of $Cu(C_2O_4)_2^{2-}$ and $Cu(H_2O)_6^{2+}$ complexes, respectively.

Table 1. Values of $\ln[\text{species}]_0$ and slope Y for different complexes

Complex	$-5 < \ln(c_i/c) < -3$*		$1 < \ln(c_i/c) < 2.5$**	
	$\ln[\text{species}]_0$	Y	$\ln[\text{species}]_0$	Y
$Cu(H_2O)_6^{2+}$	−19.5	1.35	−4.1	0.3
$Cu(C_2O_4)]$	−8	0.66	−3.4	−0.97
$Cu(C_2O_4)_2^{2-}$	−4	−0.02	−7.7	−2.2

* For $c_i = 2.93 \times 10^{-3}$ mol/dm^3. ** For $c_i = 3.52 \times 10^{-2}$ mol/dm^3.

In the concentration ratio range $0 < \ln(c_i/c) < 2.5$, the concentrations of $Cu(C_2O_4)$ and $Cu(C_2O_4)_2^{2-}$ complexes increase with decrease in c_i/c at the expense of the concentration of $Cu(H_2O)_6^{2+}$ complex. The concentration of $Cu(H_2O)_6^{2+}$ complex decreases with decrease in c_i/c (i.e. the slope $Y > 0$). However, in the concentration ratio range $-5 < \ln(c_i/c) < 0$, the concentration of $Cu(C_2O_4)_2^{2-}$ complex alone increases with decreasing c_i/c at the expense of the concentrations of $Cu(C_2O_4)$ and $Cu(H_2O)_6^{2+}$ complexes, which decrease with decrease in c_i/c. These observations are consistent with the fact that, for a given value of Cu(II) concentration c_i, addition of AO leads to the substitution of pairs of water molecules of the $Cu(H_2O)_6^{2+}$ complex by $C_2O_4^{2-}$ ligands, forming $Cu(C_2O_4)$ complex first for $c_i/c = 1$ and subsequently $Cu(C_2O_4)_2^{2-}$ complex for $c_i/c = 0.5$. Therefore, it is better to represent the data on the concentration of different copper complexes as ratio of concentrations of $Cu(C_2O_4)$ and $Cu(C_2O_4)_2^{2-}$ complexes to the concentration of $Cu(H_2O)_6^{2+}$ complex as a function of c_i/c. Fig. 3b shows the plots of $\ln\{[Cu(C_2O_4)]/[Cu(H_2O)_6^{2+}]\}$ and $\ln\{[Cu(C_2O_4)_2^{2-}]/[Cu(H_2O)_6^{2+}]\}$ against $\ln(c_i/c)$ for different Cu(II) concentration c_i. As seen from the slopes of these plots, the increase in the concentration of $Cu(C_2O_4)_2^{2-}$ complex with decrease in $\ln(c_i/c)$ is much higher than that of $Cu(C_2O_4)$ complex.

$Cu(H_2O)_6^{2+}$ complex forms deformed octahedral spatial structure with two trans bonds (Fig. 4a) [17,18], while $C_2O_4^{2-}$ ligand coordinates with Cu(II) ions at two neighboring sites such that substitution of one pair of water molecules by it forms $Cu(C_2O_4)$ complex of deformed trigonal structure (Fig. 4b). However, when two $C_2O_4^{2-}$ ligands substitute two pairs water molecules, Cu(II) atom is coordinated by four O atoms from two oxalate ligands and two O atoms from water molecules (Fig. 4c) [19,20]. This $Cu(C_2O_4)_2^{2-}$ complex shows a distorted octahedral coordination geometry, with Cu–O distances showing a typical Jahn–Teller distortion [19,20]. The Cu–O bond length lies between 0.1951 and 0.1914 nm [20]. The symmetry of this complex is close to that of $Cu(H_2O)_6^{2+}$ complex. Therefore, it may be argued that the symmetry of complexes decreases in the sequences: $Cu(H_2O)_6^{2+}$, $Cu(C_2O_4)_2^{2-}$ and $Cu(C_2O_4)$.

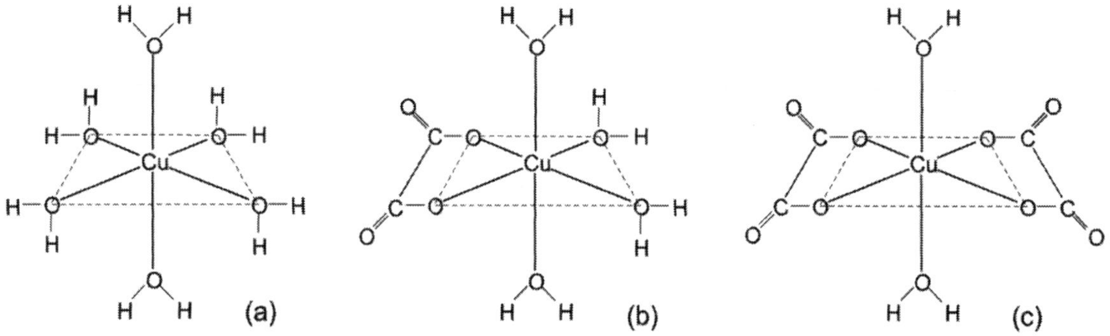

Figure 4. Schematic illustration of structure of different octahedral complexes formed in aqueous AO solutions containing Cu(II) additive: (a) $Cu(H_2O)_6^{2+}$, (b) $Cu(C_2O_4)$ and (c) $Cu(C_2O_4)_2^{2-}$; distortion: (a) tetragonal, (b) trigonal and (c) tetrahedral.

4.3. Ultraviolet region of absorption spectra

Aqueous AO solutions containing different concentrations (i.e. 5.87×10^{-3}, 1.17×10^{-2}, 1.76×10^{-2}, 2.35×10^{-2} and 3.52×10^{-2} mol/dm^3) of Cu(II) impurity exhibit characteristic bands in the wavelength intervals: 200–300 nm and 650–850 nm, which correspond to ultraviolet and visible regions, respectively. Typical example of the UV–Vis absorption spectrum of AO solution containing 1.17×10^{-2} mol/dm^3 Cu(II) impurity is presented in Fig. 5.

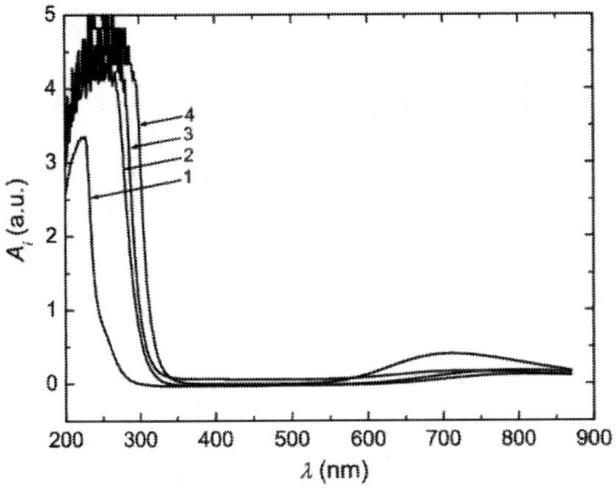

Figure 5. Typical example of the UV-Vis absorption spectrum of AO solution containing 1.173 × 10^{-2} mol/dm^3 Cu(II) impurity; AO concentration c: (1) 0, (2) 3.52 × 10^{-3}, (3) 1.407 × 10^{-2} and (4) 2.815 x 10^{-2} mol/dm^3.

Assuming that the additive rule is valid for absorption (see Eq. (5)), from the $A_i(\lambda)$ spectra of aqueous solutions of known AO concentrations containing different concentrations of Cu(II) impurity and the $A(\lambda)$ spectra of aqueous solutions of the AO concentrations without the impurity, the difference (A_i–A) was determined. Here the difference (A_i–A) refers to the absorbance A of Cu(II) impurity and c refers to the concentration c_i of Cu(II) impurity. Then from the difference (A_i–A) the values of ε_i were calculated by using Eq. (4). The results of the difference (A_i–A) are presented in Fig. 6 in the form of plots of lnε_i on λ for different concentrations of AO and Cu(II) impurity.

Fig. 6a and b shows the change in the values of lnε_i in the wavelength range 200 nm < λ < 380 nm as functions of AO concentration c and Cu(II) impurity concentration c_i in the solutions. Fig. 6a illustrates ln$\varepsilon_i(\lambda)$ curves for aqueous solutions containing Cu(II) impurity alone of three different concentrations (viz. 1.17 × 10^{-2}, 2.35 × 10^{-2}, and 3.52 × 10^{-2} mol/dm^3) as well as ln$\varepsilon_i(\lambda)$ curves for mixtures of low AO concentrations, while Fig. 6b presents ln$\varepsilon_i(\lambda)$ curves for mixtures of high AO concentrations.

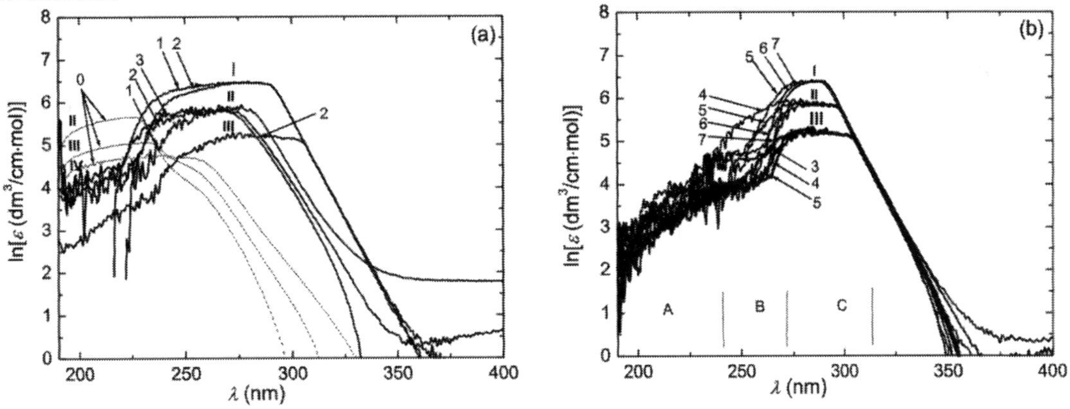

Figure 6. Plots of lnε_i against wavelength λ for different concentrations of Cu(II) impurity: (I) 5.87 × 10^{-3}, (II) 1.17 × 10^{-2}, (III) 2.35 × 10^{-2}, and (IV) 3.52 × 10^{-2} mol/dm^3. AO concentration in the solutions: Fig. 6a: (0) 0, (1) 0.007, (2) 0.014 and (3) 0.049; Fig. 6b: (4) 0.070, (5) 0.141, (6) 0.211 and (7) 0.281 mol/dm^3.

Pure aqueous solutions containing Cu(II) impurity alone exhibit maximum absorption in the wavelength region between 200 and 240 nm, whereas its molar extinction coefficients have values lying in the range: 4.7 < lnε_i < 5.5. The value of lnε_i decreases with an increase in impurity concentration c_i. In the case of AO solutions containing Cu(II) impurity, three regions of λ may be observed on the $\ln\varepsilon_i(\lambda)$ plots in which $\ln\varepsilon_i$ takes different values, as shown in Fig. 6b. These regions of λ are: (A) 200 nm $< \lambda <$ 240 nm, (B) 240 nm $< \lambda <$ 270 nm, and (C) 270 nm $< \lambda <$ 320 nm. In region A one observes a decrease in the value $\ln\varepsilon_i$ in comparison with pure Cu(II) solutions without AO. In region B there is an increase in $\ln\varepsilon_i$ for all impurity and AO concentrations. Shift of the absorption band to the higher wavelengths is a characteristic feature of this region (see Figs. 6a and b). The shift in the band increases with increasing AO concentration. In region C the values of $\ln\varepsilon_i$ are practically constant but are higher than those in regions A and B. However, increase in Cu(II) impurity concentration leads to a decrease in the value of $\ln\varepsilon_i$ in the entire range of wavelength λ (see Figs. 6a and b).

Pure aqueous copper chloride solutions of concentrations 1.17×10^{-2}, 2.35×10^{-2}, and 3.52×10^{-2} mol/dm^3 have their molar extinction coefficient ε_i in the wavelength interval between 200 and 240 nm (see Fig. 6a), and have the values 300, 150 and 110 dm^3/mol·cm, respectively. Charge-transfer bands for several hydrated metal ions lie in the interval between 200 and 250 nm of the UV region [7,11] and the values of their molar extinction coefficients ε are usually greater than 100 dm^3/mol·cm [7]. Therefore, the bands observed from pure aqueous solutions of Cu(II) impurity of the investigated concentrations c_i may be attributed to the formation of $Cu(H_2O)_6^{2+}$ aquocomplexes involving interactions between individual Cu(II) ions with H_2O molecules.

The UV absorption bands for most of the oxalate complexes are similar to the bands of dimethyl oxalate and molecular oxalic acid [12], and appear in the wavelength range between 250–270 nm and the position of their maximum changes with the cation. A characteristic feature of these bands is that the band position shifts to the direction of longer wavelength with increasing Coulomb attractive interaction between metal cation and coordinated oxalate group [12].

It has been observed [7,8,12] that a shift in the absorption band occurs to the higher wavelength region of the spectrum with increasing number of ligands in a given complex, but, in the case of a constant number of ligands, the molar extinction coefficient ε attains a constant value in the given wavelength region. The value of the molar extinction coefficient ε of a complex containing two ligands is higher than that of the complex with one ligand. This occurs due to the fact that the probability of electronic transitions, and the value of the molar extinction coefficient associated with it, increases with an increase in the number of ligands localized around the cation.

From the above considerations, it may be argued that the shift in adsorption bands to higher wavelengths, observed in Fig. 6, is caused by the entrance of oxalate ions as ligands in the coordination sphere of the Cu(II) ion; the greater the shift in the band, the higher is the concentration of complexes of higher coordination number. This means that, if the dominating complexes are $Cu(H_2O)^{2+}$ and $Cu(C_2O_4)$ in solutions composed of low AO concentrations, an increase in the AO concentration leads to a shift in the equilibrium of chemical reactions to the formation of complexes such as $Cu(C_2O_4)$ and $Cu(C_2O_4)_2^{2-}$. However, since one also observes that the band shifts are relatively large in solutions of higher AO concentrations, it may be believed that $Cu(C_2O_4)_2^{2-}$ complexes dominate in these solutions. A decrease in the values of ε with an increase in Cu(II) concentration is due to the opposite situation and is associated with the type and concentration of the complexes formed.

From the dependence of the difference $(A_i–A)$ on wavelength λ the values of half width $\Delta\nu_{1/2}$ of the peaks for all Cu(II) and AO concentrations were calculated. The data so obtained are presented as the dependence of $\Delta\nu_{1/2}$ on $\ln c$ and $\Delta\nu_{1/2}$ on $\ln(c_i/c)$ in Fig. 7a and b, respectively.

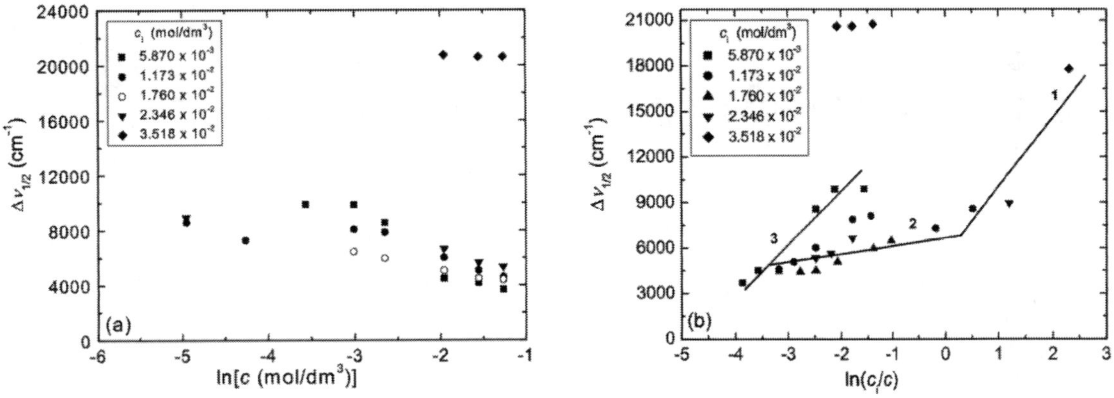

Figure 7. Dependences of half width $\Delta\nu_{1/2}$ of absorption bands on (a) ln*c* and (b) ln(c_i/*c*) for different Cu(II) concentrations c_i.

It may be seen from Fig. 7a that an increase in the AO concentration *c* leads to a decrease in $\Delta\nu_{1/2}$ in the entire range of concentration *c* used, but an increase in the concentration c_i of Cu(II) impurity leads either to a decrease or increase in the value of $\Delta\nu_{1/2}$, depending on the range of AO concentration *c*. In the range of low AO concentration corresponding to $-5 < \ln c < -2.7$, the band FWHM decreases with an increase in impurity concentration c_i. In the high AO concentration interval corresponding to $-2 < \ln c < -1$, increase in c_i leads to an increase in $\Delta\nu_{1/2}$. It is generally accepted that the lower the value of $\Delta\nu_{1/2}$, the more homogeneous is the solution in terms of the type of complexes [7]. In our case, this means that an increase in AO concentration shifts the reaction equilibrium towards the formation of $Cu(C_2O_4)$ first, followed by $Cu(C_2O_4)_2^{2-}$ complex. However, increase in impurity concentration c_i affects in this way the equilibria of formation of complexes but only in the AO concentration region corresponding to $-5 < \ln c < -2.7$. In the concentration interval $-2 < \ln c < -1$, an increase in impurity concentration c_i causes the solution to become less homogeneous, implying that $Cu(C_2O_4)_2^{2-}$ complexes dissociate and more complexes of $Cu(C_2O_4)$ type are formed.

The results of the values of half width $\Delta\nu_{1/2}$ as a function of ln(c_i/*c*) may be represented by three characteristic linear plots 1, 2 and 3. These results suggest that each of these linear plots 1, 2 and 3 represents particular types of chemical equilibria between the complexes being formed (i.e. $Cu(H_2O)_6^{2+}$ and $Cu(C_2O_4)$, and $Cu(C_2O_4)$ and $Cu(C_2O_4)_2^{2-}$) and also provides information about the dominating complexes in different ln(c_i/*c*) intervals. Thus, it may be concluded that in the three ln(c_i/*c*) regions: (1) $0 < \ln(c_i/c) < 2.5$, (2) $-3 < \ln(c_i/c) < 0$, and (3) $-4 < \ln(c_i/c) < -3$, the dominating complexes are $Cu(H_2O)_6^{2+}$, $Cu(C_2O_4)$ and $Cu(C_2O_4)_2^{2-}$, and $Cu(C_2O_4)_2^{2-}$, respectively. It should be emphasized again that, in the case of solutions of different AO and Cu(II) concentrations and the same value of ln(c_i/*c*) in the region $-3 < \ln(c_i/c) < 0$, either $Cu(C_2O_4)$ or $Cu(C_2O_4)_2^{2-}$ complex can be dominant. However, which of these complexes is predominantly formed determines the solution pH.
Displacement of peak wavelength λ_{max} to higher or lower wavelengths is usually associated with structural changes in the complex as a result of entering of different ligands in its coordination sphere [7]. Fig. 8a presents the dependence of λ_{max} on ln(c_i/*c*). It may be seen from the figure that λ_{max} increases with a decrease in the value of ln(c_i/*c*) in the entire investigated region of c_i/*c*. Here as in the case of $\Delta\nu_{1/2}$[ln(c_i/*c*)] dependence, the λ_{max}[ln(c_i/*c*)] data fit around three linear plots 1, 2 and 3 (see Fig. 6a). It should be noted that the linear plots 1, 2 and 3 describe the results of λ_{max}, the values of which lie in the intervals 250–260, 260–285, 285–290 nm, respectively. This means that the three intervals of λ_{max} are, respectively, associated with the dominating groups of the following complexes: $Cu(H_2O)_6^{2+}$, $Cu(C_2O_4)$ and $Cu(C_2O_4)_2^{2-}$, and $Cu(C_2O_4)_2^{2-}$.

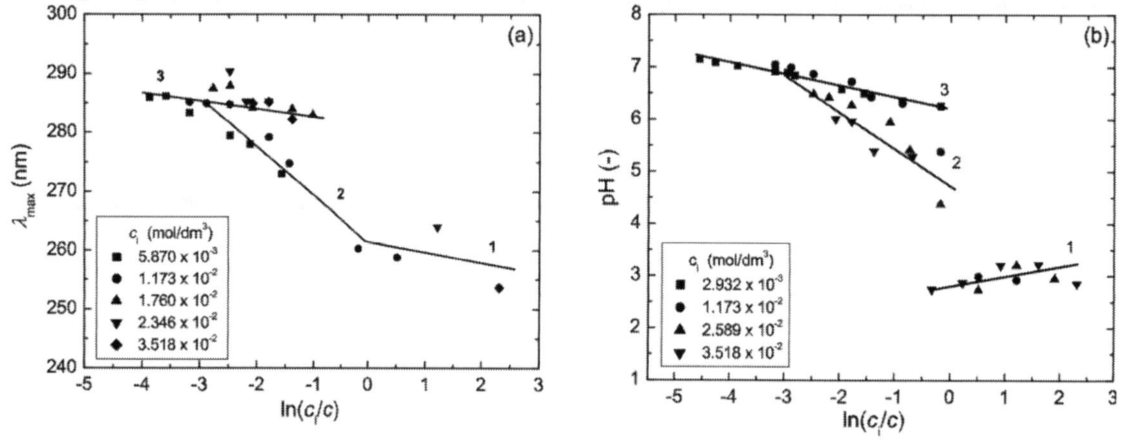

Figure 8. Dependence of (a) peak wavelength λ_{max} and (b) solution pH on $\ln(c_i/c)$ for different concentrations c_i of Cu(II) impurity.

Fig. 8b illustrates the dependence of solution pH on $\ln(c_i/c)$ for different Cu(II) concentrations c_i. This figure shows that the trends of the pH[$\ln(c_i/c)$] dependence are similar to those of the plots of $\Delta\nu_{1/2}$ and λ_{max} against $\ln(c_i/c)$, illustrated in Figs. 7b and 8a, respectively. Assuming that linear plots 1, 2 and 3 are due to the dominating complexes in Figs. 7b and 8a, the dominating complexes corresponding to plot 1 in the range $2.5 < pH < 3.5$, plot 2 in the range $4 < pH < 6.6$, and plot 3 for $pH > 6.6$ are: $Cu(H_2O)_6^{2+}$, $Cu(C_2O_4)$ and $Cu(C_2O_4)_2^{2-}$, and $Cu(C_2O_4)_2^{2-}$, respectively.

4.4. Visible region of absorption spectra

Figs 9a and b show the dependence of half width $\Delta\nu_{1/2}$ and peak wavelength λ_{max}, obtained from the spectra in the wavelength region 650–850 nm, on $\ln(c_i/c)$ for unsaturated aqueous AO solutions containing different c_i/c ratios, respectively. Fig. 9a reveals that $\Delta\nu_{1/2}$ increases with decreasing values of $\ln(c_i/c)$ in the concentration ratio defined by $0 < \ln(c_i/c) < 2.5$. At $\ln(c_i/c) = 0$, the value of $\Delta\nu_{1/2}$ abruptly decreases from the highest value of 6000 cm^{-1} to a constant value of about 4400 cm^{-1} for $\ln(c_i/c) < 0$. In the case of the dependence of λ_{max} on $\ln(c_i/c)$ (see Fig. 9b), the value of λ_{max} shifts to lower wavelengths with decreasing $\ln(c_i/c)$ in the concentration ratio given by $0 < \ln(c_i/c) < 2.5$. However, λ_{max} abruptly decreases from a value of about 790 nm at $\ln(c_i/c) = 0$ to a constant value of about 710 nm for $\ln(c_i/c) < 0$.

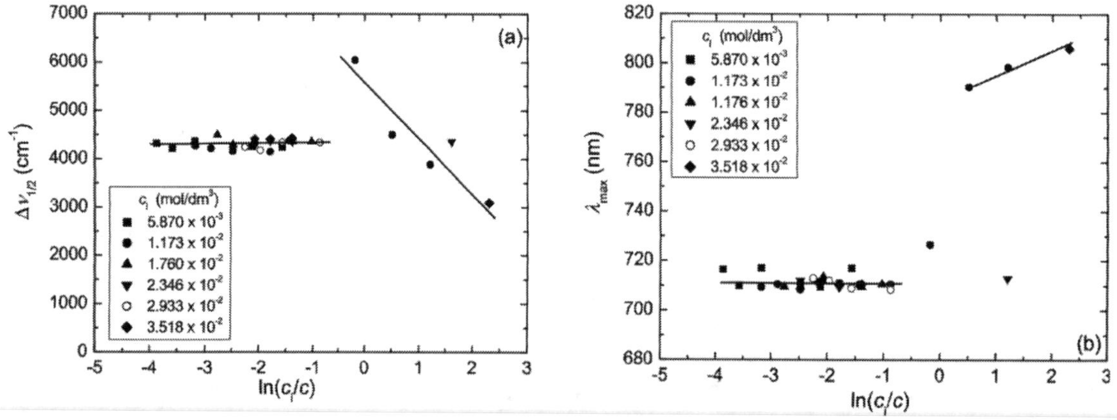

Figure 9. Dependences of (a) half width $\Delta\nu_{1/2}$ and (b) peak wavelength λ_{max} on $\ln(c_i/c)$ in the visible spectral region for unsaturated aqueous AO solutions containing different impurity concentration c_i.

As reported earlier [13], the molar extinction coefficient ε_{max} for ammonium oxalate solutions containing Cu(II) impurity increase with decreasing value of $\ln(c_i/c)$, attaining values of 12–16 $dm^3 \cdot mol^{-1} \cdot cm^{-1}$ and 33–37 $dm^3 \cdot mol^{-1} \cdot cm^{-1}$ in $0 < \ln(c_i/c) < 2.5$ and $\ln(c_i/c) < 0$, respectively. It is known [12] that $Cu(C_2O_4)_2^{2-}$ complexes exhibit their band at 700 nm and their molar extinction coefficient $\varepsilon_{max} = 33.5$ $dm^3 \cdot mol^{-1} \cdot cm^{-1}$. However, $Cu(H_2O)_6^{2+}$ aquocomplexes show their band at 794 nm and their molar extinction coefficient $\varepsilon_{max} = 12$ $dm^3 \cdot mol^{-1} \cdot cm^{-1}$ [7].

From Figs. 9a and b it may be noted that the data of the basic band parameters $\Delta\nu_{1/2}$ and λ_{max} on $\ln(c_i/c)$ are arranged around two linear plots covering the regions of concentration ratios $-5 < \ln(c_i/c) < 0$ and $0 < \ln(c_i/c) < 2.5$. This suggests that the linear plots in the two $\ln(c_i/c)$ regions are due to two types of chemical equilibria of the forming complexes in the solutions, i.e. $Cu(H_2O)_6^{2+}$ and CuC_2O_4, and CuC_2O_4 and $Cu(C_2O_4)_2^{2-}$.

4.5. Oscillator strength and structure of complexes

Figs. 10a and b show the plots of oscillator strength *f*, obtained from spectra in the ultraviolet and visible regions of aqueous solutions, against $\ln(c_i/c)$, respectively. From Fig. 10a it may be noted that the value of oscillator strength *f*, obtained from UV spectra, for copper complexes forming in aqueous AO solutions in the concentration ratio range $-4 < \ln(c_i/c) < -1$ increases with increasing $\ln(c_i/c)$ and, depending on the value of $\ln(c_i/c)$ and impurity concentration c_i, *f* changes from 0.003 to 0.03. However, in this spectral region, the oscillator strength *f* is practically constant equal to about 0.01 in the range $0 < \ln(c_i/c) < 2.5$. In contrast to the ultraviolet region, Fig. 10b reveals that the values of *f* in the visible spectral region lie in the ranges of $2 \times 10^{-4} - 4 \times 10^{-4}$ and $6 \times 10^{-4} - 7 \times 10^{-4}$ for $\ln(c_i/c) > 0$ and $\ln(c_i/c) < 0$, respectively. With an increase in $\ln(c_i/c)$, the value of *f* in this spectral region somewhat decreases in the range $\ln(c_i/c) < 0$, while it decreases significantly in the range $\ln(c_i/c) > 0$.

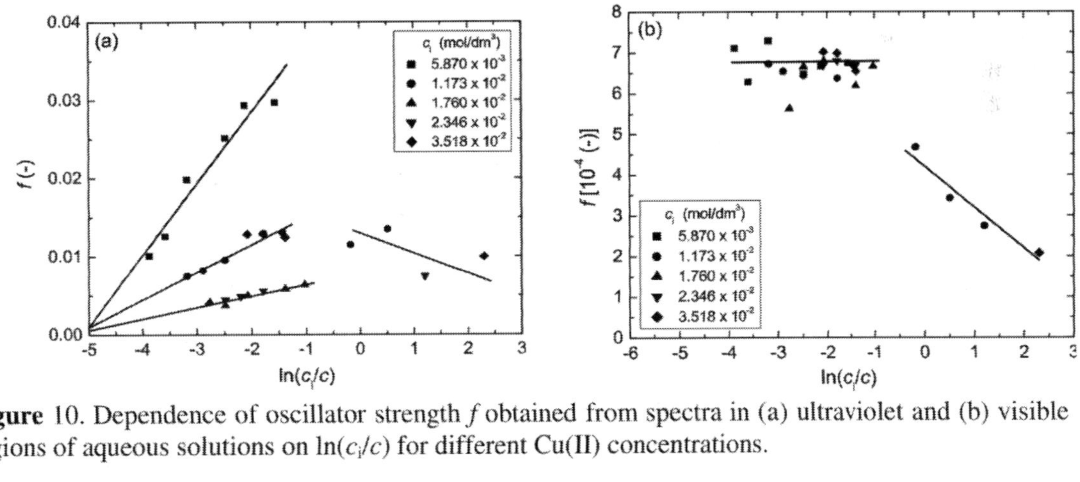

Figure 10. Dependence of oscillator strength *f* obtained from spectra in (a) ultraviolet and (b) visible regions of aqueous solutions on $\ln(c_i/c)$ for different Cu(II) concentrations.

The behavior of the plots of oscillator strength *f* against $\ln(c_i/c)$ in Figs. 10a and b is similar to that of the plots of $\Delta\nu_{1/2}$ and λ_{max} against $\ln(c_i/c)$. As before, the dependence of oscillator strength *f* on $\ln(c_i/c)$ may be attributed to the formation of metal complexes containing different ligands. However, the complexes formed are of low or high symmetries, depending on the impurity–solute concentration ratio c_i/c.

The plots of oscillator strength *f* against $\ln(c_i/c)$ are very similar to the plots of $\ln\{[Cu(C_2O_4)]/[Cu(H_2O)_6^{2+}]\}$ and $\ln\{[Cu(C_2O_4)_2^{2-}]/[Cu(H_2O)_6^{2+}]\}$ against $\ln(c_i/c)$, as shown in Fig. 3b. These observations may be attributed to the formation of increasing concentrations of $Cu(C_2O_4)$ and $Cu(C_2O_4)_2^{2-}$ complexes with decrease in $\ln(c_i/c)$ in the solution as a result of substitution of water molecules of the $Cu(H_2O)_6^{2+}$ species by $C_2O_4^{2-}$ ligands. More the reaction shifts to the formation of $Cu(C_2O_4)$ and $Cu(C_2O_4)_2^{2-}$ complexes, the greater is the oscillator strength. This conclusion is consistent with the fact that the probability of electronic transitions, and the oscillator strength associated with it, increases with an increase in the number of ligands around a metal cation [7,8]. The greater the binding force, the greater is the oscillator strength.

From Fig. 10a the following features may be noted:

(1) In the range $-2 < \ln(c_i/c) < 0$, f is practically constant or insignificantly decreases with increasing $\ln(c_i/c)$; see $f[\ln(c_i/c)]$ data for $c_i = 1.17 \times 10^{-2}$ and 3.52×10^{-2} mol/dm^3.

(2) For $-4 < \ln(c_i/c) < -2$, the value of f corresponding to a given c_i linearly decreases with an increase in $\ln(c_i/c)$. The extrapolation of the $f[\ln(c_i/c)]$ plots shows a tendency to intersect at $\ln(c_i/c) = -5$.

(3) For $-4 < \ln(c_i/c) < -2$, the slope of the $f[\ln(c_i/c)]$ plot decreases with an increase in c_i such that the slope $\Delta f/\Delta[\ln(c_i/c)]$ approaches zero for the highest value of c_i.

(4) Above a certain value of $\ln(c_i/c)$ equal to about -2, the oscillator strength has a constant value f_c. For example, for $c_i = 1.17 \times 10^{-2}$ mol/dm^3, $f_c \approx 0.012$; but for $c_i = 2.35 \times 10^{-2}$ mol/dm^3, $f_c \approx 0.007$ but for $c_i = 3.52 \times 10^{-2}$ mol/dm^3, $f_c \approx 0.01$.

(5) Except in the case of the highest impurity concentration $c_i = 3.52 \times 10^{-2}$ mol/dm^3 used in this study, the products $c_i\Delta f/\Delta[\ln(c_i/c)]$ and $c_i f_c$ are practically constant quantities equal to about 4×10^{-5} mol/dm^{-3} and 1.6×10^{-4} mol/dm^{-3}, respectively. However, for $c_i = 3.52 \times 10^{-2}$ mol/dm^3, $c_i\Delta f/\Delta[\ln(c_i/c)] \approx 0$ mol/dm^{-3} and $c_i f_c \approx 3.5 \times 10^{-4}$ mol/dm^{-3}.

Feature (1) is associated with the chemical equilibria between $Cu(C_2O_4)_2^{2-}$ and $Cu(C_2O_4)$ complexes forming in this concentration ratio range. Here an increase in c_i shifts the chemical equilibrium to the formation of $Cu(C_2O_4)$ complex, while an increase in c shifts the equilibrium to the formation of $Cu(C_2O_4)_2^{2-}$ complex. In terms of its absolute value, an increase in f means that the concentration of the $Cu(C_2O_4)$ complex increases with an increase in impurity concentration c_i and a decrease in AO concentration c. The practically constant value of f in the range $-2 < \ln(c_i/c) < 0$ suggests that $Cu(C_2O_4)$ complex dominates here. However, in comparison with the f value in $-2 < \ln(c_i/c) < 0$ range, a somewhat lower value of f for $\ln(c_i/c) > 0$ implies that $Cu(H_2O)_6^{2+}$ complex is dominant.

Feature (2) of a linear decrease in f and Feature (3) of a decrease in the slope of $f[\ln(c_i/c)]$ plot with an increase in $\ln(c_i/c)$ is associated with the capability of formation of changing copper-oxalate complexes at a given impurity concentration c_i. Feature (4) of a constant value of f is associated with the constant constitution of the complex formed above the transition value of $\ln(c_i/c)$. Feature (5) of the constant values of the products $c_i\Delta f/\Delta[\ln(c_i/c)]$ and $c_i f_c$ in the regions of $\ln(c_i/c)$ below and above its transition value, respectively, is a consequence of formation of different types of complexes in these regions.

5. Conclusions

(1) Changes in the absorption of light by aqueous ammonium oxalate solutions containing Cu(II) impurity in the concentration ratio range $-5 < \ln(c_i/c) < 2.5$ lead to decrease or increase in the intensity of bands of the UV–vis spectral region. These changes may be expressed by full width at half maximum $\Delta\nu_{1/2}$, molar extinction coefficient ε, peak wavelength λ_{max}, and oscillator strength f. The changes are caused by the coordination of $C_2O_4^{2-}$ ligand with $Cu(H_2O)_6^{2+}$ aquocomplex, and are related with the concentration ratio range $\ln(c_i/c)$. The coordination of $C_2O_4^{2-}$ ligand with $Cu(H_2O)_6^{2+}$ aquocomplex in the range $0 < \ln(c_i/c) < 2.5$ leads to the formation of $Cu(C_2O_4)$ complexes, but the coordination of the $C_2O_4^{2-}$ ligand with $Cu(C_2O_4)$ complex in the concentration ratio range $-5 < \ln(c_i/c) < 0$ results in the formation of predominantly $Cu(C_2O_4)_2^{2-}$ complex.

(2) The effect of successive coordination of the $C_2O_4^{2-}$ ligand is well–defined in the UV spectral region but poor in the visible region. This difference is associated with nature of overlapping of absorption bands corresponding to different types of complexes. The overlapping effect is pronounced in the case of visible spectral region, but no such effect occurs in the UV region.

(3) Analysis of the UV–vis spectra of aqueous ammonium oxalate solutions containing Cu(II) impurity in the UV region enables to identify two regions of concentration ratios ln(c_i/c) in which $Cu(H_2O)_6^{2+}$, $Cu(C_2O_4)$ and $Cu(C_2O_4)_2^{2-}$ are dominating complexes. Analysis of the visible region of the spectra, however, enables to establish the two ranges of ln(c_i/c) in which the complexes $Cu(H_2O)_6^{2+}$ and $Cu(C_2O_4)$, and $Cu(C_2O_4)$ and $Cu(C_2O_4)_2^{2-}$ predominantly coexist in the solutions.

Acknowledgements

The authors express their gratitude to Mr A. Holi and Dr K. Wójcik for their skillful technical assistance in this work. This work was financially supported by Ministry of Education and Science under research project PB 1041/P03/2004/26.

References

[1] Sangwal K 1996, Prog. Cryst. Growth Charact. Mater. **32**, 3–43.
[2] Kubota N 2001, Cryst. Res. Technol. **36**, 749–769.
[3] Sangwal K 2007, Additives and Crystallization Processes: From Fundamentals to Applications,Wiley, Chichester.
[4] Rossotti H 1983, Równowagi jonowe (Ionic Equilibria), PWN, Warsaw.
[5] Pataki L and Zapp E 1980, Basic Analytical Chemistry (Akademia Kiado, Budapest)
[6] Prapaipong P, Shock E.L and Koretsky C.M., 1999, Geochim. Cosmochim. Acta **63,** 2547–77.
[7] Nowicka-Jankowska T, Wieteska E ,Gorczyńska K andMichalik A 1988, Spektrofotometria UV/Vis w analizie chemicznej (UV–Vis Spectrophotometry in Chemical Analysis), (PWN, Warsaw)
[8] Lever A B P 1968, Inorganic Electronic Spectroscopy, (Elsevier, Amsterdam)
[9] Perkampus H -H 1992, UV-VIS Spectroscopy and Its Applications, (Springer, Berlin) Chap. 8.
[10] Golcuk K, Altun A and Kumru M 2003, J. Mol. Struct. **627**, 385–393.
[11] Stanila A, Marcu A, Rusu D, Rusu M, and David L 2007 J. Mol. Struct. **834/836**, 364–368.
[12] Krishnamurty K V and Harris G M 1961, Chem. Rev. **61**, 213–246.
[13] Sangwal K and Mielniczek-Brzóska E 2007, Fluid Phase Equil. **258,** 199–205.
[14] Sangwal K and Mielniczek-Brzóska E 2007, Cryst. Res. Technol. **42**, 531–543.
[15] Alderighi L, Gans P, Ienco A, Peters D, Sabatini A and Vacca A 1999, Coord. Chem. Rev. **184,** 311–318.
[16] Atkins P W 2001, Chemia fizyczna (Physical Chemistry),(PWN, Warsaw) Chap. 17.
[17] Kettle S F A 1996, Physical Inorganic Chemistry – A Coordination Chemistry Approach (Oxford University Press) Oxford.
[18] Lee C-W, Chan .-K, Chen T-R and Chen J –D 2003, Inorganica Chimica Acta **348**, 135–142.
[19] Li C-B, and Gao Y-C 2007, Acta Cryst. **E63**, m1337–m1338.
[20] Jordanovska V and Trojko R 1995, Thermochimica Acta **258**, 205–217.

15th International Seminar on Physics and Chemistry of Solids IOP Publishing
Journal of Physics: Conference Series **289** (2011) 012034 doi:10.1088/1742-6596/289/1/012034

New sources and instrumentation for neutron science

Alina Gil

Faculty of Mathematical and Natural Sciences, JD University,
Al. Armii Krajowej 13/15, 42-200 Częstochowa, Poland

e-mail: a.gil@ajd.czest.pl

Abstract: Neutron-scattering research has a lot to do with our everyday lives. Things like medicine, food, electronics, cars and airplanes have all been improved by neutron-scattering research. Neutron research also helps scientists improve materials used in a multitude of different products, such as high-temperature superconductors, powerful lightweight magnets, stronger, lighter plastic products etc.
Neutron scattering is one of the most effective ways to obtain information on both, the structure and the dynamics of condensed matter.
Most of the world's neutron sources were built decades ago, and although the uses and demand for neutrons have increased throughout the years, few new sources have been built.
The new construction, accelerator-based neutron source, the spallation source will provide the most intense pulsed neutron beams in the world for scientific research and industrial development.
In this paper it will be described what neutrons are and what unique properties make them useful for science, how spallation source is designed to produce neutron beams and the experimental instruments that will use those beams. Finally, it will be described how past neutron research has affected our everyday lives and what we might expect from the most exciting future applications.

1. Neutrons and their unique properties

A neutron is one of the fundamental particles that make up matter. It was identified in 1932 and exists in the nucleus of a typical atom along with its positively charged counterpart, the proton. Protons and neutrons each have about the same mass, and both can exist as free particles away from the nucleus.
Neutrons are released in nuclear fission. They have no electrical charge and penetrate materials more effectively than X-rays. This ability makes neutrons an especially useful tool in industrial materials analysis.
Neutrons penetrate most materials to depths of several centimeters. In comparison, X-rays and electrons probe only near the surface. X-rays and electrons are scattered by atomic electrons whereas neutrons are scattered by atomic nuclei. This results in a number of differences, perhaps the most important being in the scattering from light elements. Whereas one electron on a hydrogen atom can be hard to find by X-ray or electron diffraction, the hydrogen nucleus scatters neutrons strongly and is easily found in a neutron diffraction experiment.
Neutrons, though electrically neutral, act as small magnets, and are uniquely powerful in the atomic scale study of magnetism. Neutrons are also uniquely suited to the study of the dynamic processes (e.g. thermal vibrations) in solids.

2. Neutron methods

Just like a beam of light, X-rays or electrons a neutron beam can be reflected, scattered and absorbed. The special feature of neutrons is their deep penetration into materials, their interaction with nuclei

and their sensitivity to magnetic sources in condensed matter due to their magnetic dipole. What is more, neutrons - unlike X-rays - can distinguish light elements (e.g. H). Neutron scattering is a means to study the structure, dynamics (movement of atoms) and the compositions of materials.
The structure of material can be studied by **diffraction** or **small angle scattering**, while the dynamics can be determined by **spectroscopy**. The most common spectroscopical methods are in the order of increasing energy resolution: **time of flight, backscattering, spin echo and 3-axis spectroscopy.** Surfaces and interfaces are analysed by **reflectometry** and **grazing incidence diffraction**. **Neutron Radiography and Tomography** allows imaging that shows quite different information than X-ray imaging does. Precise composition analysis is done by **neutron activation analysis**. Last, but not least, **particle and nuclear physics** use neutrons in several ways - in this field, measurements are done directly at the source or in special setups, not at conventional instruments [11].

3. Neutron production

Neutron scattering facilities throughout the word generate neutrons either with nuclear reactors or with high-energy particle accelerators. The neutron produced have energies up to tens or even hundreds MeV, and the corresponding neutron wavelengths are far too short for investigating condensed matter.
Neutrons are produced in a nuclear reactor by the fissioning (Fig. 1a) of atoms in the reactor fuel, which, for research reactors, is invariably uranium. The neutrons are moderated and allowed to emerge from the reactor in a continuous stream. Other neutron facilities use accelerators to produce spallation neutrons. When a high-energy proton bombards a heavy atomic nucleus (Fig. 1b), causing it to become excited, and 20 to 30 neutrons are expelled. The energy spectrum from the spallation source is quite different from that produced by reactor because there are a greater percentage of high-energy neutrons [13]. Neutrons from the spallation source arrive in pulses rather than continuously as they do at a reactor. It means that the monochromator crystal needed at reactors can here be avoided and all the neutrons can be used, and relies on the measurement of the time it takes for each detected neutron to traverse the distance between the moderator and the detector. From this *time of flight*, the neutron velocity can be determined, which gives its wavelength and generating a monochromatic beam is unnecessary.

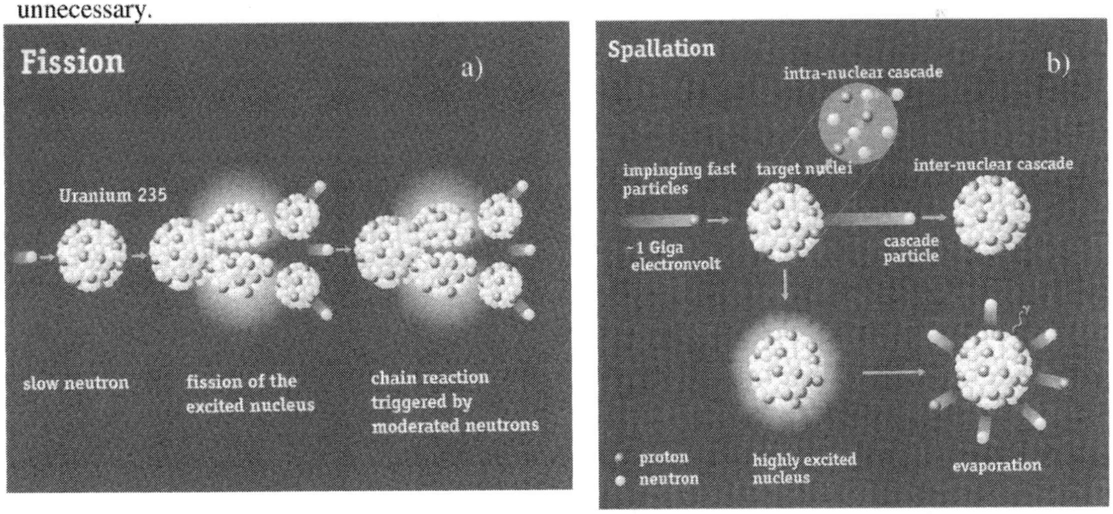

Figure 1. Neutron production: a) fission, b) spallation process [4] (with kind publisher agreement)

4. Neutron Scattering Facilities

Most of the world's neutron sources were built decades ago, and although the uses and demand for neutrons have increased throughout the years, few new sources have been built and few is projected.
There are already 6 neutron sources in Asia and Australia, 15 in North and South America and 17 in Europe [12].

Journal of Physics: Conference Series **289** (2011) 012034 doi:10.1088/1742-6596/289/1/012034

European Neutron Sources:

- Budapest Neutron Centre, AEKI, Budapest, Hungary
- Berlin Neutron Scattering Center, Helmholtz-Zentrum, Berlin, Germany
- Center for Fundamental and Applied Neutron Research (CFANR), Rez nr Prague, Czech Republic
- Frank Laboratory of Neutron Physics, Joint Institute of Nuclear Research, Dubna, Russia
- FRJ-2 Reactor, Forschungzentrum Jülich, Germany
- FRM-II Research Reactor, Garching, Germany
- GKSS Research Center, Geesthacht, Germany
- Institut Laue Langevin, Grenoble, France
- Interfacultair Reactor Instituut, Delft University of Technology, Netherlands
- ISIS Pulsed Neutron and Muon Facility, Rutherford-Appleton Laboratory, Oxfordshire, UK
- JEEP-II Reactor, IFE, Kjeller, Norway
- Laboratoire Léon Brillouin, Saclay, France
- Ljubljana TRIGA MARK II Research Reactor, J. Stefan Institute, Slovenia
- Risø National Laboratory, Denmark
- St. Petersburg Nuclear Physics Institute, Gatchina, Russia
- Studsvik Neutron Research Laboratory (NFL), Studsvik, Sweden
- Swiss Spallation Neutron Source (SINQ), Villigen Switzerland

New Projects:

- Austron Spallation Neutron Source, Vienna, Austria
- Canadian Neutron Facility, Chalk River, Ontario, Canada
- China Advanced Research Reactor (CARR), Beijing, China
- Chinese Spallation Neutron Source (CSNS), Dongwan, Guangdong, China
- European Spallation Source (ESS)
- Japan Proton Accelerator Research Complex (J-PARC), Tokai, Japan

Table 1. Spallation Sources in Europe

Centre	Organisation	Location	Type	Since	Source
ISIS	Rutherford Appleton Laboratory	Oxford, UK	spallation source (pulsed)	1985	cold, thermal und hot neutrons
SINQ Swiss Spallation Neutron Source	Paul Scherrer Institute	near Villigen, Schwitzerland	Continous spallation neutron source, 590 MeV isochronous proton cyclotron	1996	cold and thermal neutrons
European Spallation Source (**ESS**)		Lund, Sweden	spallation source	planned	

Table 2. Spallation Sources outside Europe

Centre	Organisation	Location	Country	Power (MW)
LANSCE	Los Alamos National Laboratory	Los Alamos	USA	0,16
IPNS	Argonne National Laboratory	Argonnne	USA	0,007
KENS	High Energy Accelerator Organisation KEK	Tsukuba	Japan	0,003
SNS	Oak Ridge National Laboratory	Oak Ridge	USA	1,4

4.1. New sources:

SNS (Spallation Neutron Source) is an accelerator-based neutron source in Oak Ridge, Tennessee, USA. This one-of-a-kind facility at full power will provide the most intense pulsed neutron beams in the world for scientific research and industrial development.

SNS was built by a partnership of six U.S Department of Energy laboratories. Along with its sister facility, the High Flux Isotope Reactor, SNS makes Oak Ridge a mecca for neutron-scattering research.

How SNS work? Negatively charged hydrogen ions are produced by an ion source. Each ion consists of a proton orbited by two electrons. The ions are injected into a linear accelerator, which accelerates them to very high energies. The ions are passed through a foil, which strips off each ion's two electrons, converting it to a proton. The protons pass into a ring where they accumulate in bunches. Each bunch of protons is released from the ring as a pulse. The high-energy proton pulses strike a heavy-metal target, which is a container of liquid mercury. Corresponding pulses of neutrons freed by the spallation process will be slowed down in a moderator and guided through beam lines to areas containing special instruments such as neutron detectors. Once there, neutrons of different energies can be used in a wide variety of experiments.

At full power, SNS will deliver 1.4 million watts (1.4 MW) of beam power onto the target, and it has been designed with the flexibility to provide additional scientific achievements in the future. This approach is intended to provide a facility that will meet the neutron intensity needs of the science community well into the next century [9].

The instruments at the SNS, such as neutron spectrometers, will be used to determine the positions, or arrangements, of atoms in crystals, ceramics, superconductors, and proteins – a pulse of neutrons generated by the spallation source follows a flight path to the sample. Because the neutrons have varying energies and wavelengths, they spread out in time, presenting continuous spectra to the sample. When the distance between atoms in a crystal matches the wavelength of an incident neutron, that neutron is scattered into a multidetector that records the position (scattering angle) and time of arrival of the scattered neutron. The result is a pattern of peaks showing the different positions and arrival times of various numbers of neutrons reaching each point in the multidetector. This pattern tells scientists how different atoms are arranged in the crystal.

Instrumentation based on the same principles can be used to determine the atomic structure of glasses and complex fluids or the residual stresses in industrial parts. Instruments to measure inelastic scattering will require measurement of the time of neutron travel over paths leading to and from the sample. In this way, instruments can determine the excitation spectra of materials of importance and thus the nature of the forces that hold the atoms in place. The time-of-flight technique makes it possible to collect a large number of data points for each neutron pulse. The efficiency of instruments that measure neutron time of flight and the ability of accelerator-based spallation neutron sources to produce pulsed beams of increasing intensity promise to provide continuously improved neutron sources in the future.

ISIS is the world's leading spallation neutron source [8], providing UK and international researchers access to the best scientific facilities of their kind. ISIS has contributed significantly to many of the major breakthroughs in materials science, physics and chemistry since it was commissioned in 1985. ISIS Pulsed Neutron and Muon Source is located at the Rutherford Appleton Laboratory. Their notable successes include:

- the determination of the structure of high temperature superconductors and fullerenes;
- the study of excitations in highly correlated electron systems and quantum fluids;
- the behaviour of complex fluids, such as surfactants, polymers and proteins, at interfaces;
- high pressure studies of crystalline materials;
- the structure of super-critical fluids;
- the microscopic understanding of the operation of catalysts.

Expansion of ISIS through the building of a Second Target Station was announced in April 2003, as a key part of the UK investment strategy in major facilities.

The ISIS Second Target Station started the experimental programme in 2008. It represents a new generation of neutron production target and instrument suite and open up new opportunities in

Journal of Physics: Conference Series **289** (2011) 012034 doi:10.1088/1742-6596/289/1/012034

technologically significant areas, particularly in the fields of soft condensed matter, bio-molecular science, advanced materials and nanoscale science.
It will offer unique instrumentation and unrivalled potential for structural and dynamical studies of matter, using cold neutrons and high resolution spectroscopy. The high yield of long-wavelength, low-energy neutrons over a broad spectrum with high resolution will enable a significant step forwards in the exploitation of neutrons to solve problems in different areas.
The Second Target Station Project will keep the UK at the forefront of neutron research and enable scientists to continue to make breakthroughs in materials research for the next generation of super-fast computers, data storage, sensors, pharmaceutical and medical applications, materials processing, catalysis, biotechnology and clean energy technology.

The spallation neutron source **SINQ** [10] (Swiss Spallation Neutron Source) is a continuous source – the first of its kind in the world – with a flux of about 10^{14} n/cm^2/s. Beside thermal neutrons, a cold moderator of liquid deuterium (cold source) slows neutrons down and shifts their spectrum to lower energies. These neutrons have proved to be particularly valuable in materials research and in the investigation of biological substances. SINQ is a user facility.

On Thursday evening (28 may 2009) at a meeting of research ministers in Brussels, the Swedish candidacy to build the European research facility **ESS** [11] (European Spallation Source) received support from a clear majority. It is now clear that the ESS will be built in Lund, in Sweden.
Sweden was supported by seven countries – Germany, France, Poland, Denmark, Norway, Estonia and Latvia – with a further two, Italy and Switzerland, voting with the majority. One country supported another candidate country.
The European Spallation Source, ESS, will be the world-leading research facility using neutrons for materials research and life science. It will be one of Europe's largest research centres, and co-located with the next generation synchrotron MAX IV.

Japan projects the construction of a High Intensity Proton Accelerators: the **J-PARC** (Japan Proton Accelerator Research Complex) Project [12]. It is proposed jointly by the High Energy Accelerator Research Organization (KEK) and the Japan Atomic Energy Research Institute (JAERI). The powerful pulsed spallation neutron source will be one of the major facilities of the J-PARC. The facility is being constructed at the JAERI/Tokai site, about 130 km north-east of Tokyo. Construction budget started on April 1, 2001, with an anticipated first beam in April 2008. This multipurpose facility will produce neutrons, pions, kaons antiprotons, together with muons and neutrinos.
The accelerator complex consists of following accelerators:

- 400-MeV normal-conducting Linac,
- 600-MeV superconducting Linac to increase the energy from 400 to 600 MeV will be constructed in the 2nd phase of J-PARC,
- 3-GeV synchrotron ring, which provides proton beams at 333micoA (1MW), and
- 50-GeV synchrotron ring, which provides proton beams at 15microA (0.75MW).

The 3-GeV ring will be used as a booster synchrotron for the 50-GeV main ring. In addition, it is designed to provide beam power of 1 MW.

5. Neutron Science

The diverse applications of neutron scattering research can provide opportunities for studies in practically every scientific and technical field. Neutron scattering allows scientists to count scattered neutrons, measure their energies and the angles at which they scatter, and map their final positions. Such information can reveal the molecular and magnetic structure and behavior of materials, such as high-temperature superconductors, polymers, metals, and biological samples.
Although not obvious to most people, the fruits of neutron-scattering research are improvements in the range and quality of products used in our everyday lives. A few examples are:

- cars and airplanes
- electronics such as computers and cell phones

- medicines and other health care-related products
- paints, shampoos, and other fluids

Neutrons are also an essential tool for researchers studying ways of improving materials used in high-temperature superconductors, powerful lightweight magnets, aluminum bridge decks, and stronger, lighter plastic products.

Neutrons have been used to learn how bones mineralize during development and how they decay during osteoporosis, and they make it possible for us to devise and to test remedies for demineralizing diseases. Of the people who enter hospitals, one in three (or about 100 million each year) benefit from isotopes produced by neutrons.

On the web site *http://neutron.neutron-eu.net/n_about/n_what_can_you_do_with_neutrons* you can see a simple illustration of how neutron source may form the centre of a complex of activities. Around the central core, there is a ring of specific spheres that are more or less dependent on the activities that are conducted at the neutron spallation source. Outside the ring of specific spheres, there are several examples of technical applications and industries that can be considered to be more or less directly related to these spheres.

Some Benefits of Neutron Science:

Structure of water around DNA - Hydrogen atoms constitute the skin of the majority of biologically active compounds and knowledge of their precise location and behaviour is crucial if we are to fully understand, for example, how medicines act within the body.

Neutrons are particularly good at locating the positions of hydrogen atoms. For example, the structure of the water (H_2O) molecules that surround the double helical structure of DNA, the molecule of life, can be visualised. Using neutrons, the changes in DNA structure can be measured as water is removed from around the strands by lowering the humidity around the sample.

High Temperature Superconductivity - Electricity transmission from power stations to homes and industry via metal cables is quite an inefficient process with a lot of power lost along the way in the form of heat and even sound. The power loss vanishes if the cables are cooled down to about -273°C when metals lose all resistance to the flow of electricity and it becomes a superconductor. Cooling thousands of miles of power cables to this temperature is not practical, and so there is significant interest in discovering materials that can superconduct at higher temperatures.

Neutron experiments allow us to examine the structure of materials known as high temperature superconductors which superconducts at -140°C. There is also a strong increase in the transition temperature with applied pressure accompanied by major structural changes.

Experiments such as these provide an insight into the mechanism of superconductivity, and suggest changes that can be made to the material to move closer to the technologically important goal of room temperature superconductivity.

Chemistry at the interface - The use of detergent (surfactant) molecules spans a range of applications from washing up liquid to the sophisticated stabilizers used in the formation of emulsions and colloids such as hand cream and shower gels. Their effectiveness depends greatly upon their molecular arrangement at the air-liquid / liquid-solid / liquid-liquid interfaces. The reflection of neutron beams from each interface produces patterns analogous to the beautifully color patterns seen from oil films in puddles, and provides information about the organization of detergent molecules at interfaces. These experiments are providing a fresh insight into the function of these surfactant molecules.

Understanding the atomic and molecular architecture of materials - The properties of neutrons are ideally suited to look at the precise way in which the atoms and molecules are arranged inside materials, which often form beautiful structures such as the buckyballs - units of 60 carbon atoms arranged in enclosed cages of hexagons and pentagons. The process of determining the structures is known as crystallography, and it plays a crucial role in understanding the properties of the material, allowing insights into improvements that can lead to better materials for the future.

Neutrons look at materials differently to other forms of radiation such as X-rays and light. Like X-rays, they see individual atoms, but they see each different variety of atom equally clearly, independent of size. Neutrons also penetrate deeply inside most materials, whether they are clear or opaque, unlike X-rays which only see the surface, or light which can only see deep inside transparent materials.

Geophysics - science under pressure - Hydrogen is the most abundant element in the Universe. Recent developments are making it possible to locate hydrogen atoms in structures at very high pressures, of up to 250,000 atmospheres, for key molecular materials like ice, ammonia and methane - and, eventually, hydrogen itself. This is important for the understanding of inter-molecular bonding and structures, and these materials are also constituents of many planets - such as the solid (metallic) hydrogen core of Jupiter. A study of ammonia at 65,000 atmospheres has revealed the unexpectedly complex structure and bonding.
Neutron scattering can also be used to determine magnetic structures under high pressure.

6. Summary

Neutron scattering is a unique and powerful way of studying the properties of materials at the atomic level. Neutron scattering experiments reveal where atoms are and what they are doing, enabling the spacing of atoms and the forces between them to be measured. Innovations in technique and improved instrument performance over the last twenty years have made a huge contribution to our understanding of materials, and the number of disciplines where neutron scattering has made an impact has steadily increased.
Spallation neutron sources, which are designed with the future in mind, will be the leading neutron research facility for many years to come. They allow measurements of greater sensitivity, higher speed, higher resolution, or in more complex sample environments than have been possible at existing neutron facilities.

7. References

[1] Pynn R, 1990, *Neutron Scattering – A Primer*, Los Alamos Science.
[2] Liang L; Rinaldi and R; Schober H (Eds.); 2009, Neutron Applications in Earth, Energy and Environmental Sciences, Series:Neutron Scattering Applications and Techniques,Vol. XVIII, 638 p.
[3] Jones B, King P, Sivia D, Clements D and Whitehead S, The ISIS 2007 Annual Report. Science and Technology Facilities Council.
[4] Tindemans P et al, 2002, ESS Council, The ESS Project, Europeaen Source of Science, Vol. I.
[5] Krämer K and Zolliker M, 2008, Swiss Neutron News, Paul Scherer Institute.
[6] Neutron News, Published by: Taylor & Francis 2008, 2009.
[7] Journal of Neutron Research, Published by: Taylor & Francis 2008.
[8] http://www.isis.stfc.ac.uk/ Science & Technology Facilities Council
[9] http://neutrons.ornl.gov/facilities/SNS/works.shtml
[10] http://www.psi.ch/info/facility-news, PSI
[11] http://neutron.neutron-eu.net/, European Neutron Portal, Neutron Tools for Science
[12] http://www.neutron.anl.gov/facilities.html
[13] http://www.fas.org/sgp/othergov/doe/lanl/pubs/00326651.pdf

Effects of oxygen supply on the biodegradation rate in oil hydrocarbons contaminated soil

I Zawierucha[1], G Malina[2]

[1] Institute of Chemistry and Environment Protection, Jan Dlugosz University of Czestochowa, Waszyngtona 4/8, 42-200 Czestochowa, Poland

[2] Faculty of Hydrogeology and Geology Engineering, Department of Geology, Geophysics and Environment Protection, AGH University of Science and Technology, Mickiewicza 30, 30-059 Cracow, Poland

E – mail: iwona_zawierucha@o2.pl

Abstract. Respirometry studies using the 10-chamber Micro-Oxymax respirometer (Columbus, Ohio) were conducted to determine the effect of biostimulation (by diverse ways of O_2 supply) on enhancing biodegradation in soils contaminated with oil hydrocarbons. Soil was collected from a former military airport in Kluczewo, Poland. Oxygen was supplied by means of aerated water, aqueous solutions of H_2O_2 and $KMnO_4$. The biodegradation was evaluated on the basis of O_2 uptake and CO_2 production. The O_2 consumption and CO_2 production rates during hydrocarbons biodegradation were estimated from the slopes of cumulative curve linear regressions. The pertinent intrinsic and enhanced biodegradation rates were calculated on the basis of mass balance equation and O_2 uptake and CO_2 production rates. The biodegradation rates of 5-7 times higher as compared to a control were observed when the aqueous solution of $KMnO_4$ in concentration of 20 g L^{-1} was applied. Permanganate is known to readily oxidize alkene carbon – carbon double bonds; so it can be successfully applied in remediation technology for soils contaminated with oil hydrocarbons. While hydrocarbons are not completely mineralized by permanganate oxidation reactions, their structure is altered by polar functional groups providing vast improvements in aqueous solubility and availability for biodegradation. The 3% aqueous solution of H_2O_2 caused significant improvement of the biodegradation rates as compared to a control (on average about 260%). Aerobic biodegradation of hydrocarbons can benefit from the presence of oxygen released during H_2O_2 decomposition. Adding of aerated water resulted in an increase of biodegradation rates (about 114 - 229%) as compared to a control. The aerated water can both be the source of oxygen for microorganisms and determine the transport of substrate to bacteria cells.

1. Introduction

Dynamic development of transportation causes that more of petroleum substances may penetrate natural environment, resulting in degradation of soils, surface water and groundwater [1]. Nearly 2.5 million tons of oil products is estimated to enter the environment yearly [2]. Since late 70-ties intensive studies have been carried out in Poland to estimate the degree of soil and groundwater contamination. A number of locations have been identified and documented with very high contamination of oil hydrocarbons [3]. Oil-hydrocarbon contaminants may appear locally and periodically as a consequence of leaks of crude oil from mining shafts, damage pipelines, cisterns, tankers, and storage and distribution stations of petroleum products. They can also appear permanently on definite areas and derive from refineries or engine industry [4]. Moreover, floods are a serious factor of contamination spreading in the environment [5]. Oil products in soils affect vegetation, threaten quality of surface water and groundwater, and are also potentially hazardous for humans and the environment [2].

Under natural conditions soils contaminated with oil hydrocarbons may undergo self purification, mainly due to biodegradation with the participation of indigenous microorganisms.

The activity of native microflora decides about efficiency of biodegradation, and it can be enhanced by providing optimal conditions for the microbial growth [6,7].

In the case when O_2 is considered as a limiting factor, biostimulation by oxygen supply is applied to stimulate microbial activity and enhance aerobic biodegradation rates. Commonly used oxygen supply techniques may include tilling, forced aeration and chemical methods [8,9,10]. Tilling has been recommended as a physical method to accelerate biodegradation in landfarming. This technique is effective only for topsoils. Forced aeration techniques, including injection of aerated water, air and pure oxygen, are commonly used for enhancing biodegradation in subsurface soils and groundwater contaminated with petroleum hydrocarbons [9,10]. Chemical methods involve addition of alternative oxygen sources, such as oxygen releasing compounds ORC®, or agents (e.g. $KMnO_4$). Hydrogen peroxide (H_2O_2) and ozone (O_3) are also applied, but they can have toxic effects to microorganisms [10].

In this paper, the respirometry studies were conducted to determine the effects of biostimulation (by diverse ways of oxygen supply) on enhancing intrinsic biodegradation of oil hydrocarbon, both in case of fresh leakage and aged contamination. Oxygen was supplied by means of aerated water, aqueous solutions of hydrogen peroxide (H_2O_2) or potassium permanganate ($KMnO_4$). The intrinsic and enhanced biodegradation rates were evaluated by the mean O_2 uptake and CO_2 production rates obtained using a linear regression of the cumulative curves of O_2 uptake and CO_2 production.

2. Materials and methods

2.1. *Soil samples*

Soils samples contaminated with oil hydrocarbons ("historical" - aged contamination) originated from the former military airport in Kluczewo, Poland. Contaminated soil samples were collected at the depths of 1.5 m (G1) and 2.0 m (G2). Uncontaminated soil samples were also collected at the depths of 1.5 m (G1R) and 2.0 m (G2R).

2.1.1. *Soil characteristics*

pH and Eh of soil suspension were determined according to PN-ISO 10390:1997, soils moisture – PN-ISO 11465:1999, organic matter (OM) content – PN-78/C-04541, content of nitrogen – PN–ISO 11261:2002, content of phosphorus – PN–ISO 11263:2002.

2.1.2. *Number of bacteria*

Microbial enumeration of bacteria was measured as the number of colony – forming units (CFU) per gram, using the standard agar-plate technique. This method relies on executing of dilution series of soil suspension, and inoculating stable volumes of the dilutions on the agar medium. The dilution series of soil suspension were made with sterile distilled water. Bacteria were incubated at 28oC for 48 hours.

2.1.3. *Contents of hydrocarbons in soil*

Soil samples were extracted with mixture of methanol:dichloromethane (4:1) by ultrasounds. Contents of aliphatic and aromatic hydrocarbons were determined using a gas chromatograph with mass detector - Varian Star 3400 CX.

2.2. *Respirometer test*

Biodegradation of oil hydrocarbons under aerobic conditions was analyzed using a 10-chamber Micro-Oxymax respirometer (Columbus Instruments, OH, USA), in conjunction with an Ultra IBM-compatible computer to collect and record all data. This is a closed-circuit respirometer capable to measure on-line the O_2 consumed and CO_2 produced during hydrocarbons biodegradation. Experiments were carried out using batch tests at a temperature of $25\pm5^{O}C$. Soil samples of 30 g were placed into 500 mL measuring chambers of the respirometer, to which were added: 8 mL of aqueous solutions of $KMnO_4$ (concentrations of 5, 10, 20 g L^{-1}), H_2O_2 (in dilution of 1:10), and aerated distilled water with the O_2 concentrations of 6.67, 7.36, 7.83 mg L^{-1}, which corresponded to the O_2 saturation (under existing conditions i.e. t= 27oC and barometric pressure = 1013 hPa) of 83, 92 and 98%, respectively. The initial concentration of H_2O_2 was of 300 g kg^{-1}. All tests were done in triplicate. As a control, uncontaminated soil samples G1R and G2R (control 1), as well as contaminated soil samples G1 and G2 without any enhancement (control 2) were used. Control 1 determined the basal respiration rate, which is the result of soil organic matter degradation, whereas control 2 determined an overall respiration, including basal respiration and hydrocarbons biodegradation rates.

Biodegradation of oil hydrocarbons was evaluated based on the cumulative curves of O_2 consumption and CO_2 production. Cumulative curves were plotted as O_2/CO_2 content values in the headspace versus time. Rates of O_2 uptake and CO_2 production were estimated from the slopes of linear regressions of cumulative curves. The coefficients of the linear regression equations represented the mean rates of O_2 consumption and CO_2 production during hydrocarbons biodegradation. The linear regression model was applied because it was found to fit with enough accuracy the experimental data plotted as accumulative curves. It was indicated by the regression coefficients (R^2) within the range of 0.94-0.99.

Having data on mean O_2 uptake and CO_2 production rates, and assuming that the only products of hydrocarbons biodegradation are biomass, carbon dioxide and water, pertinent biodegradation rates were calculated.

Mass balance equation is as following:

$$C_nH_m + a\,O_2 = Y\,CH_2O + b\,CO_2 + c\,H_2O \qquad (1)$$

where: m – number of hydrogen atoms, n – number of carbon atoms, a, b, c – stechiometric coefficients of reaction, Y – microbial yield

Assuming that hydrocarbons formula is $CH_{1,5}$ [11,12], hence m/n =1,5; whereas microbial yield Y=0,5.

Hydrocarbons biodegradation rates ($k'_{NA/ENA}$, $k''_{NA/ENA}$) were calculated according to formulas [12,13]:

$$k'_{NA/ENA} = 2{,}144\frac{4\left(12+\frac{m}{n}\right)}{4(1-Y)+\frac{m}{n}}k_{O_2} \qquad \left[\frac{mg\ of\ hydrocarbons}{kg\ of\ soil\quad day}\right] \qquad (2)$$

$$k''_{NA/ENA} = 58{,}090\,k_{CO_2} \qquad \left[\frac{mg\ of\ hydrocarbons}{kg\ of\ soil\quad day}\right] \qquad (3)$$

where: $k'_{NA/ENA}$ – natural/enhanced biodegradation rate calculated on the basis of O_2 consumption rate, $k''_{NA/ENA}$ - natural/enhanced biodegradation rate calculated on the basis of CO_2 production rate, k_{O_2} – O_2 consumption rate [µl/min], k_{CO_2} – CO_2 production rate [µl/min], 2,144 – a conversion coefficient for O_2 consumption rate to [mmol of O_2/kg of soil*day]; 58,090 - conversion coefficient for CO_2 production rate to [mg of hydrocarbons/kg of soil*day]

The values of O_2 consumption and CO_2 production rates obtained for contaminated soil samples (without and with enhancement by O_2 supply) minus the values obtained for uncontaminated soil samples were placed to formulas (2) and (3). It provided to calculate the "net" hydrocarbons biodegradation rates.

3. Results and discussion

The characteristics of soils are presented in Table 1. The obtained soil physical-chemicals parameters are considered optimal for biodegradation of oil hydrocarbons under aerobic conditions [11,14,15]. Sufficient number of microorganisms (105 CFU g-1) [16] for biodegradation was also present in impacted soils.

Soil samples were mainly contaminated with aliphatic hydrocarbons. Aliphatic hydrocarbons are well susceptible to microbiological decomposition [17,18,19].

Tab. 1. Soil characteristics

Soil sample	pH	Eh (mV)	Moisture (% w/w)	OM content (% w/w)	Number of bacteria (CFU/g)	N (mg/kg $_{s.m.}$)	P (mg/kg $_{s.m.}$)	Σ aromatic hydrocarbons (mg/kg $_{d.w.}$)	Σ aliphatic hydrocarbons (mg/kg $_{d.w.}$)
G1	7,4	451	18,18	2,76	20×10^5	452	314	30	155
G1R	6,7	447	26,71	3,92	1×10^5	462	345	-	-
G2	7,6	457	19,23	3,20	10×10^5	303	305	-	82
G2R	8,1	449	13,07	1,99	3×10^5	330	332	-	-

The exemplary cumulative curves of O_2 consumption and CO_2 production during biostimulation tests are presented in Fig. 1. The mean rates of O_2 uptake and CO_2 production during intrinsic and enhanced biodegradation (obtained from linear regressions of cumulative curves) are presented in Fig.2.

In our experimental set-up the air in the respirometer is circulated in a closed loop, and soil samples are continuously aerated. The addition of O_2 could cause reading clutter of O_2 contents at the top of measuring chamber, therefore the evaluation of biodegradation enhancement by O_2 supply was based on the biodegradation rate calculated on the basis of CO_2 production rate. Only for contaminated soil samples without enhancement (control), pertinent biodegradation rates were calculated on the basis of O_2 consumption rates and then used to comparative analysis. The effectiveness of biodegradation enhancement (biostimulation) by diverse ways of O_2 supply was presented in Fig. 3.

A

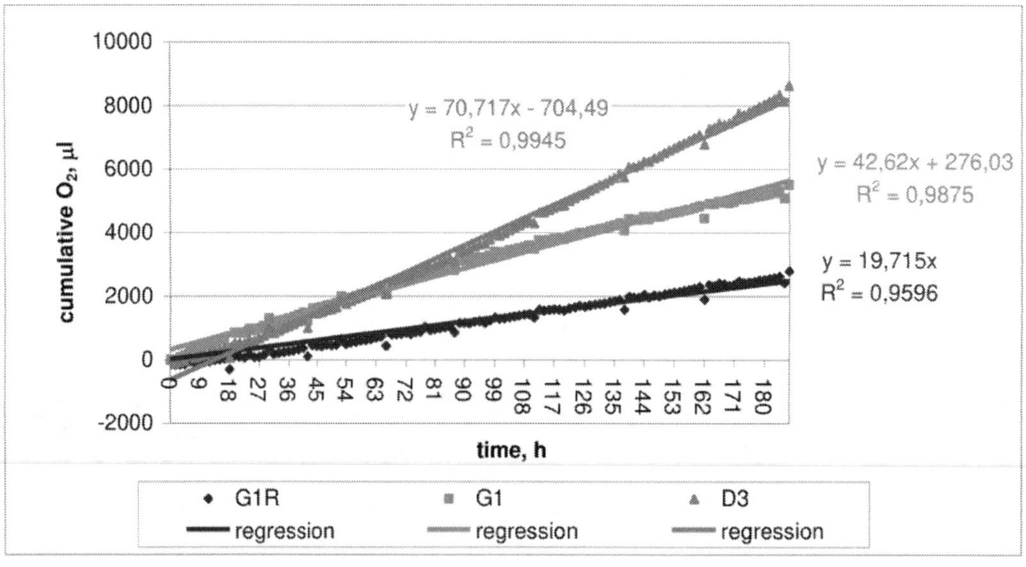

B

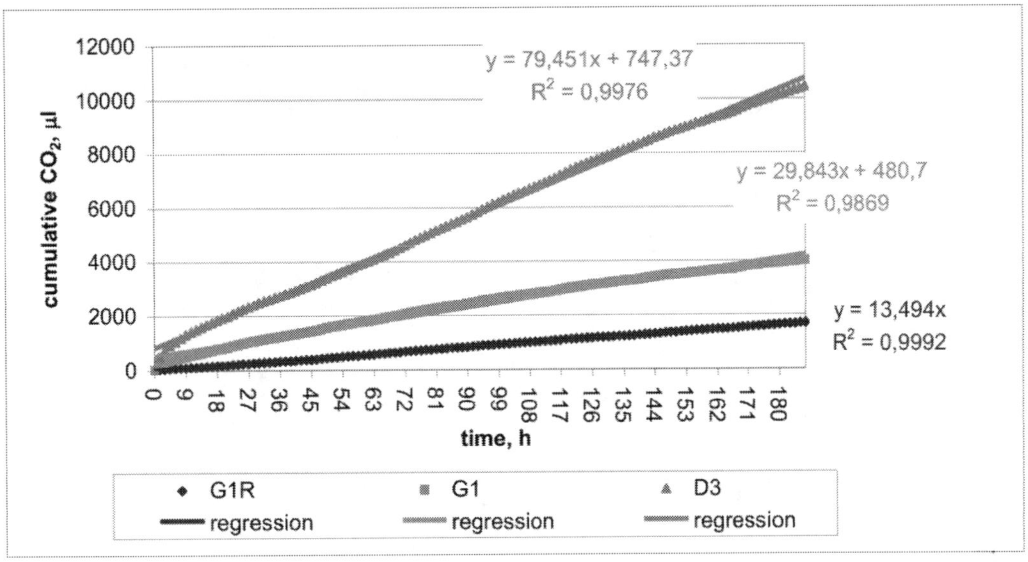

Figure 1. Cumulative curves of O_2 uptake (A) and CO_2 production (B) during biostimulation tests
Notation: G1R – control 1 (uncontaminated soil - depth of 1.5 m), G1 – control 2 (contaminated soil without enhancement - depth of 1.5 m), D3 – addition of aqueous solution of $KMnO_4$ (concentration of 20 g L^{-1})

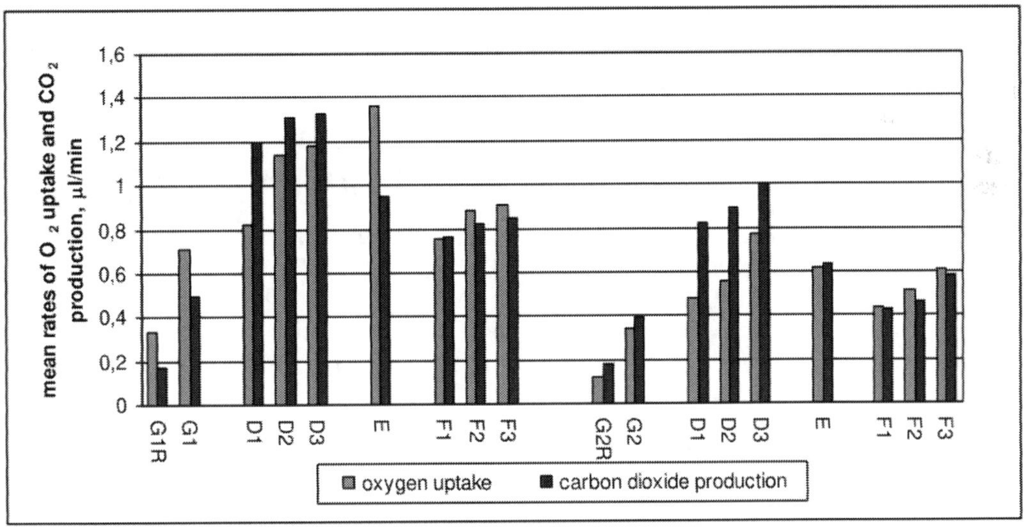

Figure 2. Mean rates of O_2 uptake and CO_2 production during intrinsic and enhanced biodegradation (biostimulation tests)

Notation: G1R, G2R – control 1 (uncontaminated soil - depths of 1.5 and 2.0 m); G1, G2 – control 2 (contaminated soil without enhancement - depths of 1.5 and 2.0 m); D_1, D_2, D_3 – addition of aqueous solution of $KMnO_4$ (concentration of 5, 10 and 20 g L^{-1}); E – addition of aqueous solution of hydrogen peroxide in dilution of 1:10; F_1, F_2, F_3 – addition of aerated distilled water with oxygen saturation of 83, 92 and 98%, respectively

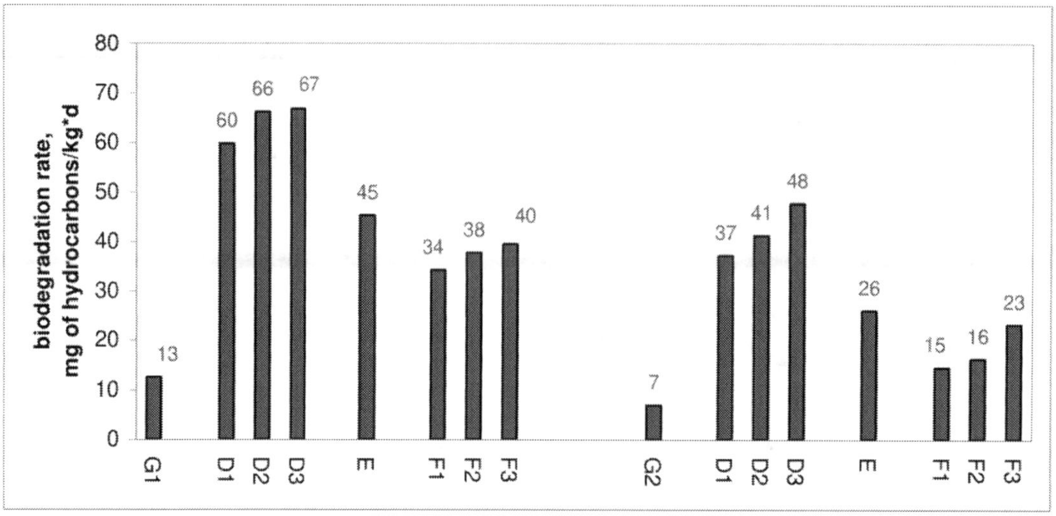

Figure 3. The effect of biostimulation (by diverse ways of O_2 supply) on the biodegradation rate

The biodegradation rates of 5-7 times higher as compared to a control were observed when the aqueous solution of $KMnO_4$ in concentration of 20 g L^{-1} was applied. Permanganate is known to readily oxidize alkene carbon – carbon double bonds [20,21], so it can be probably used also for oxidation of aromatic hydrocarbons. Brown et al. [22] observed the reduction of contents in soil for benzo(a)pyrene (of 72.1%), pyrene (of 64.2%), phenanthrene (of 56.2%) and anthracene (of 53.8%) after 30 min oxidation at $KMnO_4$ concentration of 160 mM. The results suggested that permanganate oxidation can be successfully applied in remediation technology for soils contaminated with oil hydrocarbons. While hydrocarbons are not completely mineralized by permanganate oxidation reactions, their structure is altered by polar functional groups providing vast improvements in aqueous solubility and availability for biodegradation.

The 3% aqueous solution of H_2O_2 caused significant improvement of the biodegradation rates as compared to a control (on average about 260%). Aerobic biodegradation of hydrocarbons can benefit from the presence of oxygen released during H_2O_2 decomposition [23]. However, Goi et al. [24] noted that efficiency of H_2O_2 application was strongly dependent on the soil matrix. Treatment of shale oil and transformer oil adsorbed in peat, used as a model of organic-rich soil, resulted in lower oil removal degree, and required more H_2O_2 than the treatment of oil in sand matrix representing the mineral part of the soil.

Adding of aerated water resulted in an increase of biodegradation rates (about 114 - 229%) as compared to a control. The aerated water can both be the source of oxygen for microorganisms and determine the transport of substrate to bacteria cells. The water facilitated diffusion of substrate for microorganisms and, consequently improved biodegradation rate [25].

5. Conclusions

Biostimulation by oxygen supply can be attractive approach of enhacing biodegradation in soil contaminated with oil hydrocarbons.

The results of biostimulation show that the highest biodegradation rates were achieved when aqueous solution of $KMnO_4$ in concentration of 20 g L^{-1} was applied. Then, the biodegradation rates were about 415-585% higher compared to a control.

The adding of aerated water seems to be the most optimal way of O_2 supply. The use of aerated water causes improvement of the biodegradation rate and is safe for natural environment.

Acknowledgments

This work was financially supported by MNISW grant no. 4 T12B 019 30.

References

[1] Deska I Malina G Radło A 2003 Wpływ uziarnienia gruntu oraz paliwa na różnicę między pozorną i rzeczywistą LNAPL na zwierciadle wody podziemnej. *Inżynieria i Ochrona Środowiska* **6**(2) 203-220

[2] Gaca J 2000 Biodegradacja olejów i produktów naftowych i ich wpływ na środowisko Wpływ zanieczyszczeń naftowych i chemicznych na środowisko przyrodnicze VII Międzynarodowe Sympozjum Szkoleniowe Piła 1-8

[3] Kościelniak S Steininger M 2001 Dopuszczalne stężenia zanieczyszczeń gruntów substancjami chemicznymi (propozycje zmian) Wpływ zanieczyszczeń naftowych i chemicznych na środowisko przyrodnicze Materiały szkoleniowe cz. 1 Poznań 11-17

[4] Malina G Derda A. Derda M 1999 Ocena zagrożenia użytkowych zbiorników wód podziemnych substancjami ropopochodnymi w rejonie Częstochowy *Inżynieria i Ochrona Środowiska* **2**(3-4) 279-295

[5] Różański H 1999 Zaburzenia składu chemicznego roślin leczniczych, pastwiskowych i warzywnych pochodzących z gleb i wód zanieczyszczonych ropą naftową oraz produktami ropopochodnymi Techniczne i społeczne aspekty gospodarki odpadami III Międzynarodowe Forum Gospodarki Odpadami Poznań 365-373

[6] NRC 2000 Natural attenuation for Groundwater Remediation (Washington D.C. National Academy Press)

[7] Dechema 2002 (Proceedings of 5. Symposium on Natural Attenuation Heidelberg)

[8] Atlas RM 1991 Microbial hydrocarbon degradation-bioremediation of oil spills. *J. Chem. Tech. Biotechnol.* **52** 149-156

[9] Brown RA Crosbie JR 1994 Oxygen sources for in situ bioremediation Bioremediation – Field Experience (Lewis Publishers) 311-332

[10] Riser-Roberts E 1998 Remediation of Petroleum Contaminated Soil: Biological, Physical, and Chemical Processes (CRC Press LLC Boca Raton FL)

[11] Malina G 1999 Biowentylacja strefy aeracji zanieczyszczonej substancjami ropopochodnymi. (Monografia 69 Wydawnictwo Politechniki Częstochowskiej Częstochowa)

[12] Płaza G Ulfig K Worsztynowicz A Malina G Krzemińska B Brigmon L 2005 Respirometry for assessing the biodegradation of petroleum hydrocarbons *Environmental Technology* **26** 161-169

[13] Berry CJ Story S Altman DJ Upchurch R Whitman W Singleton D Płaza G Brigmon RL 2006 Biological Treatment of Petroleum in Radiologically Contaminated Soil (WSRC-MS- 2005-00341)

[14] Karkulski K Zieńko J 1997 Substancje ropopochodne w środowisku przyrodniczym (Wyd. Uczel. Polit. Szczecińskiej Szczecin)

[15] Kołwzan B 2000 Biodegradacja produktów naftowych Zanieczyszczenia naftowe w gruncie (Oficyna Wydawnicza Politechniki Wrocławskiej Wrocław) 207-235

[16] Turco RF Sadowski MJ 1995 The Microflora of Bioremediation Bioremediation, Science and Applications Soil Science Society of America Special Publication **43** 87-103

[17] Galas E Kwapisz E 1997 Charakterystyka wybranych szczepów bakterii degradujących węglowodory ropy naftowej *Biotechnologia* 1 **36** 145-157

[18] Grabas K Kołwzan B 1999 Remediacja gruntów skażonych produktami naftowymi na terenie lotniska „Bagicz" VI Ogólnopolskie Sympozjum Naukowo – Techniczne „Biotechnologia Środowiskowa" Wrocław 253-263

[19] Łebkowska M Muszyński A 1997 Mikrobiologiczne oczyszczanie gruntów ze składników ropopochodnych I Konferencja Naukowo-Techniczna, Technologie odolejania gruntów, odpadów i ścieków Gorlice-Wysowa Zdrój 115-118

[20] Wolfe S Ingold CF 1981 Oxidation of olefins by potassium permanganate, mechanism of α-ketol formation. *Journal of the American Chemical Society* **103** 938-939

[21] Walton J Labine P Reidies A 1992 The chemistry of permanganate in degradative oxidations Chemical Oxidation Technologies for the Nineties Technomic Publishing Company New York 205-221

[22] Brown GS Barton LL Thomson BM 2003 Permanganate oxidation of sorbed polycyclic aromatic hydrocarbons *Waste Management* **23** 737-740

[23] Sturman PJ Stewart PS Cunningham AB Bouwer EJ Wolfram JH 1995 Engineering scale-up of in situ bioremediation processes: a review *Journal of Contaminant Hydrology* **19** 171-203
[24] Goi A Kulik N Trapido M 2006 Combined chemical and biological treatment of oil contaminated soil *Chemosphere* **63** 1754-1763
[25] Malina G 2007 Likwidacja zagrożenia środowiska gruntowo-wodnego na terenach zanieczyszczonych (Monografia 132 Wydawnictwo Politechniki Częstochowskiej Częstochowa)

PN-ISO 10390:1997 Jakość gleby Oznaczanie pH
PN-ISO 11465:1999 Jakość gleby Oznaczanie zawartości suchej masy gleby i wody w glebie w przeliczeniu na suchą masę gleby Metoda wagowa
PN-78/C-04541 Woda i ścieki Oznaczanie suchej pozostałości, pozostałości po prażeniu, straty przy prażeniu oraz substancji rozpuszczonych i lotnych
PN – ISO 11261:2002. Jakość gleby Oznaczanie azotu ogólnego Zmodyfikowana metoda Kjeldahla
PN – ISO 11263:2002. Jakość gleby Oznaczanie fosforu Spektrometryczne oznaczanie fosforu rozpuszczalnego w roztworze wodorowęglanu sodu

Institute of Physics
Dirac House, Temple Back
Bristol BS1 6BE UK